CASE STUDIES IN ULTRASOUND

ROYAL J. BARTRUM, JR., M.D. **HARTE C. CROW, M.D.**

Department of Diagnostic Radiology,
Mary Hitchcock Memorial Hospital and
Hitchcock Clinic,
Dartmouth Medical School,
Hanover, New Hampshire

1979
W. B. SAUNDERS COMPANY
Philadelphia London Toronto

W. B. Saunders Company: West Washington Square
Philadelphia, PA. 19105

1 St. Anne's Road
Eastbourne, East Sussex BN21 3UN, England

1 Goldthorne Avenue
Toronto, Ontario M8Z 5T9, Canada

Case Studies in Ultrasound ISBN 0-7216-1553-8

Last digit is the print number: 9 8 7 6 5 4 3 2 1

This book is dedicated to the individuals —physicians, administrators, and patients— who with imagination and vision encouraged and facilitated the introduction of ultrasound into the North Country—among whom Frank W. Lane, Marchant E. Tulloh, and especially Peter K. Spiegel played major roles.

Preface

A book of this type can be many things—a textbook, an atlas of pathologic conditions, a reference source, or a means of self-testing. We envision this book as containing bits and pieces of all these but feel that its most important characteristic must be "enjoyability." (Pardon the neologism; there just does not seem to be another word which expresses this concept as succinctly as "enjoyability.") Above all, we have tried to make the following pages pleasant as well as instructive reading.

Abdominal and pelvic ultrasound is not well suited for the traditional radiology teaching format. Because of the tremendous variability in scanning techniques and the great influence the ultrasonographer has on the appearance of the final image, there is nothing which could be called a "standard" ultrasound picture. (It took 15 years for ultrasonographers to agree on an orientation system for scans; they are still quibbling over whether the scans should be displaced as white echoes on a black background or vice versa.) Like radiographic images, the ultrasound scan shows gross anatomy, and, equipped with a knowledge of human anatomy and pathophysiology, the ultrasonographer in most cases should be able to convert the ultrasonic picture into a reasonably specific diagnosis. This requires considerable input on the physician's part; he cannot glance idly through ultrasound scans waiting for "Aunt Minnie" to appear—she rarely shows up in this business. Even more than in conventional radiology, the satisfactory application of ultrasound requires that the practitioner know the clinical history and relevant laboratory and physical findings and have a well-conceived idea of the differential diagnostic possibilities and specific clinical questions to be answered. So prepared, the ultrasonographer can tailor the examination so that the differential diagnostic possibilities are sorted out and the clinical questions answered.

On the following pages we have assembled selected ultrasound scans from over 300 patients, all with proven diagnoses either by biopsy, surgery, or autopsy. All the common pathologic conditions encountered in a busy clinical ultrasound practice are included and, because all the diagnoses are proven, each scan can be considered "typical" of the specific pathologic process. It will soon become apparent, however, that there are frequently more similarities than differences between the scans and you will find that you must rely heavily on the clinical data provided as well as on your own deductive powers to arrive at the correct diagnosis.

We feel this book will be most valuable if it is used as a quiz; therefore, we have organized the material as unknown cases or problems to be solved. On the odd-numbered (right-hand) page that begins each case we present one or more selected scans from an ultrasound examination along with any relevant clinical data and specific questions to be answered. When you have reached your conclusion turn to the following even-numbered (left-hand) page where the scans are shown again, this time with the diagnosis and our accompanying comments and teaching points along with references for further reading. We have tried to present only one or two teaching points (a.k.a. pearls) per case: sometimes this is the diagnosis, sometimes it is a point in differential diagnosis, sometimes a hint on scanning technique or suggestions for further diagnostic work-up. We have mixed the cases up so that they are not in any logical order; you should not attempt to determine what is coming next by what has preceded. (Our patients, after all, do not present in any logical diagnostic order.) We have also included some normal cases just to keep everybody honest. You can use the index to look up specific discussions and illustrations, but we urge everyone to treat the scans as unknowns initially—that is what makes the whole thing fun.

ROYAL J. BARTRUM, JR.
HARTE C. CROW

Scan Orientation

Each picture is labeled to show the scanning plane, using the standard of the American Institute of Ultrasound in Medicine for labeling of ultrasound images: a copy of the entire standard is available from the AIUM Executive Office, 6161 N. May Avenue, Oklahoma City, Oklahoma 73112. We will briefly summarize the standard as it applies to the scans in this book.

Transverse (cross-sectional) scans are oriented as if the viewer were looking up from the patient's feet; the liver is on the viewer's left. Sagittal (logitudinal) scans are presented as if we were looking at the patient from his right side, that is, the patient's head will be on the viewer's left. Unless stated otherwise, it is assumed that the patient was scanned supine. If the patient was not supine, the scanning position is *always* given, e.g., "prone," "upright," "right lateral decubitus," and so on.

All scans are referred to a standard "reference plane." The reference plane for transverse scans passes through the iliac crests; it is labeled **Transverse section 0.** The reference plane for longitudinal scans is the midline (midsagittal) plane; it is labeled **Longitudinal section 0L.** Each scan is labeled with its distance, in centimeters, from the reference plane. For transverse scans, displacements in the direction of the patient's head are considered positive, and those toward the feet are negative. (For the sake of clarity in this text we have also added a term to help you determine whether the scan is transverse or longitudinal. We want to stress that this is not necessary in everyday work; we do it only as an aid in orientation.) Thus you will find a transverse scan through the head of the pancreas labeled **Transverse section +8,** meaning it was performed 8 cm. above the reference plane (iliac crests). A transverse scan through the pelvis might be labeled **Transverse section −7** indicating it was performed 7 cm. below the iliac crests. Longitudinal scans are labeled with the distance from the midline and either an **R** or an **L,** depending on whether the scan is to the patient's *right* or *left*. Thus a longitudinal scan through the right lobe of the liver and right kidney is labeled **Longitudinal section 6R,** meaning it was performed 6 cm. to the right of the reference plane (midline). A longitudinal scan of the aorta would be labeled **Longitudinal section 2L** indicating it was made 2 cm. to the left of midline.

Oblique scans are given two labels. The first is the distance of the scan from the transverse reference plane; this is measured at the point the scan plane intersects the midline and is considered positive if the displacement is toward the patient's head and negative if toward the feet. The second label is the number of degrees of rotation about the midline: vertical (12 o'clock) is considered 0 degrees and the angle increases in a clockwise rotation. (Thus 3 o'clock would be 90 degrees and 5 o'clock 150 degrees.) As an example, consider an oblique scan through the head and body of the pancreas. The scan plane might cross the midline 8 cm. above the iliac crests and have an orientation of about 2 o'clock; this would be labeled **Oblique section +8, 60°.**

(We have tried to keep oblique planes to a minimum in this book, and sometimes instead of using the formal notation we will simply label the scan "Oblique section through the long axis of gallbladder.")

For abdominal scans the reference planes of the iliac crests and midline are adequate. When other organs are examined it is often more convenient to use a "local" reference plane. Thyroid scans, for example, utilize the standard longitudinal reference plane (the midline or **0 section**), but for transverse scans the line through the sternal notch is used as a reference plane: all transverse sections are labeled with their displacement from this local reference plane. Prone scans of the kidney also make use of a local reference plane. Transverse scans are labeled as usual, utilizing the iliac crests for reference, but the longitudinal reference plane is that plane passing through the *middle of the kidney in its long axis*. Thus the 0 plane for a kidney is the line connecting the center of the upper pole with the center of the lower pole. Scans are labeled according to their displacement from this reference line: the letters **R** and **L** are appended, as usual, to indicate displacement to either the patient's right or left. Of course the scan also specifies which kidney, right or left, is being scanned. As an example, a longitudinal scan through the medial portion of the right kidney might be labeled **Longitudinal section 2L** *Prone scan of the right kidney.* Scans of the extremities and scrotum usually are given only a descriptive label such as *Cross-section of palpable mass on right thigh* rather than using reference planes and numbers.

All the scans are displayed as dark dots on a light background. There is presently a good deal of lather being raised over which display mode, positive or negative, is preferable. We have tried it both ways and prefer the light background. We find that it is difficult to appreciate differences in gray levels when using light dots on a dark background. Even though we have heard the bit about "the ease of seeing the stars in the night sky" cited as support for displaying the data in reverse (white dots on a black background), we suspect it was the inability of early real-time scanners to produce images with a light background that got the white dot-dark background business started. It is really a tempest in a tea-pot; a technically good scan made on a well-adjusted instrument will do the trick for the informed ultrasonographer regardless of the display format.

Contents

Case 1

CLINICAL INFORMATION

This 53-year-old man is feeling well, but on routine physical examination his blood pressure was found to be 145/95 and his urinalysis showed microscopic hematuria. An IVP was performed at another hospital and the report was ". . . mass in the lower pole of the right kidney, either cyst or tumor, recommend ultrasound examination." Since the outside hospital does not have an ultrasound machine, the patient is referred to your laboratory to have this examination performed. This longitudinal scan of the right kidney is obtained with the patient prone.

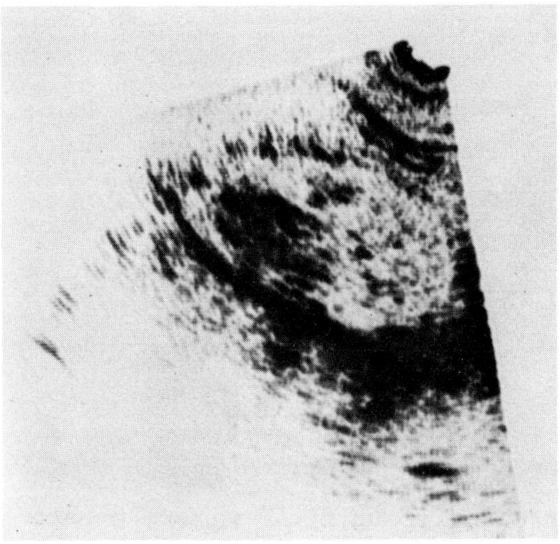

Figure 1–A. **Longitudinal section 0L.** Prone scan of right kidney.

DIAGNOSIS

Hypernephroma (renal cell carcinoma).

ULTRASOUND FINDINGS

This case nicely illustrates the typical or classic findings in a renal tumor. The lower pole of the kidney is distorted and slightly enlarged; the calyceal and renal sinus echoes are distorted and absent in this region, and there are multiple echoes of differing strengths scattered throughout the mass. Look carefully at the back wall of the tumor. Notice that the boundary is slightly irregular — there are some "step-offs" present. This is a very important point in distinguishing a tumor from a cyst; cysts (more accurately "benign fluid-filled renal masses") have a *smooth, continuous* back wall. The slight irregularity of the back wall seen in this case is sufficient grounds for calling this lesion a tumor, even if none of the other criteria are present.

There is no way to tell whether this tumor is benign or malignant from the appearance of the ultrasound scan alone. Either arteriography (Figure 1–B) or percutaneous aspiration biopsy is required to make this distinction. It has been reported that one can distinguish a hypernephroma from the much rarer angiomyolipoma on the basis of the ultrasound scan appearance alone, but we do not believe that this is possible.

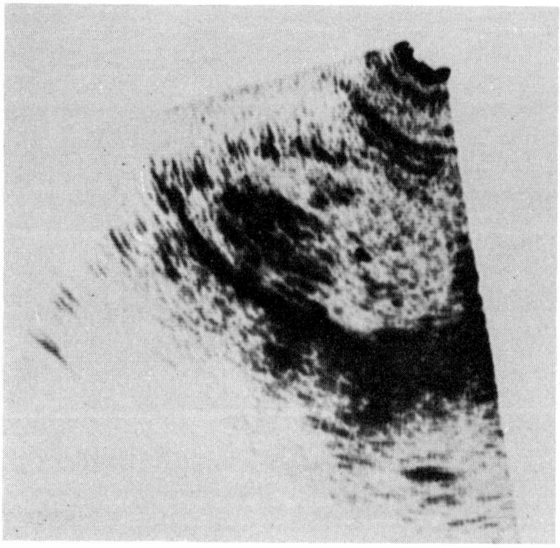

Figure 1–A. Longitudinal section 0L. Hypernephroma. Prone scan of right kidney.

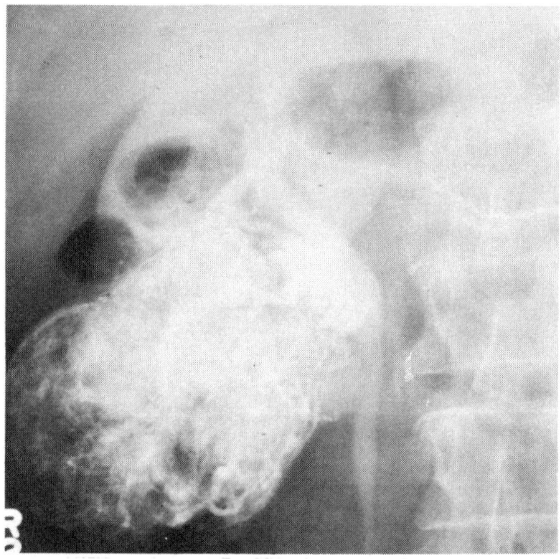

Figure 1–B. Hypernephroma. Selective renal arteriogram from patient in Figure 1–A.

REFERENCES

Maklad NF, Chuang VP, Doust BD, et al: Ultrasonic characterization of solid renal lesions. Radiology 123:733 (1977)

Pollack MM, Goldberg BB: Chapter 10, Kidney, in *Abdominal Gray Scale Ultrasonography*, Goldberg BB, ed. John Wiley & Sons, New York (1977)

Duffy P, Ryan J, Aldous W: Ultrasound demonstration of a 1.5 cm angiomyolipoma. J Clin Ultra 5:111 (1977)

Case 2

CLINICAL INFORMATION

These scans were obtained from several different patients, all of whom had epigastric pain as their complaint. The scans are all normal and illustrate the many vascular structures that can be identified in the upper abdomen. Let's see if you can identify each of the labeled vessels A to J.

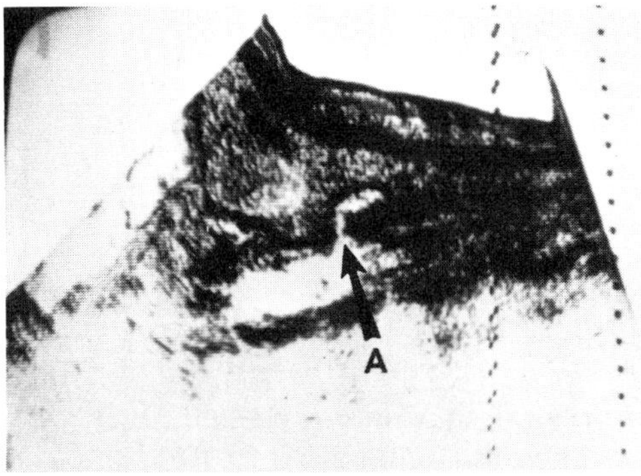

Figure 2–A. Longitudinal section 1L.

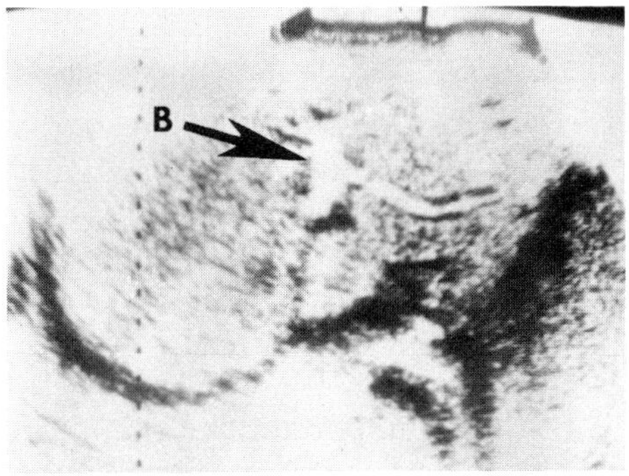

Figure 2–B. Transverse section +12.

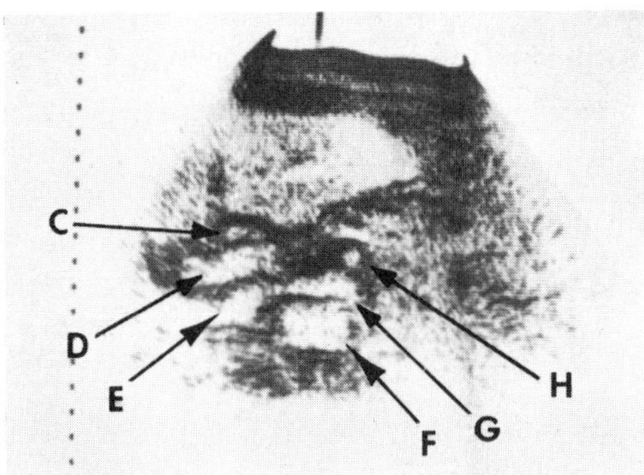

Figure 2–C. Transverse section +10.

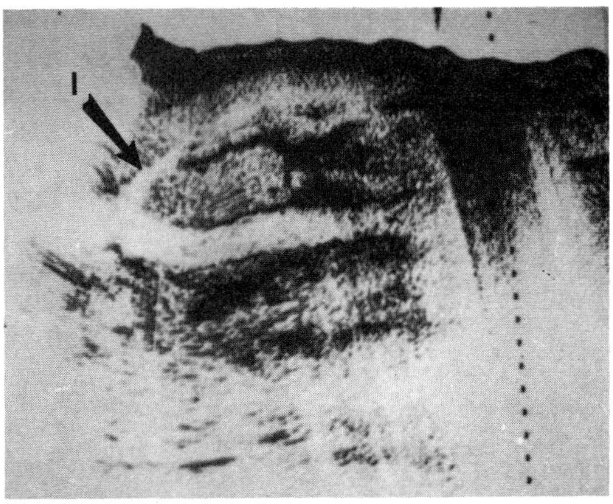

Figure 2–D. Longitudinal section 2R.

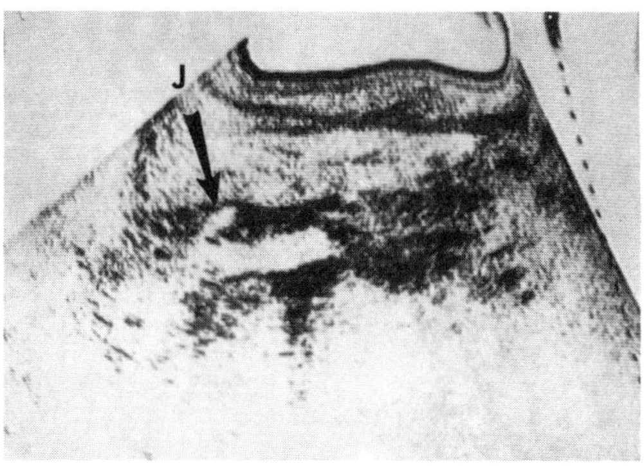

Figure 2–E. Longitudinal section 1L.

DIAGNOSIS

Normal anatomy of the upper abdomen.

ULTRASOUND FINDINGS

The various structures are:

A — Superior mesenteric artery
B — Portal venous branches in the liver
C — Main hepatic artery and common bile duct (the duct is probably the smaller of the two small, rounded echo-free structures, but we cannot be certain)
D — Portal vein
E —Inferior vena cava
F — Aorta
G — Left renal vein
H — Superior mesenteric artery
I — Hepatic vein
J — Celiac artery

It is important that the ultrasonographer become familiar with the vascular anatomy in the upper abdominal region because the major vessels serve as valuable anatomic landmarks for localizing normal organs and pathologic masses. There are several good articles and atlases on this subject. Figures 2–F through 2–H are other examples of normal vessels in this region.

sma = superior mesenteric artery; smv = superior mesenteric vein; ivc = inferior vena cava; ao = aorta; ra = right renal artery; hv = hepatic vein.

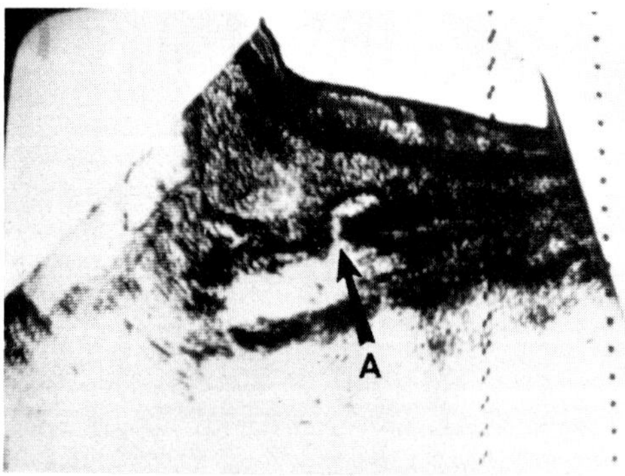

Figure 2–A. Longitudinal section 1L.

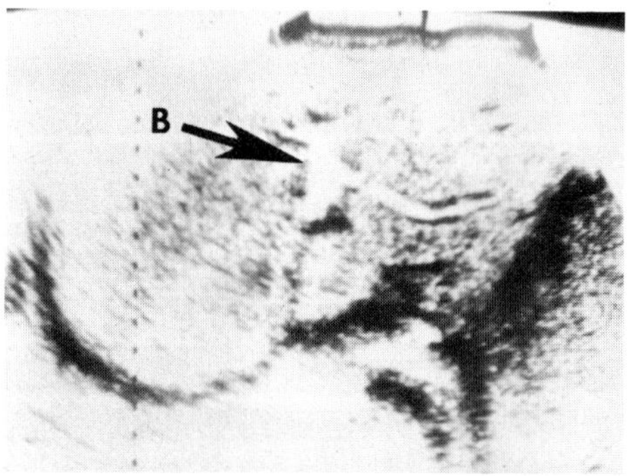

Figure 2–B. Transverse section +12.

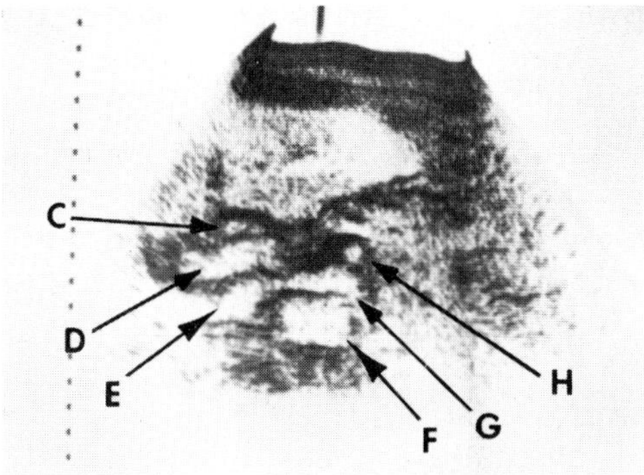

Figure 2–C. Transverse section +10.

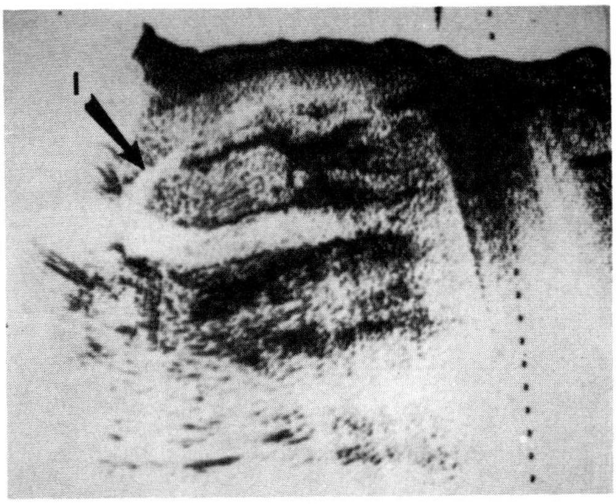

Figure 2–D. Longitudinal section 2R.

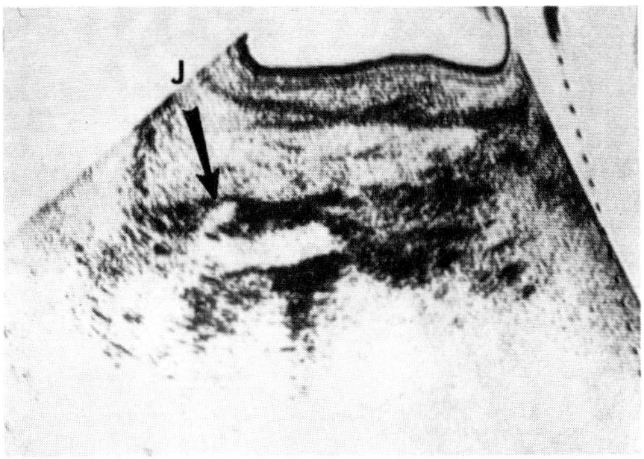

Figure 2–E. Longitudinal section 1L.

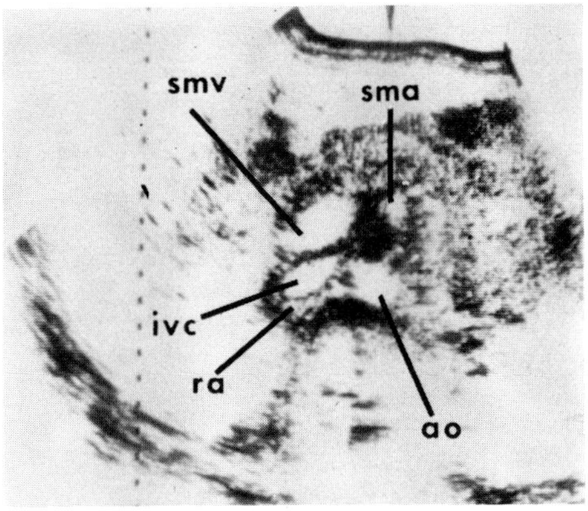

Figure 2–F. Transverse section +10.

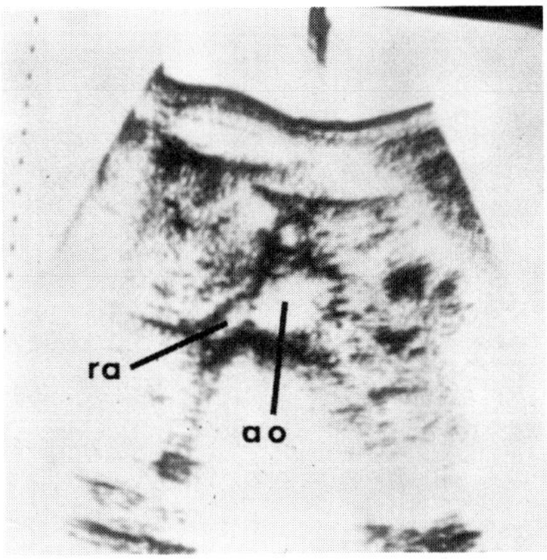

Figure 2–G. Transverse section +7.

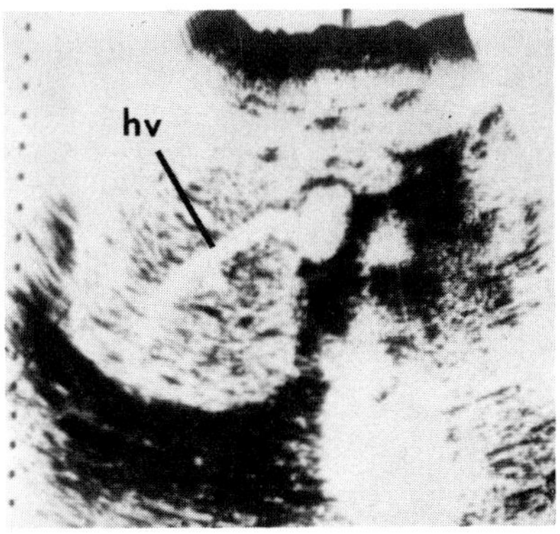

Figure 2–H. Transverse section +14.

REFERENCES

Sample WF: Techniques for improved ultrasound delineation of normal anatomy . . . with gray scale ultrasound. Radiology 124:197 (1977)

Carlsen EN, Filly R: Newer ultrasonographic anatomy in the upper abdomen. Parts I and II. J Clin Ultra 4:91 (1976)

Leopold GR: Gray scale ultrasonic angiography of the upper abdomen. Radiology 117:665 (1975)

Case 3

CLINICAL INFORMATION

This 37-year-old man has symptoms suggesting peptic ulcer disease. An upper GI series shows only edema of the duodenal mucosa and spasticity of the duodenal bulb without an ulcer niche. Although there is nothing in the clinical history, physical examination, or basic laboratory studies to suggest any other abdominal disease, his doctor wants an ultrasound scan to exclude other pathology. What is the tubular structure found in the porta hepatis (arrow)? What is underneath it?

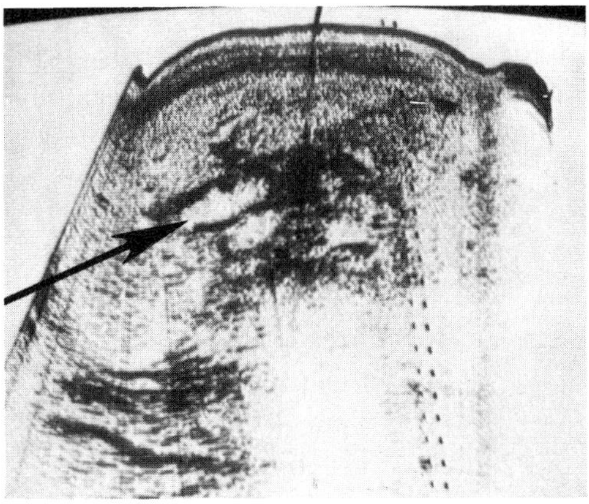

Figure 3–A. Transverse section +9.5.

DIAGNOSIS

Portal vein (normal abdomen).

ULTRASOUND FINDINGS

The arrow in Figure 3–A points to the portal vein. Tubular structures in the upper abdomen are almost always major vessels and the ultrasonographer must learn quickly to recognize the normal vascular anatomy and its variations in this region. This is important for two reasons: we do not want to be confused by normal anatomic structures and label them erroneously as pathologic masses; and, more important, the major vessels serve as valuable landmarks for locating other organs and masses in the abdomen. (The oval structure beneath the portal vein in Figure 3–A is the inferior vena cava.)

The portal venous system should be followed from its origins (in the splenic vein) to its termination (as branching portal venous radicles in the liver) in every patient. This not only verifies the identity of the vessels but also defines the upper extent of the pancreas and shows where to look for the common bile duct and major branches of the aorta. Scans at several levels are usually required to accomplish this because the portal venous system is rarely aligned in a single transverse plane. Figure 3–B is another scan, 2 cm. lower than Figure 3–A. It shows the portal vein at its formation by the confluence of the splenic and inferior mesenteric veins. Figures 3–C and 3–D are scans at different levels in another normal patient showing different portions of the portal system. Sometimes a long segment of the portal and splenic veins can be shown on a single scan if the scanning plane is properly angled (Figure 3–E).

Although a real time scanner greatly facilitates identifying the portal system, identification can also be done readily with a conventional B-scanner as long as the operator is familiar with the anatomy and makes careful serial scans at 1 cm. intervals.

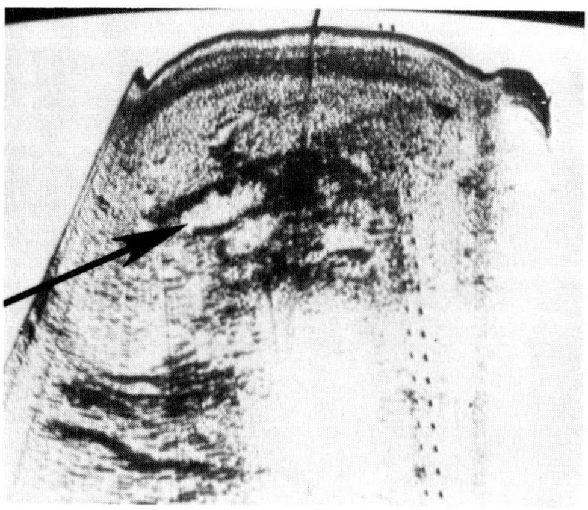

Figure 3–A. Transverse section +9.5. Portal vein (arrow) and IVC.

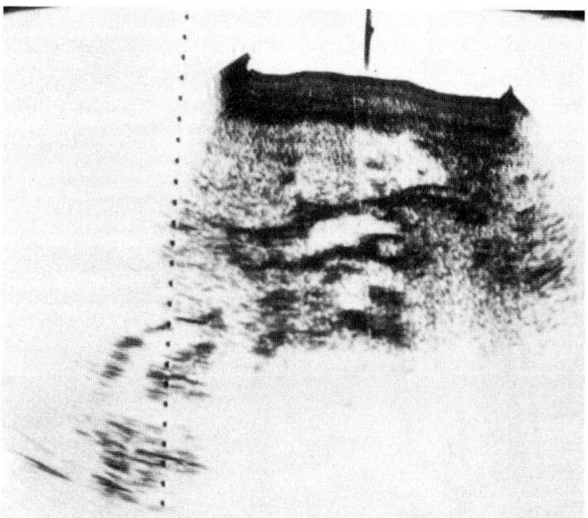

Figure 3–B. Transverse section +7.5. Portal venous system.

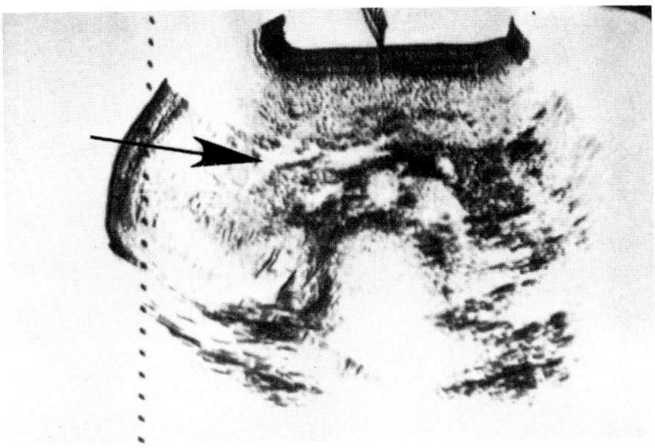

Figure 3–C. Transverse section +10. Portal venous radicles.

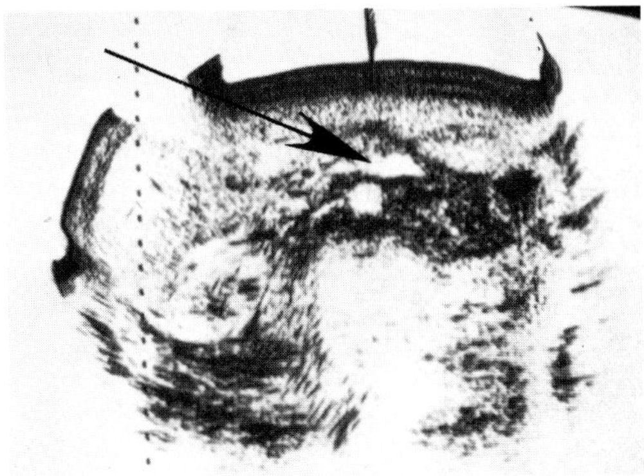

Figure 3–D. Transverse section +8.5. Splenic vein.

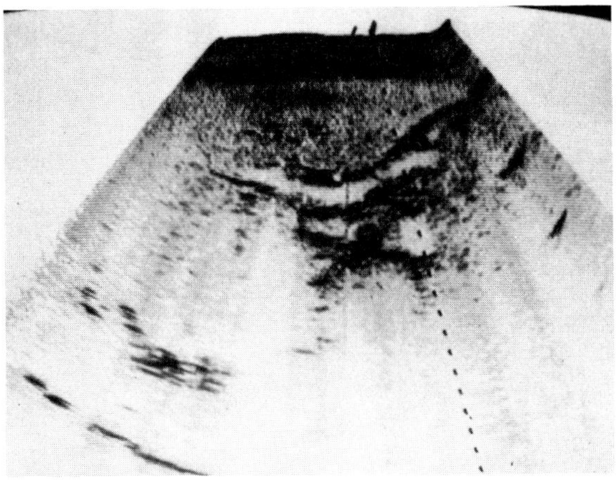

Figure 3–E. Transverse section +8.5, 80°. Portal and splenic veins.

REFERENCES

Sample WF: Techniques for improved ultrasound delineation of normal anatomy . . . with gray scale ultrasound. Radiology 124:197 (1977)

Sanders RC, Conrad MR, White RI Jr: Normal and abnormal upper abdominal venous structures as seen by ultrasound. Am J Radiol 128:657 (1977)

Carlsen EN, Filly R: Newer ultrasonographic anatomy in the upper abdomen. Parts I and II. J Clin Ultra 4:91 (1976)

Case 4

CLINICAL INFORMATION

Figures 4–A, 4–B, and 4–C are scans of a fetal head. Which, if any, are acceptable for measuring a biparietal diameter?

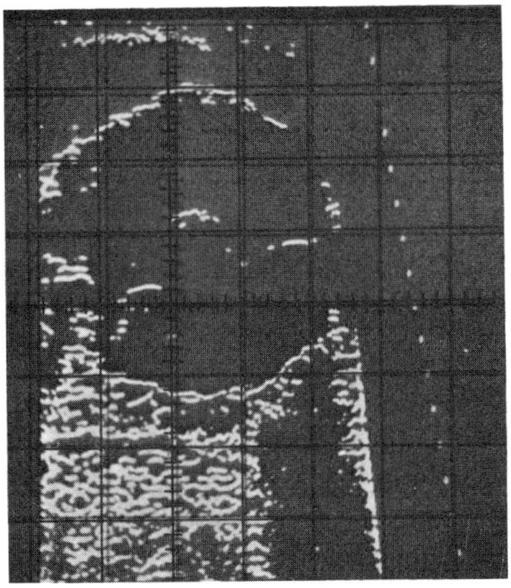

Figure 4–A.
Fetal skull.

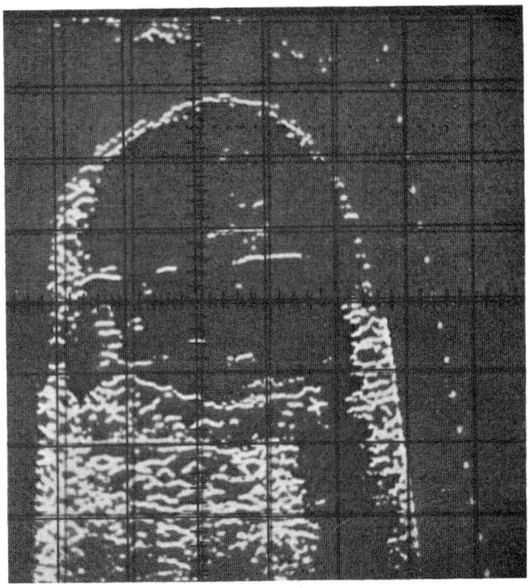

Figure 4–B.
Fetal skull.

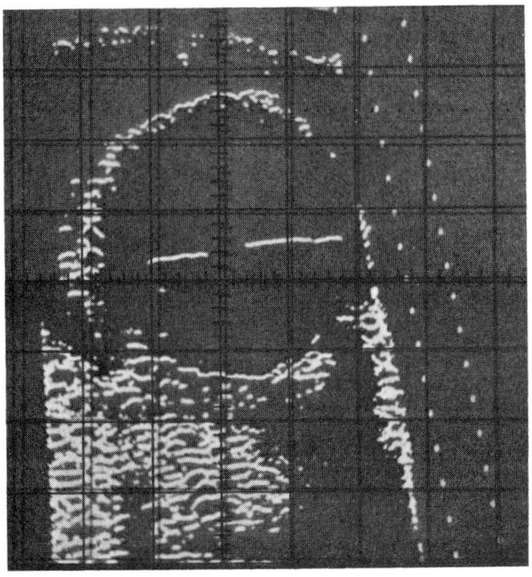

Figure 4–C.
Fetal skull.

DIAGNOSIS

Normal fetal skull — all three figures are satisfactory.

ULTRASOUND FINDINGS

We think all these scans are adequate for measuring fetal biparietal diameter (BPD). There is a voluminous literature on the topic of accurate gestational dating and scanning techniques to obtain a "true" BPD. Some writers recommend complicated scanning regimens to ensure that the skull is being sectioned in exactly the right plane while others have insisted on real time visualization of the head to obtain a correct BPD. We are certain that these methods will produce accurate BPD measurements but we are also convinced that all that work is not necessary.

The charts and tables which relate BPD to gestational ages are themselves only approximations based on data obtained from large series of patients. When using these tables it is easy to forget that they represent a mythical "average" baby and are not necessarily accurate for each individual case. If we read the original articles and analyze the data in the various tables, we find that with a BPD of 7.8 cm., for example, the predicted gestational age is 30½ weeks. This figure is either the median age or average age for the babies in that series. But what about babies who are not "average" but are otherwise normal and healthy: how far can their BPDs differ from the median before we should become concerned for their health? Continuing our example we find that the 10th and 90th percentile range for babies with a BPD of 7.8 cm. is 28½ weeks and 33 weeks. This means that 80 out of 100 normal babies with BPD of 7.8 cm. will be between 28½ and 33 weeks of age. We can look at this another way; 80 out of 100 babies who are 30½ weeks of age will have a BPD measurement between 7.3 and 8.3 cm. The point is that the ultrasonographic estimation of gestational age is just that — an estimate. There is no reason to strive for extreme accuracy in measuring the head if the improvement in the measurement is only on the order of a few millimeters. This improvement is more than offset by the normal variations between babies.

Here is how we estimate gestational age between 13 and 36 weeks.

1. We obtain six different scans of the head in at least two different planes, attempting to get the largest diameter possible on each scan.

2. All six head scans are measured and a "consensus" diameter obtained. Usually this is very easy because most, if not all, of the measurements will be identical. If more than two of the six are different we go back and get another three scans. Naturally, we disregard any obviously inadequate pictures, such as those with assymetrical or misshapen skulls or those in which the midline echoes are off to the side.

3. We report the actual measurement and the "median" or "average" age for that measurement. In addition we report the 10th and 90th percentile age ranges so that the obstetrician will have an idea of the variation in our estimate.

We like to use bi-stable scans, but gray scale pictures are acceptable, providing the machine can be adjusted to resolve the skull table as a thin, easily measurable line. There are many different BPD tables available and the difference between them is minor. The one we use is shown in Table 1.

There is some controversy about measuring the BPD. The problem arises because the line of the skull in gray-scale scans often has some thickness and can introduce several millimeters' variation. The secret is to make the image of the skull as thin as possible by decreasing the gain (and using the "bi-stable" gray-scale presentation). We then put the tips of the calipers at the same point on both sides — usually the middle of the line proves to be easiest. If your scanner is equipped with electronic calipers, use them, of course. If not, measuring from any type of hard copy will do but don't forget to include a centimeter scale on the image.

TABLE 1. BPD CHART

BPD (cm.)	Median (wks.)	Range (wks.)	BPD (cm.)	Median (wks.)	Range (wks.)
2.5	12½	11 –13½	6.3	25	23½–27
2.6	13	11½–14	6.4	25½	24 –27
2.7	13½	12 –14	6.5	26	24 –27½
2.8	13½	12½–14½	6.6	26	24½–28
2.9	14	12½–15	6.7	26½	25 –28½
3.0	14	13 –15½	6.8	27	25 –28½
3.1	14½	13 –15½	6.9	27	25½–29
3.2	15	13½–16	7.0	27½	26 –29½
3.3	15	14 –16½	7.1	28	26 –29½
3.4	15½	14 –16½	7.2	28	26½–30
3.5	16	14½–17	7.3	28½	26½–30½
3.6	16	15 –17½	7.4	29	27 –31
3.7	16½	15 –18	7.5	29	27½–31½
3.8	16½	15½–18	7.6	29½	28 –32
3.9	17	16 –18½	7.7	30	28 –32½
4.0	17½	16 –19	7.8	30½	28½–33
4.1	17½	16½–19	7.9	31	29 –33½
4.2	18	17 –19½	8.0	31	29 –34
4.3	18½	17 –20	8.1	32	29½–34½
4.4	19	17½–20	8.2	32	30 –35
4.5	19	17½–20½	8.3	32½	30 –35½
4.6	19½	18 –21	8.4	33	31 –36
4.7	20	18½–21½	8.5	34	31½–36½
4.8	20	19 –21½	8.6	34½	32 –37
4.9	20½	19 –22	8.7	35	32½–37½
5.0	20½	19½–22	8.8	35½	33 –38½
5.1	21	19½–22½	8.9	36½	33½–39
5.2	21½	20 –23	9.0	37	34 –40
5.3	21½	20½–23½	9.1	37½	34½–41
5.4	22	20½–24	9.2	38	35 –42
5.5	22	21 –24	9.3	39	36 –>42
5.6	22½	21½–24½	9.4	39½	36½–>42
5.7	23	21½–25	9.5	40	37 –>42
5.8	23	22 –25	9.6	41	37½–>42
5.9	23½	22 –25½	9.7	42	38 –>42
6.0	24	22½–26	9.8	>42	38½–>42
6.1	24½	23 –26	9.9	>42	40 –>42
6.2	24½	23 –26½	10.0	>42	41 –>42

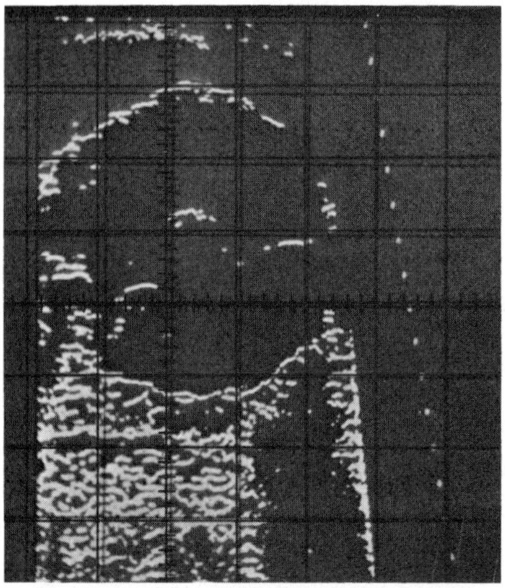

Figure 4–A.
Fetal skull; satisfactory biparietal diameter (BPD).

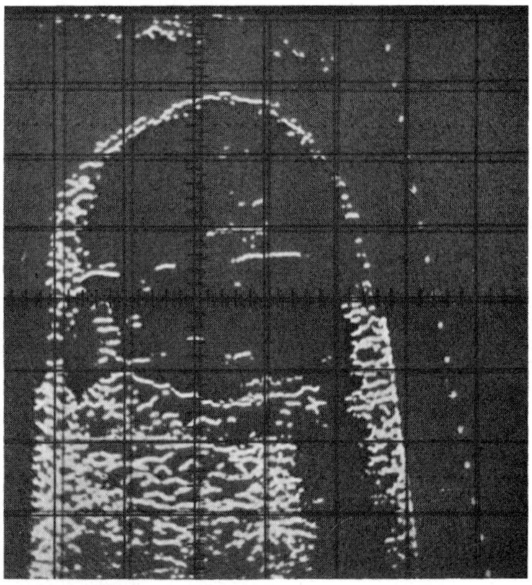

Figure 4–B.
Fetal skull; satisfactory BPD.

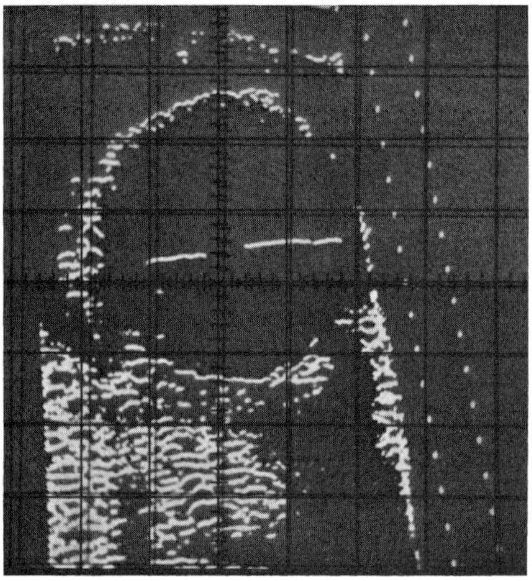

Figure 4–C.
Fetal skull; satisfactory BPD.

REFERENCES

Wiener SN, Flynn MJ, Kennedy AW, et al: A composite curve of ultrasonic biparietal diameters. Radiology 122:781 (1977)

Lawson TL, Albarelli JN, Greenhouse SW, et al: Gray scale measurement of biparietal diameter. J Clin Ultra 5:17 (1977)

Case 5

CLINICAL INFORMATION

This patient is a 34-year-old man with epigastric pain and belching. Physical examination and basic laboratory studies are normal. The upper GI series shows a small hiatal hernia but no gastroesophageal reflux. The ultrasound technician is having trouble getting any picture of the left upper quadrant; there is only a line of very loud echoes near the skin and no echoes can be obtained from any deeper structures.

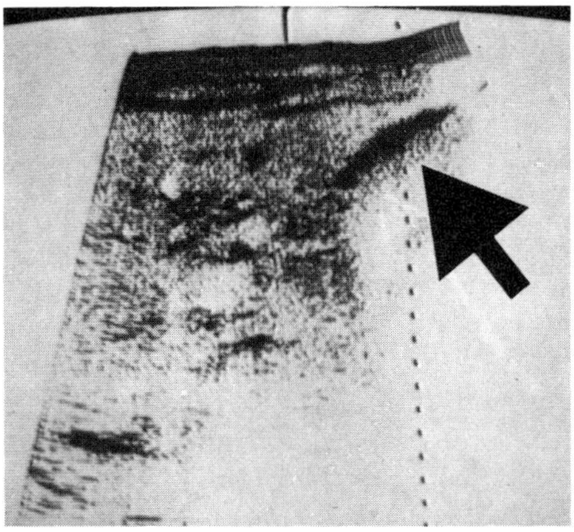

Figure 5–A. Transverse section +12.

DIAGNOSIS

Acoustical shadowing from air in the stomach (normal abdomen).

ULTRASOUND FINDINGS

At an air-soft tissue interface there is a great mismatch of acoustic impedance, which results in complete reflection of the ultrasound beam. On the picture this appears as a very high-intensity echo with "shadowing" of underlying structures. Figure 5–A illustrates this effect: air in the body of the stomach is reflecting the sound beam, causing the long, high-intensity echo, and shadowing the remainder of the left upper quadrant (arrow). Air in the stomach is a nearly universal occurrence and hampers, to a greater or lesser extent, all ultrasound scans of the upper abdomen. It does not take much air to cause a great deal of trouble because with the patient in the supine position whatever air is present will float to the most anterior portions of the stomach, usually the body and antrum.

When the amount of air is large, as in this case, it is easy to recognize the echo and shadowing phenomena. When smaller amounts of air are present, however, the scan picture can become confusing. Small air collections may interrupt only small portions of the beam; the sound which gets past generates echoes from underlying structures. Unfortunately, small air collections give rise to reverberation artifacts, which also appear as "echoes" in the underlying tissue. The ultrasonographer is thus confronted with the problem of deciding which "echoes" are real and which are artifacts (Figures 5–B and 5–C). We do not know of any easy way to make this distinction. Obviously, if a familiar organ shape such as the kidney or spleen appears, there is no problem. On the other hand if the scan seems to show a "mass" we proceed with caution. Is the "mass" well defined and can it be continually reproduced and seen in different scanning planes? Does the "mass" have a back wall? Acoustic shadowing can create false side walls but cannot create an artifactual back wall. Do the "echoes" in the "mass" appear relatively uniformly throughout the mass — as occurs in real lesions — or do they fill in from anterior to posterior — as occurs in artifact production? Do the findings correlate with the rest of the clinical picture; for example, if the "mass" is large, can we palpate it? If the answers to these questions are yes, then we feel comfortable in saying the "mass" is real. If some of the answers are no, however, we are usually reluctant to interpret the scan as abnormal.

There is a curious correlary to the phenomena of shadowing and artifact production by gastric air. If the spleen and left kidney can be easily and well seen on a *supine* scan, the ultrasonographer must be suspicious of a pathologic condition in the left upper quadrant. Normal patients have air in the stomach and this should obscure this area. If the organs of the left upper quadrant are easily demonstrated, something must be displacing the air and we must explain this.

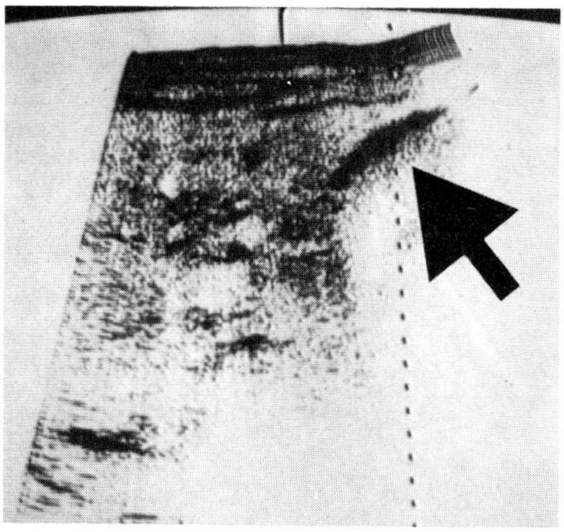

Figure 5–A. Transverse section +12. Air in stomach (normal).

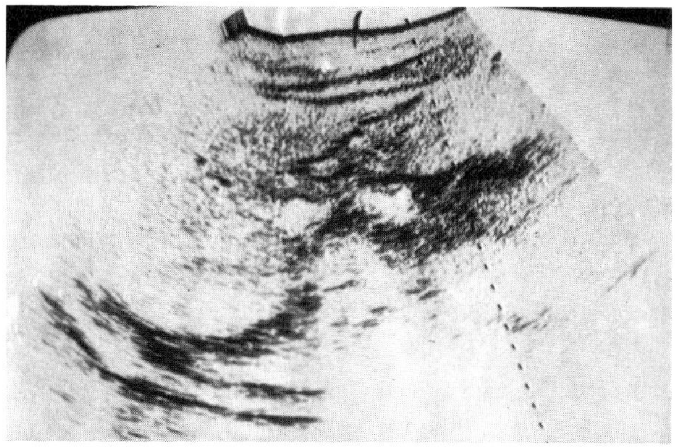

Figure 5–B. Transverse section +11.5. Normal left upper quadrant.

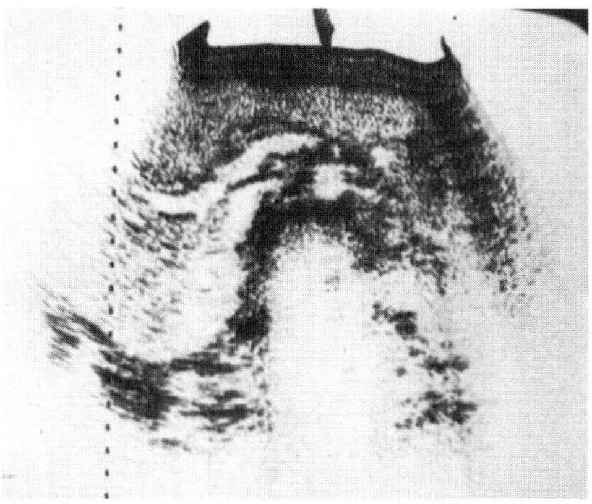

Figure 5–C. Transverse section +9. Normal left upper quadrant.

REFERENCE

Bartrum RJ Jr, Crow HC: *Gray-Scale Ultrasound: A Manual for Physicians and Technical Personnel.* Chapters 5 and 6. WB Saunders Co, Philadelphia (1977)

Case 6

CLINICAL INFORMATION

This 19-year-old woman thinks that she might be pregnant. Her periods have always been "regular as clockwork" and now she is ten days overdue.

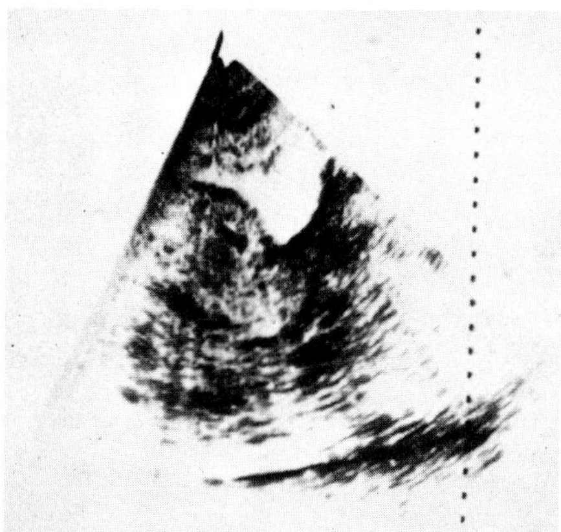

Figure 6–A. Longitudinal section 0L.

DIAGNOSIS

Intrauterine pregnancy, 5½ weeks.

ULTRASOUND FINDINGS

There is a tiny gestational sac within the uterus (arrow). This is the earliest pregnancy which we have identified; the sac is just barely discernible. Because of the reliable menstrual history we know it has been less than three weeks since conception! Ultrasound scanning is among the most sensitive detectors of early pregnancy; it is usually positive at about the same time as the standard hormonal laboratory tests. The gestational sac is visible as a tiny, echo-free area in the central uterus as early as five weeks gestational age (one week after the first missed period). How reliably ultrasound can identify such small sacs is not known because very few patients come to medical attention so early. By six weeks of gestational age a sac should be present in all cases; and if the sac is not visible by seven weeks, we can be confident that there is no intrauterine pregnancy (this does not exclude an ectopic pregnancy, of course). There is no special trick to making this diagnosis; just make sure to get serial scans through the uterus at 5 mm. intervals — if a gestational sac is there, you will see it. Figure 6–B shows a 6½-week pregnancy.

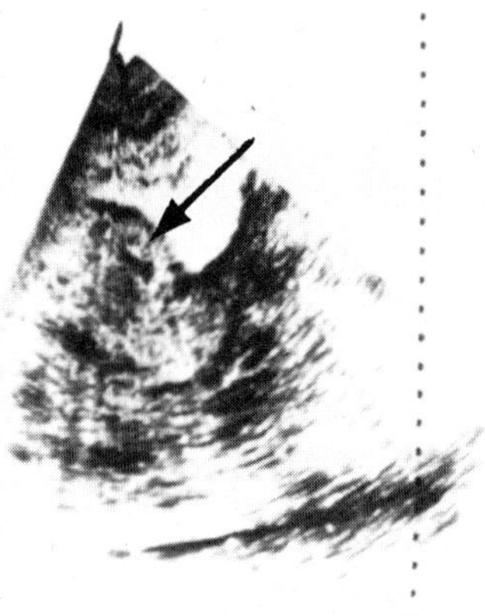

Figure 6–A. Longitudinal section 0L. 5½ week gestational sac.

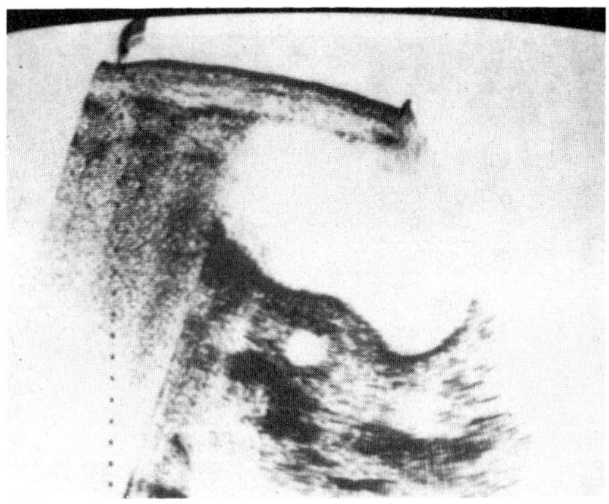

Figure 6–B. Longitudinal section 1R. 6½ week gestational sac.

REFERENCE

Ghorashi B, Gottesfeld KR: The gray scale appearance of the normal pregnancy from 4 to 16 weeks gestation. J Clin Ultra 5:195 (1977)

Case 7

CLINICAL INFORMATION

This 62-year-old woman has recently had a barium enema which showed a typical "apple-core" carcinoma of the sigmoid colon. She is undergoing a metastatic workup before surgery, which includes an abdominal ultrasound scan to evaluate the possibility of retroperitoneal adenopathy or liver metastases. What do you think about the texture of the left lobe of her liver?

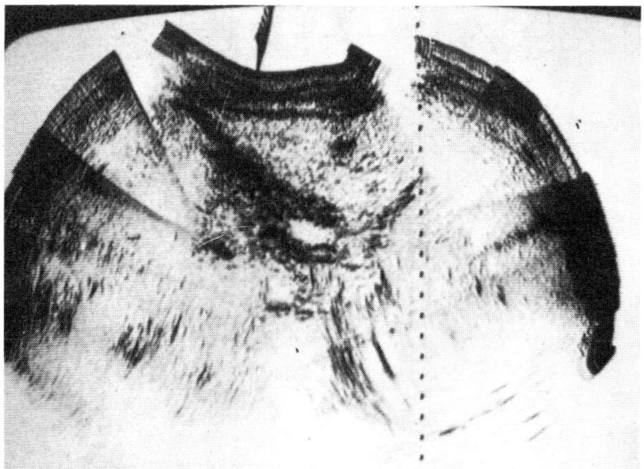

Figure 7–A. Transverse section +10.

DIAGNOSIS

Normal liver showing ligamentum teres.

ULTRASOUND FINDINGS

There is a small focal collection of high-intensity echoes near the junction of the right and left lobes of the liver (arrow). This is a normal finding and can be identified in almost all patients if carefully sought. It is important to recognize this as a normal "liver echo" and not confuse it with a focal liver lesion. The source of the echo is not completely certain, but most ultrasonographers feel it arises from the ligamentum teres, the intrahepatic portion of the falciform ligament.

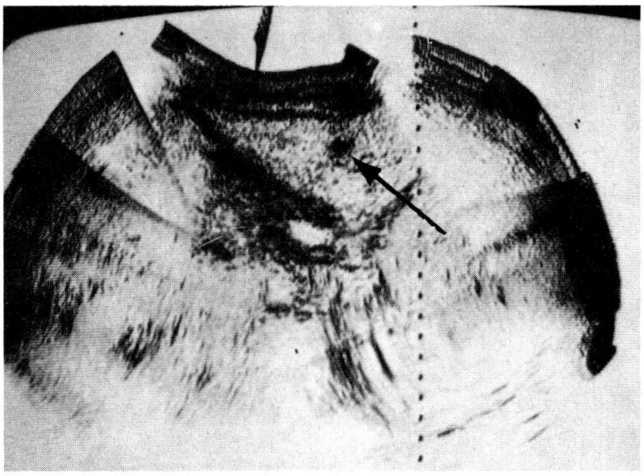

Figure 7–A. Ligamentum teres (normal). **Transverse section +10.**

REFERENCE

Hillman BJ, D'Orsi CJ, Smith EH, et al: The ultrasonic appearance of the falciform ligament. Am. J. Roentgenol (in press).

Case 8

CLINICAL INFORMATION

This 22-year-old woman thinks she is pregnant but, because of very irregular periods, she has no idea of how far along she might be. Can you help her out?

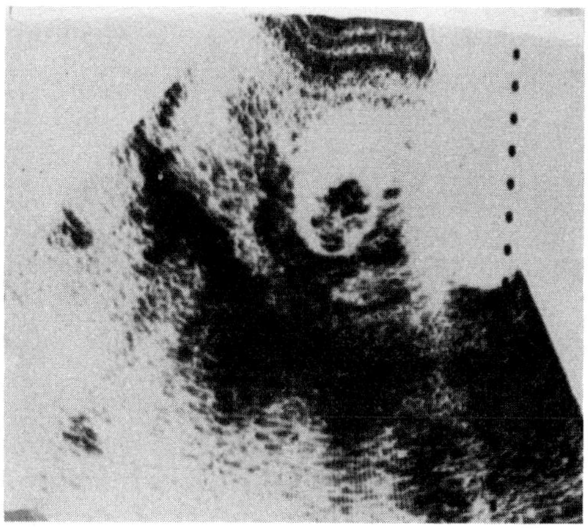

Figure 8–A. Longitudinal section 1L.

DIAGNOSIS

Ten- to 12-week intrauterine pregnancy.

ULTRASOUND FINDINGS

In the uterus there is a fluid-filled (echo-free) area, which represents the amniotic cavity; in the center of the cavity is an amorphous collection of echoes arising from the developing fetus. Because no specific fetal parts or fetal head can be identified, this pregnancy must be between 10 and 12 weeks' gestational age.

In order to date a pregnancy in the first trimester we must be familiar with the ultrasonic appearance of the developing embryo. We define four separate periods, each with easily recognizable ultrasonic landmarks. *Less than eight weeks* (Figure 8–B): there is a gestational sac (arrow) present within the uterus, but the sac does not contain any internal echoes and is almost always round. *Eight to 10 weeks* (Figure 8–C): The gestational sac (arrow) is no longer round but is changing into the amniotic cavity and adopting an elongated configuration to fill the uterine cavity. In the last part of this period faint echoes appear within the cavity. *Ten to 12 weeks* (Figure 8–D): There are now definite echoes within the amniotic cavity (arrow), but no individual fetal parts can be recognized. In particular, the fetal head cannot be precisely defined. *Twelve to 14 weeks* (Figure 8–E): The fetal head (arrow) can be identified. By 14 weeks the head is large enough to be measured, and the regular BPD charts can be used.

Some ultrasonographers have compiled measurements of the gestational sac and developing fetus in an attempt to add more objectivity to this process. If you like this sort of thing — fine. We feel comfortable applying only the anatomic criteria described above. It is easy to place the pregnancy within one of the four categories. Look at Figure 8–A again. The amniotic cavity clearly contains echoes but no parts are recognizable, so this gestation is somewhere between 10 and 12 weeks. The echoes are really quite amorphous, with no hint as to what might be head, body, placenta, and so on. Therefore, we might go further and place this baby in the first half of this period or about 10 to 11 weeks of age. In another week there will be more differentiation and we will almost be able to pick out the head. In practice we find this degree of accuracy is quite acceptable; in fact, it is better than one can do at any other stage in pregnancy!

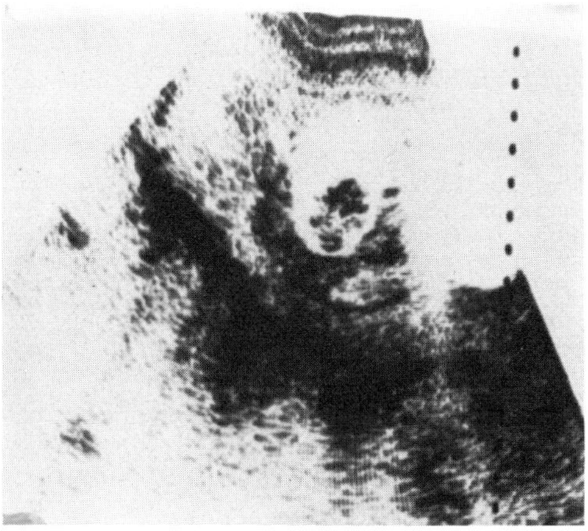

Figure 8–A. Longitudinal section 1L. Pregnancy, 10 to 12 weeks.

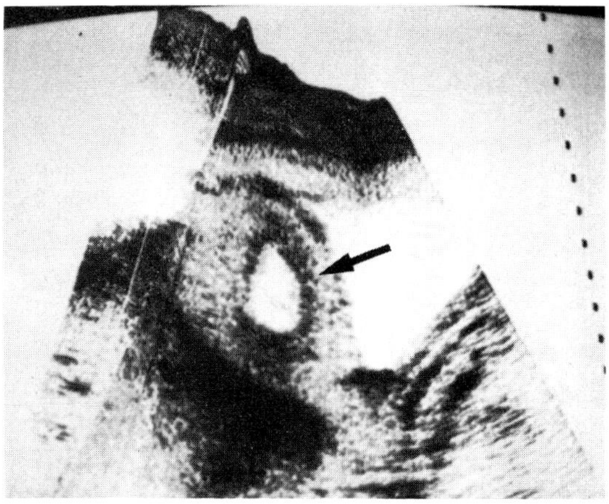

Figure 8–B. Longitudinal section 1R. Pregnancy, 8 weeks.

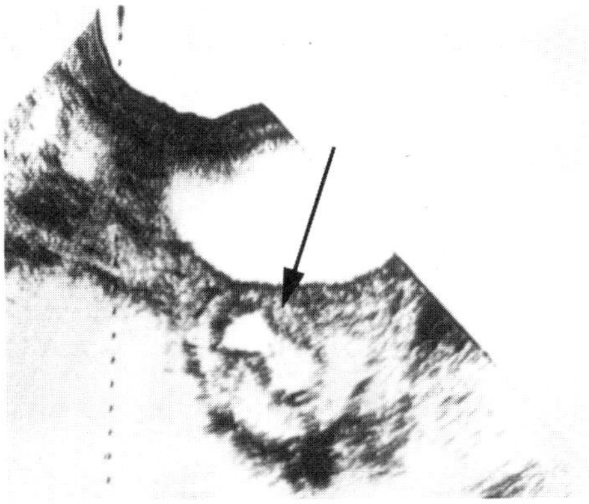

Figure 8–C. Longitudinal section 0L. Pregnancy, 9 weeks.

Figure 8–D. Transverse section –7. Pregnancy, 11 weeks.

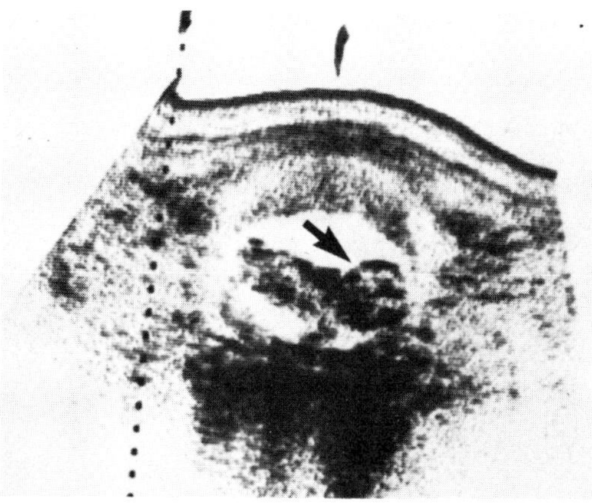

Figure 8–E. Transverse section –11. Pregnancy, 13 weeks.

REFERENCE

Ghorashi B, Gottesfeld KR: The gray scale appearance of the normal pregnancy from 4 to 16 weeks gestation. J Clin Ultra 5:195 (1977)

Case 9

CLINICAL INFORMATION

This unfortunate 32-year-old woman has adenocarcinoma of the colon which was discovered six months ago when she presented with melena and pain in the left lower quadrant of the abdomen. An "apple-core" tumor was found in the sigmoid colon on barium enema. The remainder of the colon was normal. A curative surgical procedure involving wide local excision with primary reanastomosis was attempted; local node dissection was performed, and the anastomotic site was covered with omentum and fixed to the posterior pelvis. At the time of surgery there was no evidence of metastatic tumor in the abdomen, and all the lymph nodes removed were free of tumor. The postoperative course was benign, and the patient felt well until one week ago. In the past week she has had several periodic severe abdominal cramps with increasing nausea. For the past two days she has been running a fever of 101 degrees. She now appears quite ill; and her abdomen shows a mild diffuse tenderness with no rebound pain (peritoneal irritation). There are a few scattered, weak bowel sounds. Her hematocrit is 38, and the white cell count is 14,000 with 15 per cent immature forms. She is sent now for ultrasound scanning before any x-ray studies have been obtained.

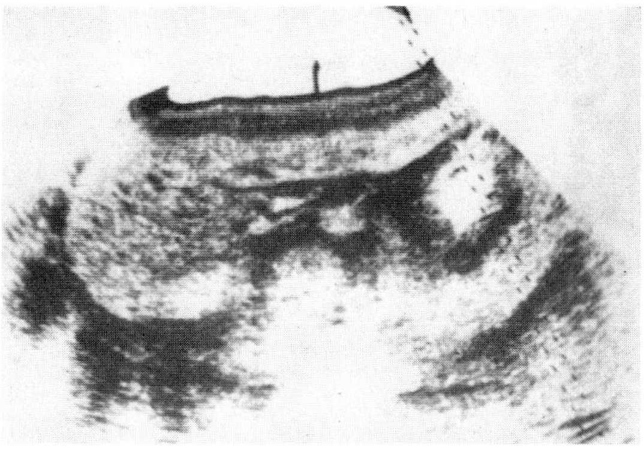

Figure 9–A. Transverse section +8.

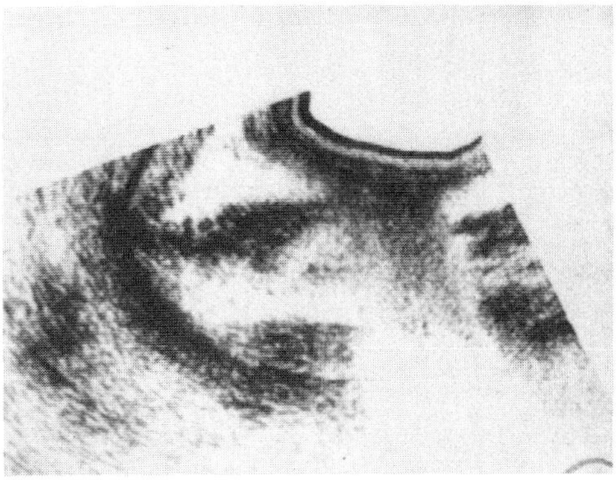

Figure 9–B. Longitudinal section 6L.

DIAGNOSIS

Fluid-filled stomach (secondary to low colonic obstruction).

ULTRASOUND FINDINGS

There is an echo-free mass in the left upper quadrant which is clearly seen on both the cross-sectional and longitudinal scans. The absence of echoes tells us that the mass is fluid-filled; but examining the boundaries, we can see that this is not a simple cyst. The contour is irregular rather than spherical, and the walls of the mass are irregular with multiple small echoes projecting into the lumen.

What are the differential diagnostic possibilities? When confronted with an abdominal mass, the first thing we must consider is whether we are looking at a normal structure, particularly the GI tract. The location of the mass in this case is just right for a fluid-filled stomach. The shape and the presence of intraluminal projections (representing mucosal folds) are also consistent with stomach. The patient says that she has not had anything to eat for 48 hours and has vomited on several occasions during that period. This information tends to be against the diagnosis of fluid-filled stomach. Could this be an intra-abdominal abscess? Perhaps; but the appearance is not typical because most abscesses do not have well-demarcated boundaries and tend to contain a few internal echoes. The lack of a palpable mass or focal tenderness in this area also argues against this diagnosis. On the other hand, the clinical setting certainly is compatible with abscess. How about ascites? We can rule this out because there is no fluid encompassing the right lobe of the liver or in the region of the porta hepatis. Maybe this is a pancreatic pseudocyst? Unlikely; serum amylase is normal, and there is nothing in the clinical setting to go along with this diagnosis.

Our differential comes down to separating a fluid-filled stomach from an intra-abdominal abscess. The simplest way to do this is to get a radiogram of the area. An upright plain film of the abdomen (Figure 9–C) confirms our suspicion of a fluid-filled stomach and also suggests bowel obstruction as the underlying problem. At surgery, a low small bowel obstruction was found secondary to herniation beneath the omentum; there was no evidence of recurrent tumor.

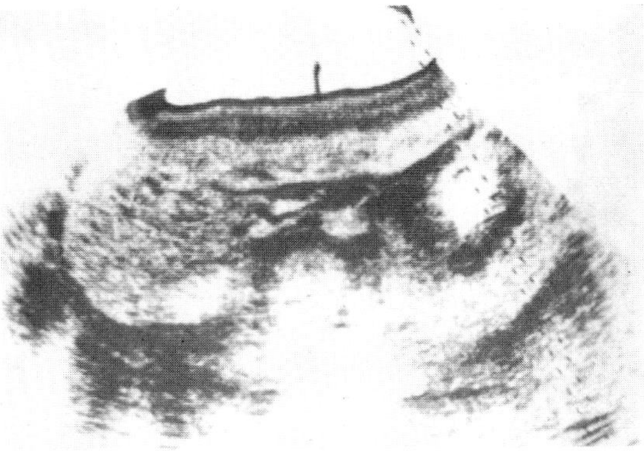

Figure 9–A. Transverse section +8. Fluid-filled stomach.

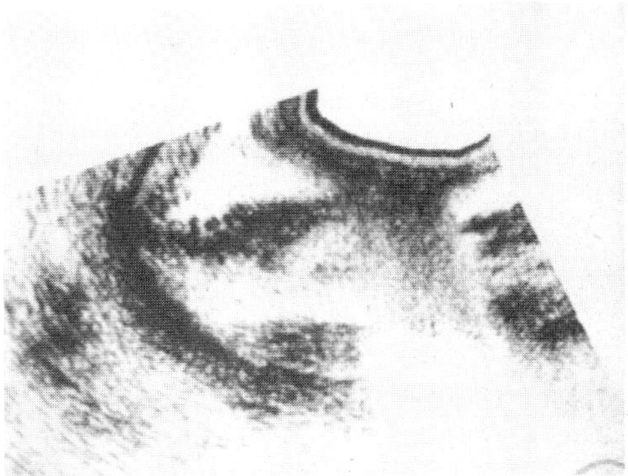

Figure 9–B. Longitudinal section 6L. Fluid-filled stomach.

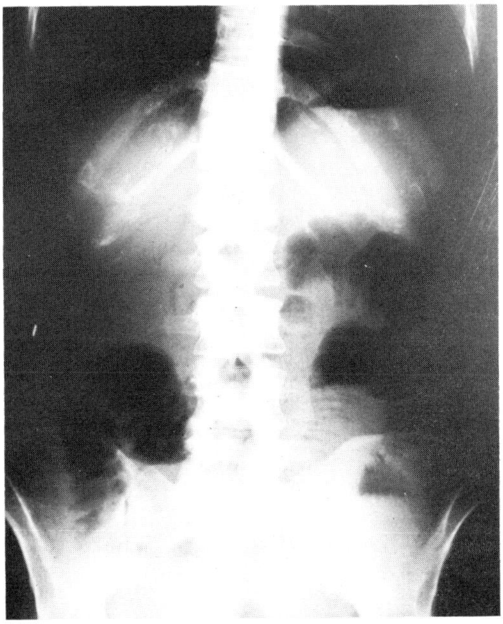

Figure 9–C. Upright abdominal film of patient in Figures 9–A and 9–B.

Case 10

CLINICAL INFORMATION

This 27-year-old woman complains of dyspareunia. The pelvic examination is normal. Figure 10–A is a longitudinal scan in the midsagittal plane; Figures 10–B, 10–C, and 10–D are cross-sections through the pelvis.

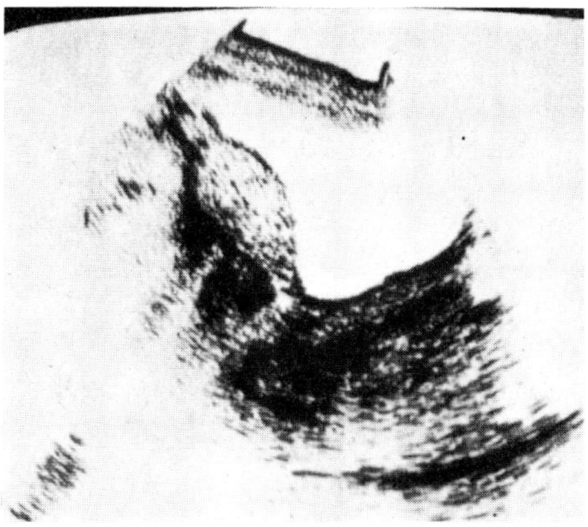

Figure 10–A. Longitudinal section 0L.

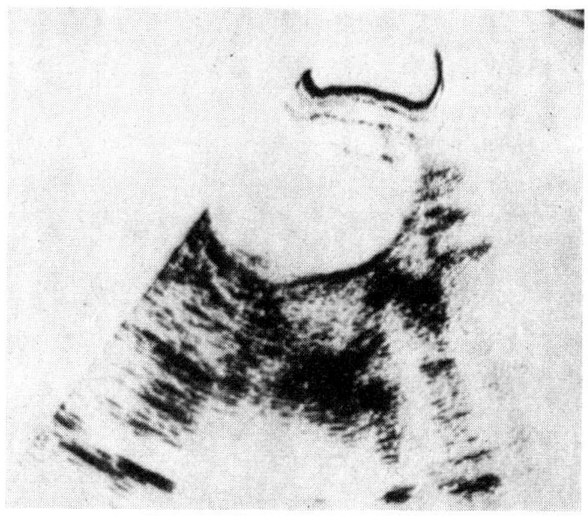

Figure 10–B. Transverse section –9.

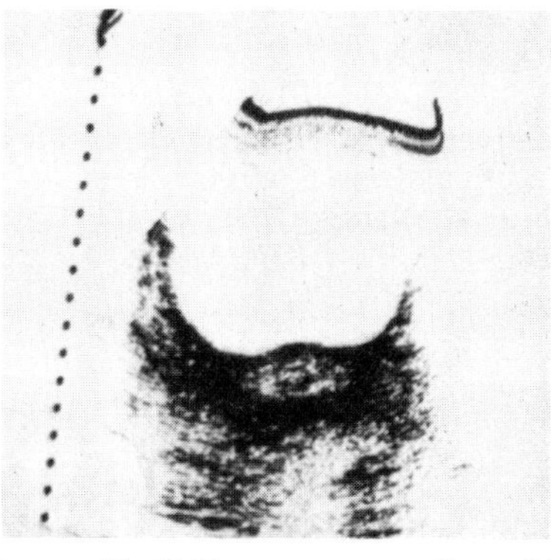

Figure 10–C. Transverse section –10.

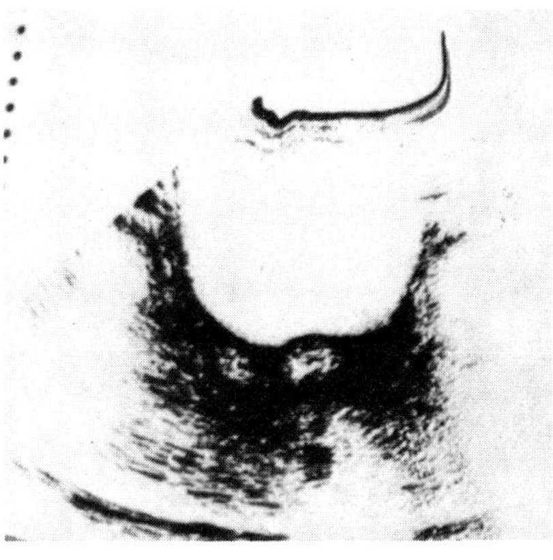

Figure 10–D. Transverse section –12.

DIAGNOSIS

Normal pelvis.

ULTRASOUND FINDINGS

The normal anatomy of the female pelvis is nicely seen in this case. The uterus is best seen on longitudinal scans (Figure 10–A), where it can easily be found beneath the bladder. The myometrium generates uniform low-level echoes: the endometrial surface may produce slightly louder echoes in the center of the uterus, but the difference in intensity should be minor. If there are prominent endometrial echoes we should inquire about an IUD or recent endometrial insult such as a D & C. The ovaries are best seen on a transverse cross-section. If serial scans are performed starting at the fundus of the uterus and working toward the cervix, the ovaries can usually be spotted as two small masses lying just underneath the bladder and lateral to the uterus (arrows in Figures 10–B and 10–D). Sometimes the ovaries of normal size are difficult to find; we don't worry about this — when a pathologic condition is present they stand out readily.

There is only one trick to examining the pelvis. *The bladder must be full.* The bladder is your acoustic window into this area; if you don't insist on having it full, you are wasting your time.

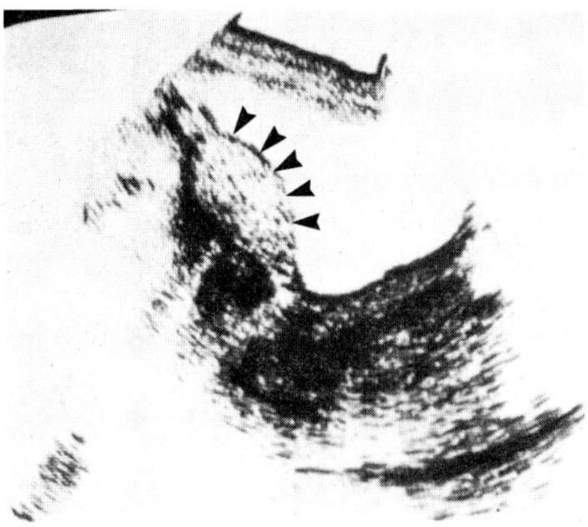

Figure 10–A. Longitudinal section 0L. Normal female pelvis.

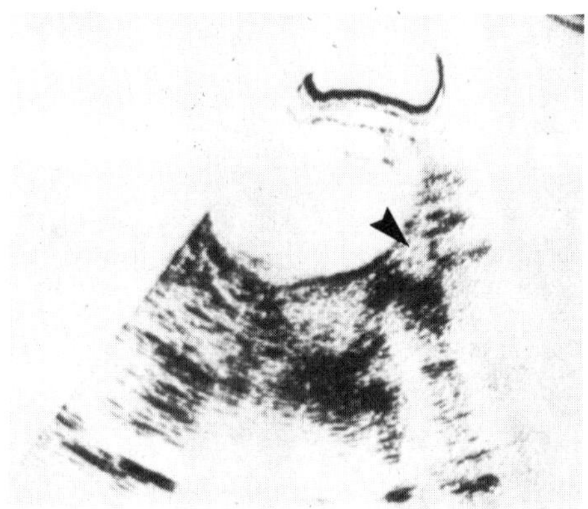

Figure 10–B. Transverse section –9. Normal female pelvis.

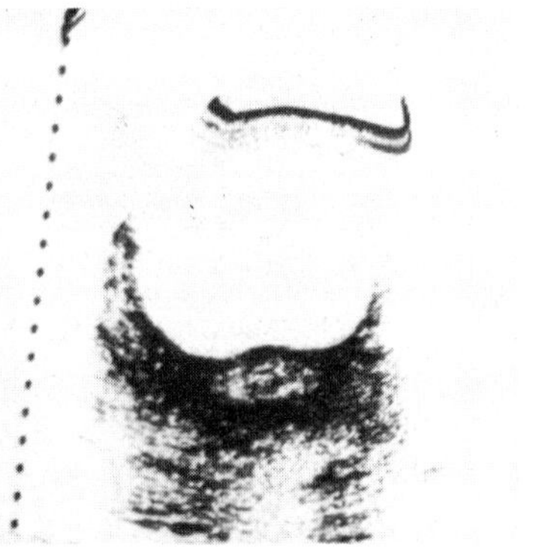

Figure 10–C. Transverse section –10. Normal female pelvis.

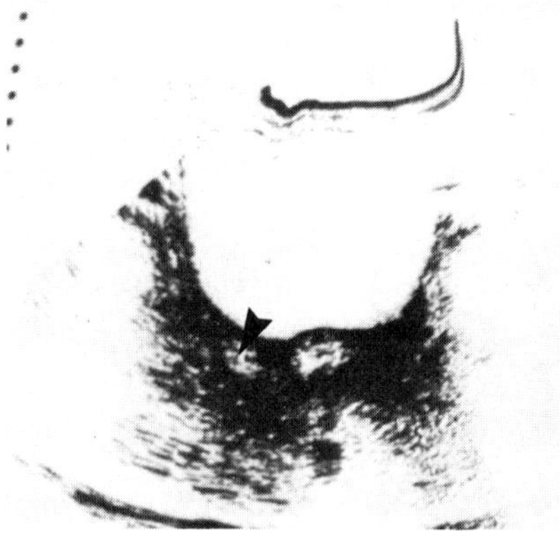

Figure 10–D. Transverse section –12. Normal female pelvis.

REFERENCES

Sample WF, Lippe BM, Gyepes MT: Gray scale ultrasonography of the normal female pelvis. Radiology 125:477 (1977)

Rubin C, Kurtz AB, Goldberg BB: Water enema: a new ultrasound technique in defining pelvic anatomy. J Clin Ultra 6:28 (1978)

Miller EI, Thomas RH, Lines P: The atrophic postmenopausal uterus. J Clin Ultra 5:261 (1977)

Case 11

CLINICAL INFORMATION

These three scans were all obtained in the same plane (3R), in the same patient, at the same time. The only difference among them is that they were made using different frequency transducers: a 1.6 MHz, a 2.25 MHz, and a 3.5 MHz. The technician carefully adjusted the gain and the TCG to get the "best" images possible. Can you tell which scans were performed with which transducer? Can you explain why the pictures look different? Which transducer do you prefer? (There is no pathologic condition in this case.)

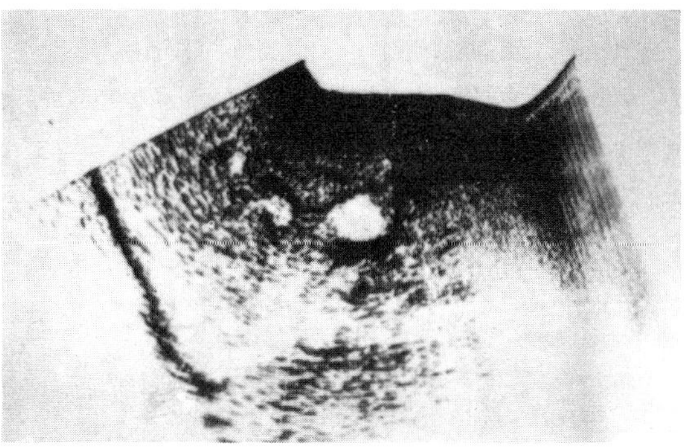

Figure 11–A. Longitudinal section 3R.

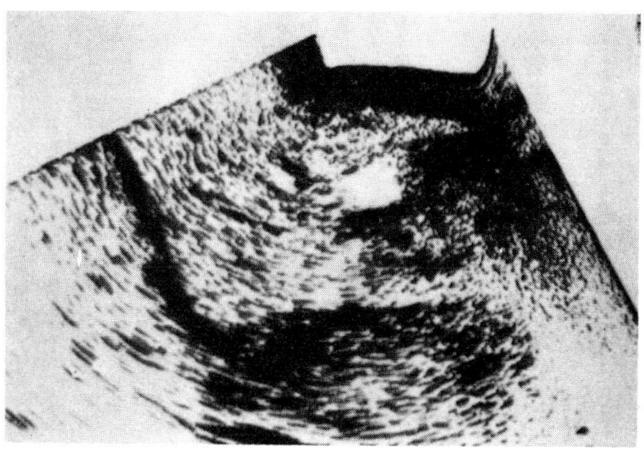

Figure 11–B. Longitudinal section 3R.

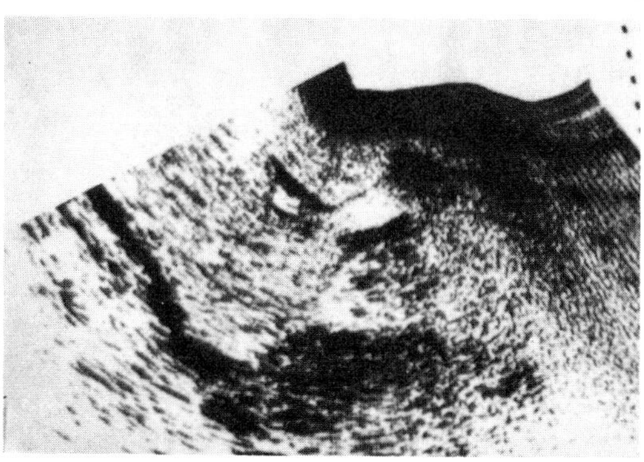

Figure 11–C. Longitudinal section 3R.

DIAGNOSIS

Normal liver and gallbladder. Figure 11–A is the 3.5 MHz, Figure 11–B is the 1.6 MHz, and Figure 11–C is the 2.25 MHz.

ULTRASOUND FINDINGS

Although subject to occasional exceptions, the following rule of thumb applies to the great majority of abdominal ultrasound transducers: As the frequency of the transducer increases, its resolution will increase and its penetrating power (ability to visualize echoes far from the transducer face) will decrease. The improvement in resolution is due to the fact that higher-frequency transducers can be more precisely focused than those of lower frequency. The loss of penetrating power is a function of increased attenuation of higher frequencies by soft tissues. We can compensate for this attenuation to a certain degree by increasing the gain and the TCG in the scanner amplifier; but eventually there is a problem of increasing noise and artifact production as the gain is increased (Figure 11–A). It is much more important to have an image that is free of noise and artifacts than it is to have slightly better resolution; therefore, we use high-frequency transducers only in small patients in whom penetration is not a problem (or when we are examining a very small superficial structure such as the thyroid, testicle, or popliteal space).

Newer scanners have made tremendous improvement in decreasing noise at high gain compared to the first generation units — many ultrasonographers have found that they can use a frequency one level higher with no loss in image quality with the second and third generation machines. Figures 11–A, 11–B, and 11–C were made on a first generation scanner; Figures 11–D, 11–E, and 11–F illustrate that the differences in frequency are not as pronounced with a later model machine.

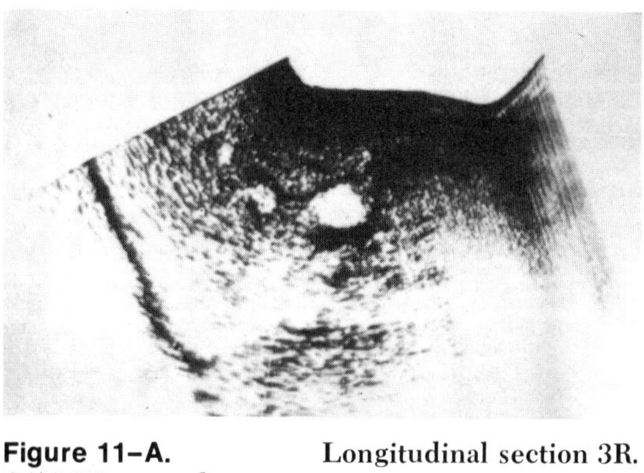

Figure 11–A. Longitudinal section 3R.
3.5 MHz transducer.

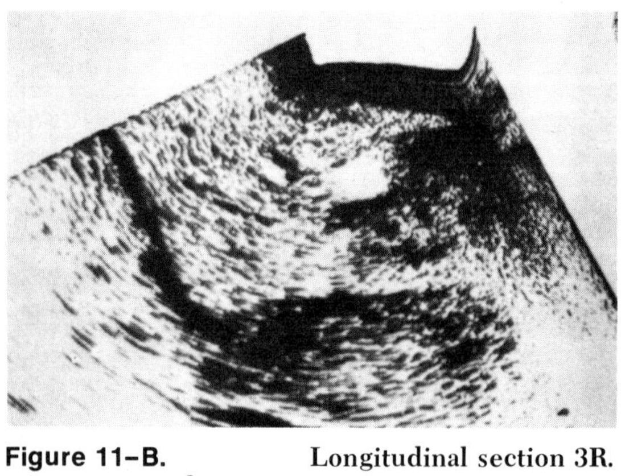

Figure 11–B. Longitudinal section 3R.
1.6 MHz transducer.

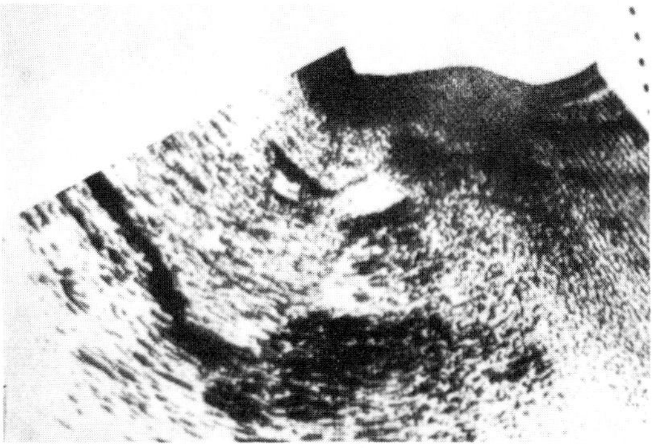

Figure 11–C. Longitudinal section 3R. 2.25 MHz transducer.

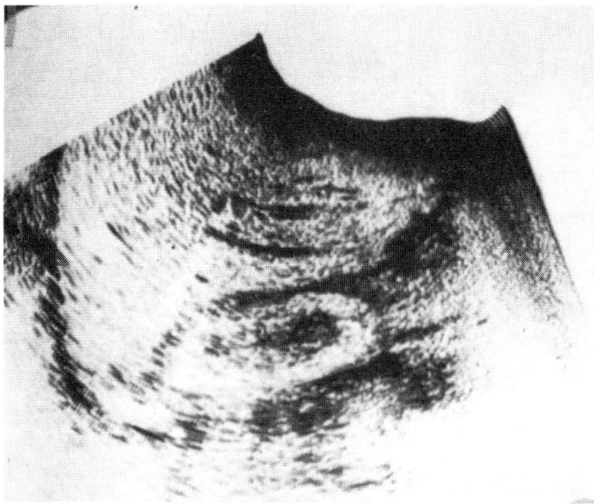

Figure 11–D. Longitudinal section 6R. 3.5 MHz transducer.

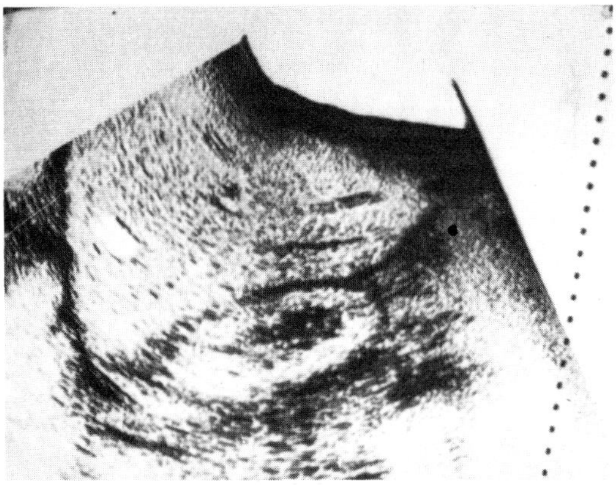

Figure 11–E. Longitudinal section 6R. 1.6 MHz transducer.

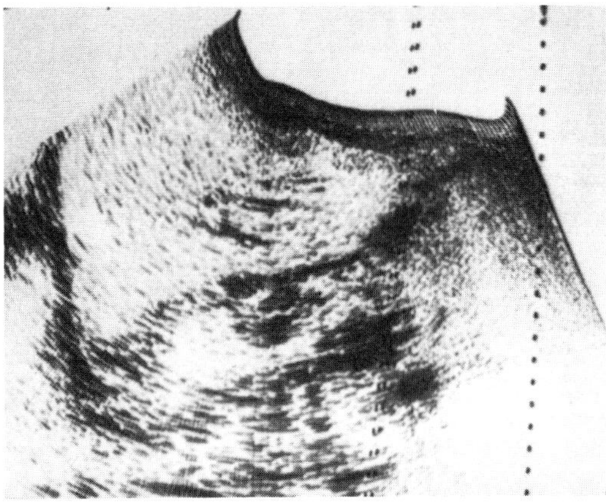

Figure 11–F. Longitudinal section 6R. 2.25 MHz transducer.

REFERENCE

Bartrum RJ JR, Crow HC: *Gray-Scale Ultrasound: A Manual for Physicians and Technical Personnel.* Chapter 2. WB Saunders Co, Philadelphia (1977)

Case 12

CLINICAL INFORMATION

This 36-year-old woman was sent to an endocrinologist because her general practitioner had run out of ideas in his efforts to relieve her complaint of lassitude and fatigue. He had been unable to document any signs of thyroid deficiency. Because she felt so strongly that there was something wrong with her "glands," however, he referred her to a "specialist." The endocrinologist was equally frustrated, but he did feel an unusual firmness in the right lobe of the thyroid and felt that this part of the gland was a bit enlarged. The radioisotope scan was also unsettling; there was an area of diminished uptake on the right which the radiologist interpreted as a "cold nodule." What about the ultrasound findings; do they help?

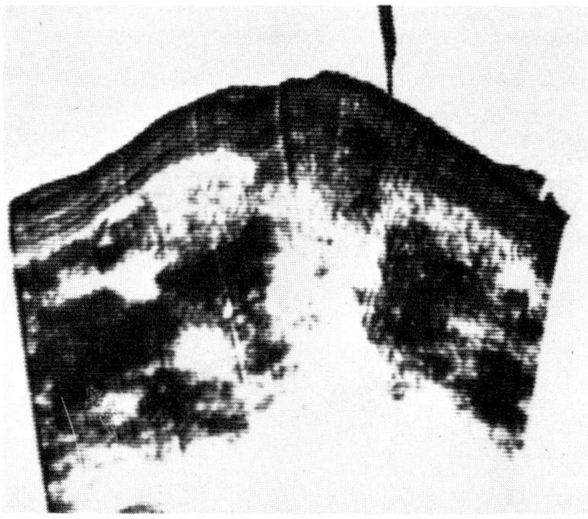

Figure 12–A. Transverse section +5.

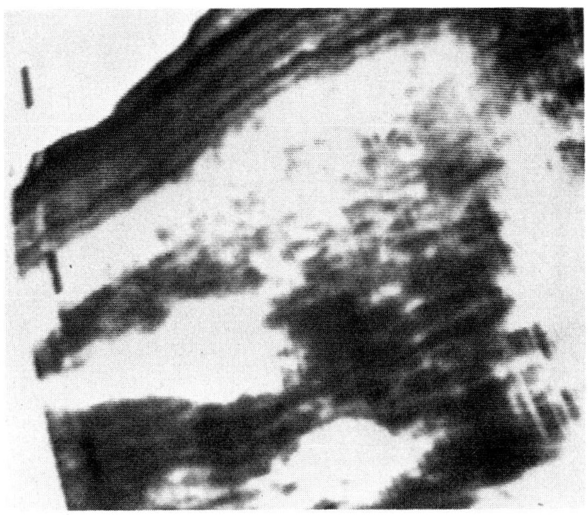

Figure 12–B. Transverse section +5. (Same section as Figure 12–A scanned on an expanded scale.)

DIAGNOSIS

Follicular carcinoma of the thyroid.

ULTRASOUND FINDINGS

There is certainly an abnormal area in the anterior portion of the right lobe of this thyroid; this part of the gland is also enlarged a bit. Even with the magnified view it is not possible to say whether or not this area is made up of many small cysts or very homogeneous solid tissue. Pathologically, this proved to be malignancy. We do not intend to imply that the histologic diagnosis can be made reliably by ultrasound; we can say, however, that the abnormality is easily recognized as different from the normal gland, and in this case that was enough to prompt the surgery. Remember, we are looking at texture with ultrasound, and subtle variations can be perceived readily. The problem in the thyroid is that both benign and malignant processes alter the texture and cannot be differentiated from each other.

Perhaps we should say a word about technique. We use a 5 MHz transducer, 6 mm. in diameter, internally focussed at 5 cm. We also use direct contact rather than a water bath, whenever possible. Scanning takes a steady hand and practice, but with care the pictures are quite satisfactory. We display the image on the 1:1 scale, and this is the *only* instance when we use the zoom feature of our scan converter. Although the combination of the high frequency and 1:1 scale makes an elegant picture, we have to resist the temptation to read more into the scan than is there. We can show only that we are dealing with a solid lesion which *could* be cancer; it could just as well be a benign process such as adenomatous hyperplasia or a small area of nontoxic nodular goiter.

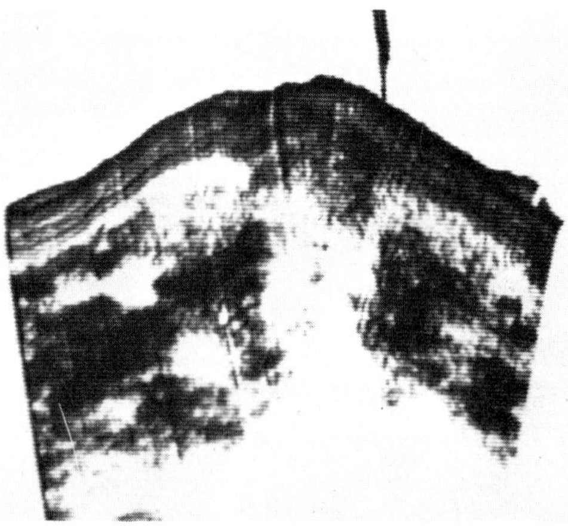

Figure 12–A. Transverse section + 5. Follicular carcinoma.

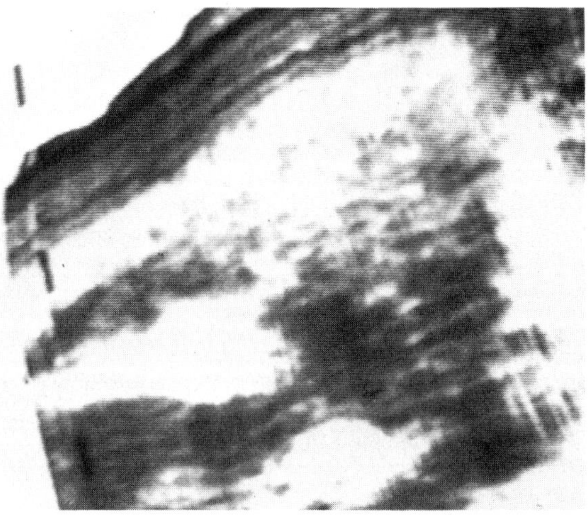

Figure 12–B. Follicular carcinoma. Magnified view of Figure 12–A.

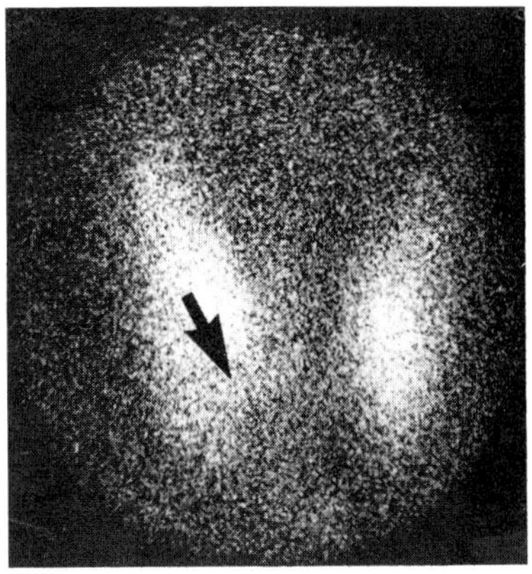

Figure 12–C.
Follicular carcinoma. Tc scan of thyroid. Area of decreased activity in lower portion of right lobe corresponds to mass in Figures 12–A and 12–B.

REFERENCES

Sackler JP, Passalaqua AM, Blum M, et al: A spectrum of diseases of the thyroid gland as imaged by gray scale water bath sonography. Radiology 125:467 (1977)

Blum M, Goldman AB, Herskovic A, et al: Clinical applications of thyroid echography. N Engl J Med 287:1164 (1972)

Case 13

CLINICAL INFORMATION

This 48-year-old woman has had persistent abdominal pain for 11 weeks. Her general health is "terrible." She has been hospitalized on several previous occasions for depression, peptic ulcer disease, and acute alcohol detoxification. Ten weeks ago she was in the hospital for 8 days with acute pancreatitis. Ever since her discharge she has had abdominal pain and her serum amylase continues to be marginally elevated.

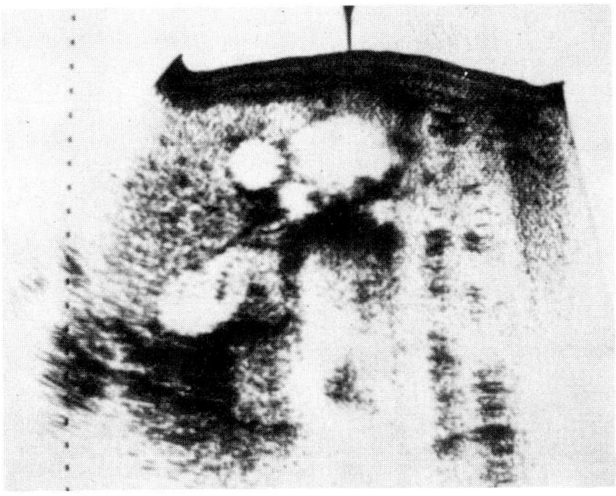

Figure 13–A. Transverse section +8.5.

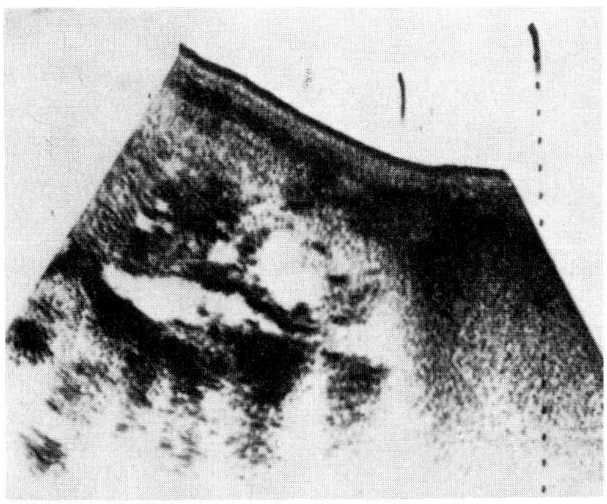

Figure 13–B. Longitudinal section 2.5R.

DIAGNOSIS

Pancreatic pseudocyst.

ULTRASOUND FINDINGS

If you missed this diagnosis go directly to jail, do not pass GO, and do not collect your ultrasound fee. The clinical setting is almost perfect for a pseudocyst, and the scans show a large echo-free mass where we would normally expect to see the body of the pancreas. On the transverse section there are small nodules of pancreatic tissue on the back wall of the cyst, and there is a small daughter cyst adjacent to the main cavity. (The gallbladder is just to the patient's right of the main cyst and somewhat anterior to the daughter cyst.) This is the classic appearance of a pseudocyst; and in this clinical setting there is really no other diagnosis to consider.

Figures 13–C and 13–D are an ultrasound and a CT scan from a different patient, again with a typical pseudocyst. The loud echo on the back wall of this cyst (arrow) is caused by pancreatic calcification.

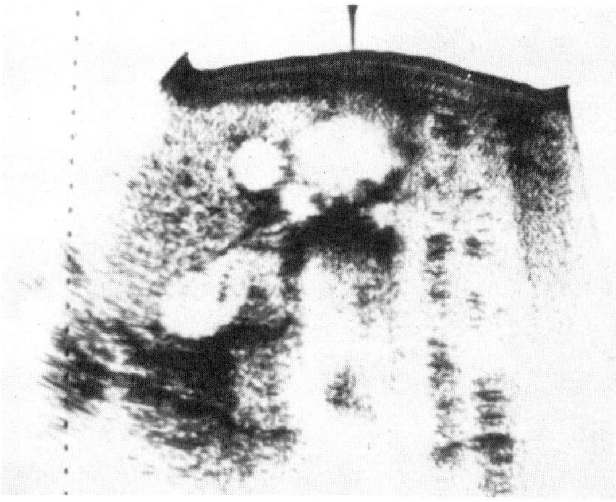

Figure 13–A. Transverse section +8.5. Pancreatic pseudocyst.

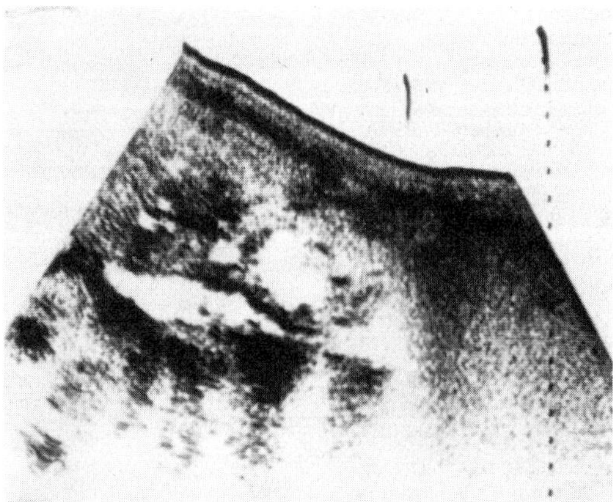

Figure 13–B. Longitudinal section 2.5R. Pancreatic pseudocyst.

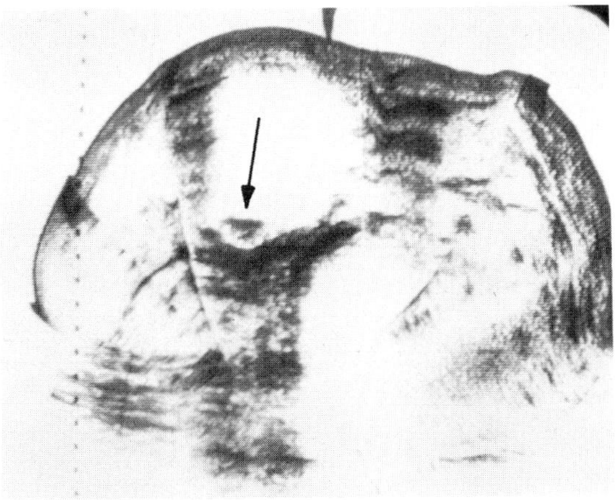

Figure 13–C. Transverse section +5.5. Pancreatic pseudocyst.

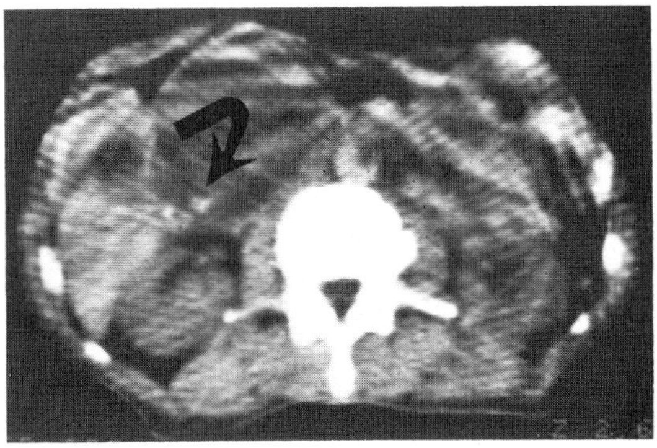

Figure 13–D. Transverse section +5.5. Pancreatic pseudocyst. CT scan of patient in Figure 13–C.

REFERENCES

Sarti DA: Rapid development and spontaneous regression of pancreatic pseudocysts documented by ultrasound. Radiology 125:789 (1977)

Van Hurden JA, ReMine WH: Pseudocysts of the pancreas: review of 71 cases. Arch Surg 110:500 (1975)

Case 14

CLINICAL INFORMATION

This is the first pregnancy for this 17-year-old girl, who has had no prenatal care. Her periods have always been regular, however, and she knows that they stopped nine months ago. She assumes she is ready to have her baby and has come to the hospital for delivery. She is not in labor and on physical examination her uterus is consistent with only a 30-week gestation. No fetal heart tones can be heard. She is sent to ultrasound for accurate dating of the pregnancy.

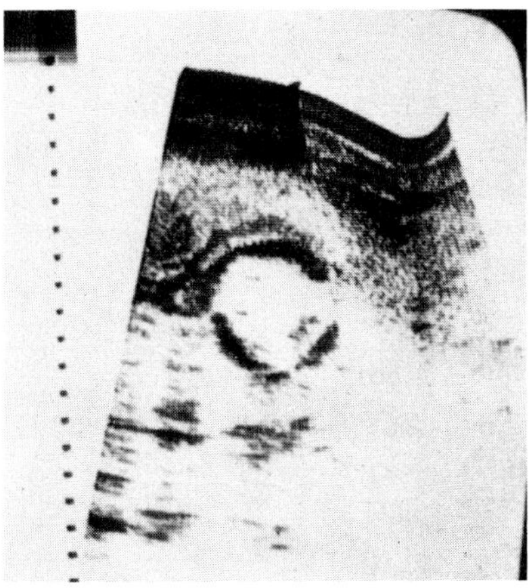

Figure 14–A.
Fetal skull—gray scale.

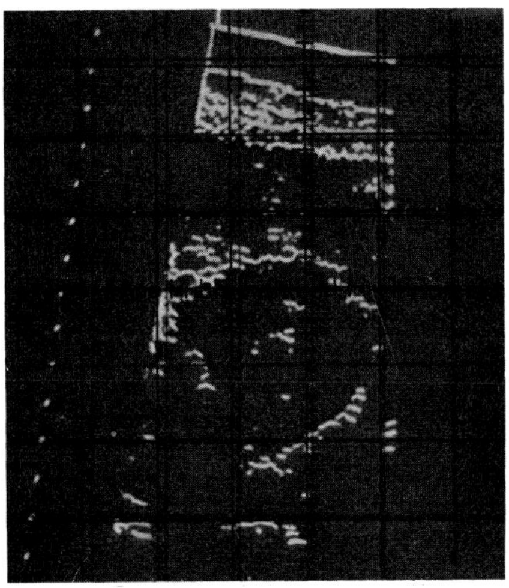

Figure 14–B.
Fetal skull—bi-stable.

DIAGNOSIS

Fetal demise.

ULTRASOUND FINDINGS

Figures 14–A and 14–B are gray-scale and bi-stable scans through the fetal head. The skull is distorted and incomplete, and there are areas of flattening and loss of the calvarial echo. The gray-scale picture suggests that there may be some overlapping of the bones. These are all signs of fetal demise and are the same as the findings on conventional radiograms in this condition. These are late changes and indicate that the fetus has been dead for at least two weeks. There are much earlier ultrasonic findings in fetal death, but they are difficult to demonstrate in static scans. From the fifteenth week on, the fetal heart should be visible by ultrasound; if it is not moving, we diagnose fetal death. With a real-time scanner we can observe fetal movements as well as see the fetal heart beat. In most institutions, however, doppler monitoring of the fetal heart (Doptone) is still the easiest method of screening for this problem.

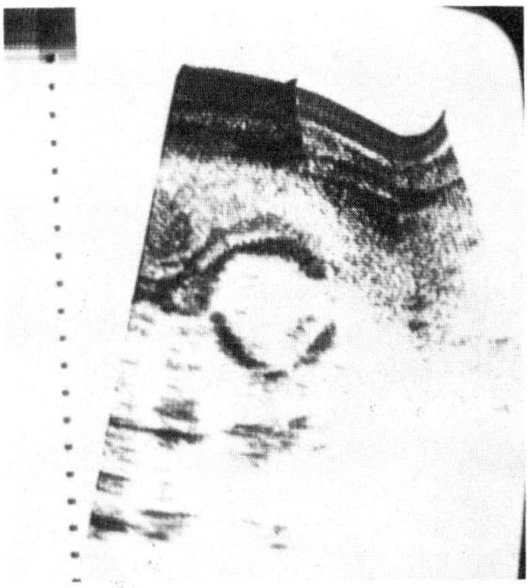

Figure 14–A.
Fetal demise. Fetal skull—gray scale.

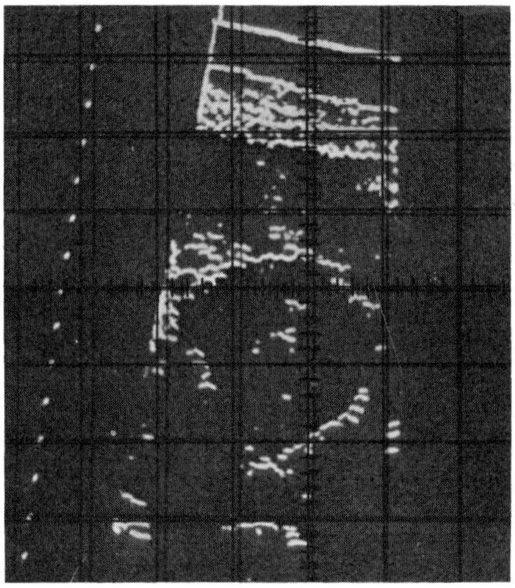

Figure 14–B.
Fetal demise. Fetal skull—bi-stable.

REFERENCES

Scheer K, Nubar JC: Rapid conclusive diagnosis of intrauterine fetal death. Am J Obstet Gynecol 128:907 (1977)

Gottesfeld KR: The ultrasonic diagnosis of intrauterine fetal death. Am J Obstet Gynecol 108:623 (1970)

Case 15

CLINICAL INFORMATION

This 84-year-old woman had an episode of midepigastric pain with nausea and vomiting three days ago. She has now recovered somewhat, but her appetite has not returned, and her entire abdomen is slightly tender. She has mild chronic congestive heart failure, well controlled with Digoxin and salt restriction. The physical examination is normal except for the presence of a prominent, non-tender, pulsatile mass in the midabdomen. She is unaware of the mass and it is not mentioned in the report of her most recent physical examination, performed 18 months ago.

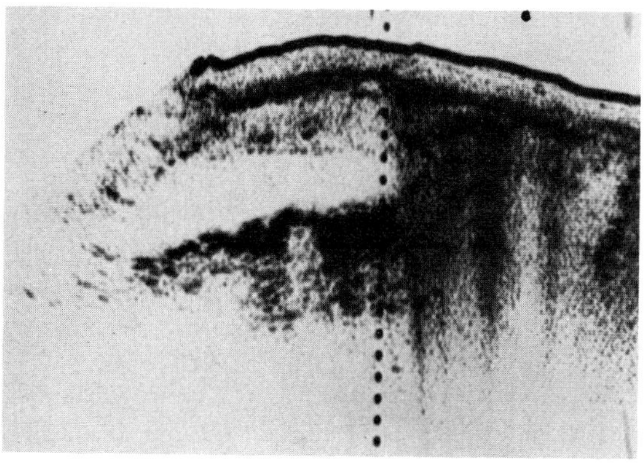

Figure 15–A. Longitudinal section 0L.

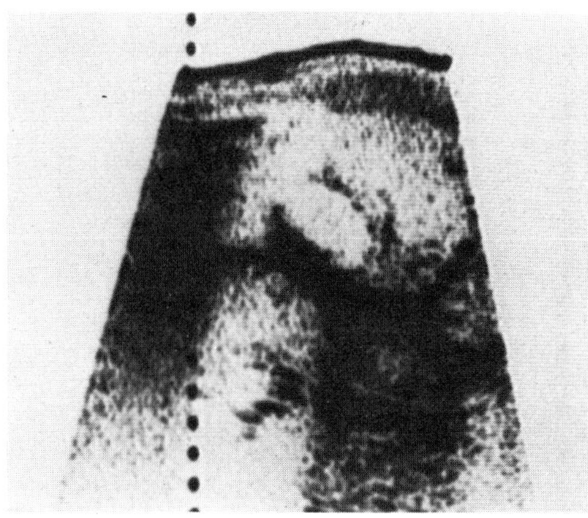

Figure 15–B. Transverse section +2.

DIAGNOSIS

Fusiform aortic aneurysm.

ULTRASOUND FINDINGS

The longitudinal and cross-section scans of the abdominal aorta (Figures 15–A and 15–B) show a fusiform dilatation characteristic of an aneurysm. The lower extent of the aneurysm is obscured by overlying bowel gas in this case; Figure 15–C shows a similar case in which the entire aneurysm was visualized. The echo-free posterior area represents the true lumen, while echoes seen anterior in the vessel are arising from clotted blood. We do not think it is possible (or important?) to tell whether the clot is within the lumen of the aneurysm or is in the wall — a so-called "false channel." Similarly, we feel it is very difficult in most cases to define the relationship of the aneurysm to the renal arteries, and we believe that every patient in whom surgery is contemplated should have an aortogram to answer these questions.

The ultrasonographer can measure the aneurysm quite accurately and this should be done carefully on both longitudinal and transverse sections. We report the greatest diameter from outside wall to outside wall and try to refrain from adding editorial comments such as "large," "significant," "dangerous," and so on. The decision to operate on such aneurysms is a difficult and controversial matter among vascular surgeons, and we suggest that the ultrasonographer confine himself to reporting only objective data such as greatest diameter and change in size or appearance since the most recent previous examination.

Figure 15–D is a longitudinal scan of another fusiform abdominal aortic aneurysm; in this case there is no clot.

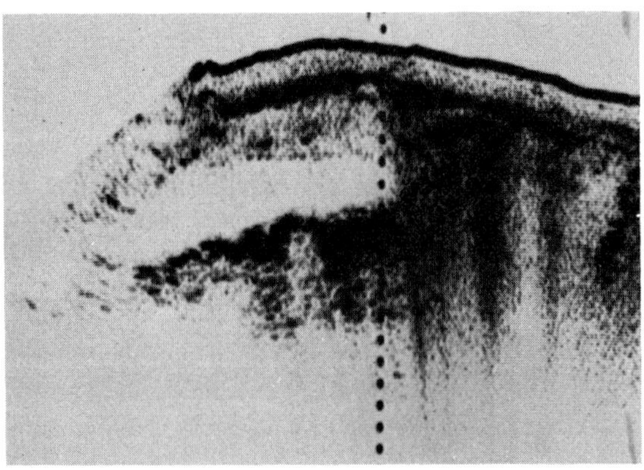

Figure 15–A. **Longitudinal section 0L.**
Aortic aneurysm.

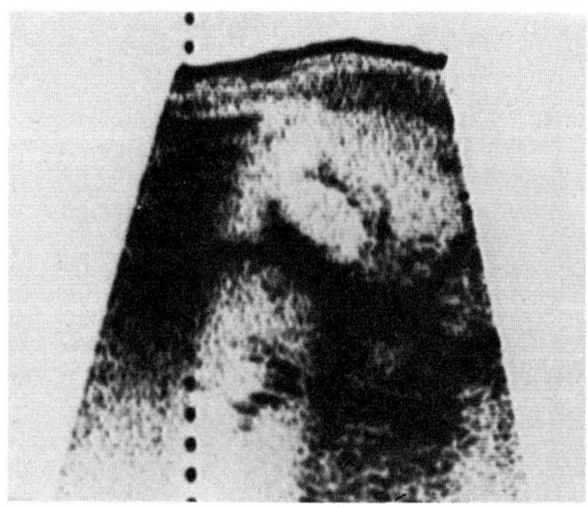

Figure 15–B. **Transverse section +2.**
Aortic aneurysm.

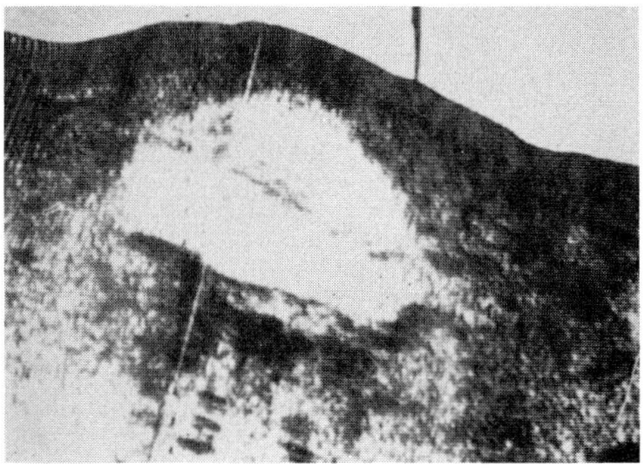

Figure 15–C. Longitudinal section 1L. Aortic aneurysm.

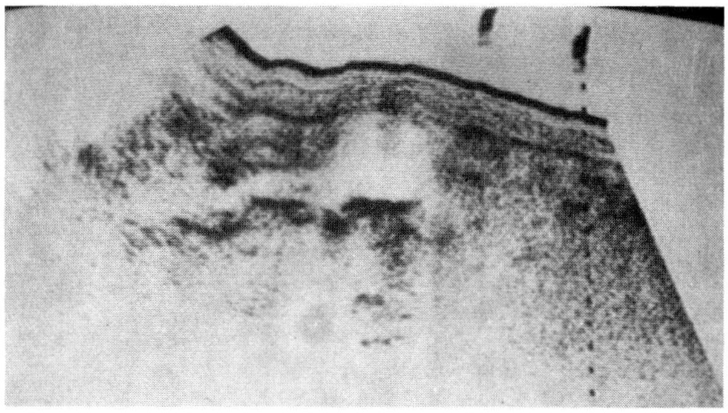

Figure 15–D. Longitudinal section 2L. Aortic aneurysm.

REFERENCES

Filly RA, Goldberg BB: Abnormal Vessels, Chapter 4 in *Abdominal Gray Scale Ultrasonography,* BB Goldberg, editor. John Wiley & Sons, New York (1977)

Leopold G, Goldberger L, Bernstein E: Ultrasonic detection and evaluation of abdominal aortic aneurysms. Surgery 72:939 (1972)

Case 16

CLINICAL INFORMATION

This 73-year-old man has been having difficulty urinating. He reports increasing frequency and says that sometimes he has to get up as many as four times during the night. He has difficulty starting urination, and the caliber of the stream is diminished. His doctor suspects prostatic enlargement but is concerned because the urinalysis shows microscopic hematuria. He obtained an IVP which showed, in addition to the large prostate, bilateral attenuation of the infundibula of the collecting systems. This ultrasound scan is of the left kidney. The right kidney has a similar ultrasonic appearance.

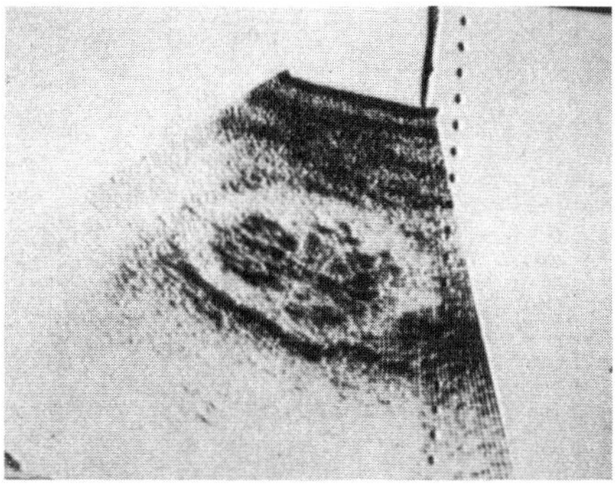

Figure 16–A. Longitudinal section 0L. Prone scan of left kidney.

DIAGNOSIS

Renal fibrolipomatosis (renal sinus lipomatosis).

ULTRASOUND FINDINGS

The renal contour and cortical echoes are normal. The striking feature of this scan is the large collection of high-level echoes in the central portion of the kidney. There are many more echoes here than are present in the normal renal sinus. The echoes are all of high intensity, so we know that we are not dealing with multiple parapelvic cysts or hydronephrosis. By the same token the echo pattern is not that of normal renal cortex, and for this reason we can exclude the presence of a column of Bertin or a solid renal tumor. Echoes of this intensity are usually due to fat or dense fibrous tissue. The combination of this hilar echo pattern, especially if bilateral, and the attenuation of the infundibula on the IVP is almost pathognomonic for renal fibrolipomatosis.

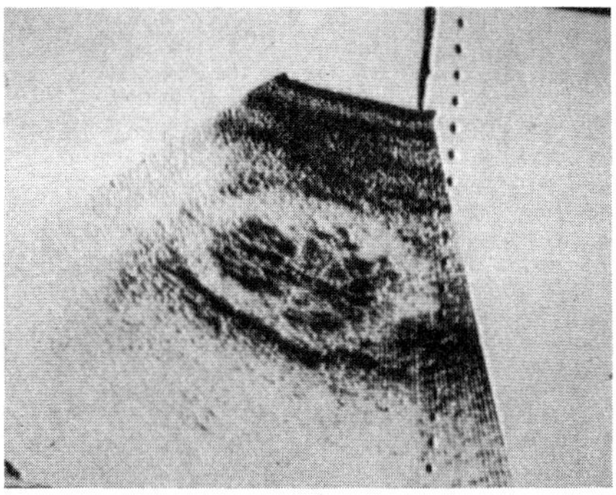

Figure 16–A. Longitudinal section 0L. Renal fibrolipomatosis. Prone scan of left kidney.

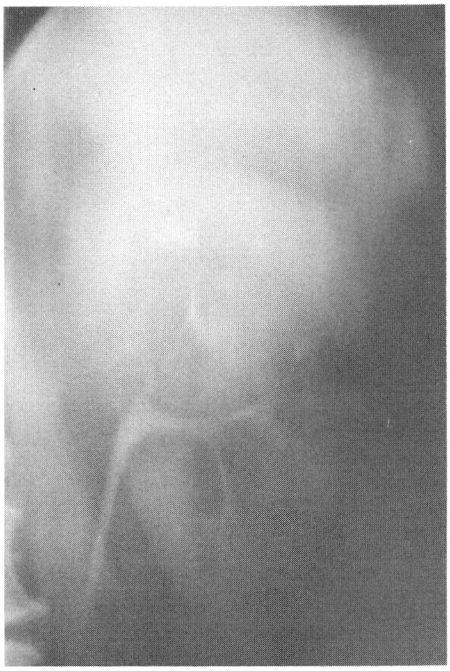

Figure 16–B. Renal fibrolipomatosis. Nephrotomogram from patient in Figure 16–A.

REFERENCES

Becker JA, Schneider M, Staiano S, et al: Renal pelvic lipomatosis: a sonographic evaluation. J Clin Ultra 2:299 (1974)

Yeh HC, Mitty HA, Wolf BS: Ultrasonography of renal sinus lipomatosis. Radiology 124:799 (1977)

Case 17

CLINICAL INFORMATION

This 72-year-old man with chronic leukemia has been doing fairly well for eight years despite the presence of slowly increasing hepatosplenomegaly and two brief episodes of pneumonia. Two days ago he had some sharp pains in his upper abdomen over a period of four or five hours and has now developed a low-grade fever (100.2) and nausea.

The technician is out sick today and the first-year resident has been recruited to do the ultrasound scanning. He brings this case to you and says that he believes that, in addition to the enlargement of the liver and spleen, there may be focal areas of either necrosis or infection because the texture of both organs is very heterogeneous. What do you think?

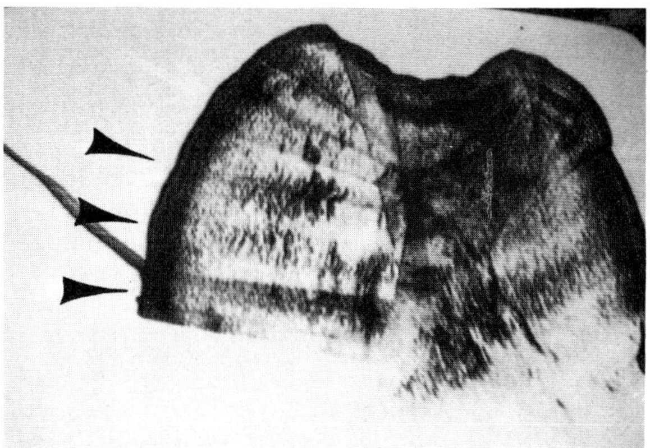

Figure 17–A. Transverse section +8.

DIAGNOSIS

Technically inadequate scan with multiple rib artifacts.

ULTRASOUND FINDINGS

The resident has fallen into the most common trap for beginning ultrasonographers: failing to recognize reverberation artifacts and interpreting them as real echoes. Whenever the sound beam passes over the strongly reflective interfaces caused by air or bone, reverberation of sound occurs and a shower of artifactual echoes is projected into the underlying tissue. In this case the transducer was moved completely across the abdomen; and every time a rib was crossed, a band of artifacts was projected into the underlying liver and spleen (arrowheads). If the resident had had more scanning experience, he would have known that the sound beam must never knowingly be passed through bone or air because this inevitably leads to excessive artifact production. Limited sector scans between the ribs are what is needed in this case if we are to evaluate the parenchymal texture of the liver or the spleen. We had the resident sit down and read the reference cited below; his scanning technique has improved considerably.

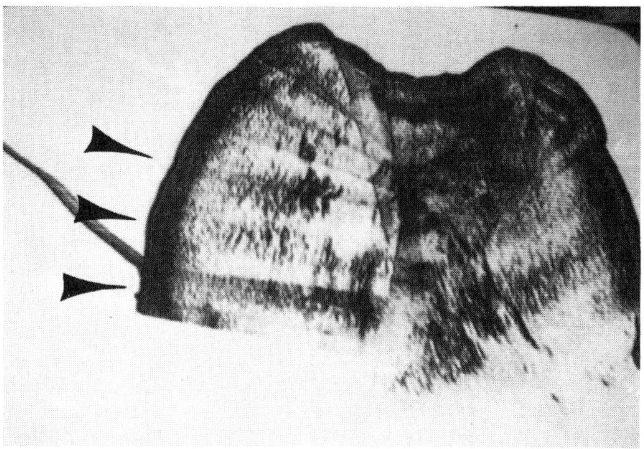

Figure 17–A. Transverse section +8.
Rib artifacts.

REFERENCE

Bartrum RJ Jr, Crow HC: *Gray-Scale Ultrasound: A Manual for Physicians and Technical Personnel.* Chapters 5 and 6. WB Saunders Co, Philadelphia (1977)

Case 18

CLINICAL INFORMATION

This 27-year-old woman cannot locate the string on her IUD. Where is it? Has she passed the IUD without knowing it? Is the IUD still in place and has only the string fallen off? Has the device perforated the uterus? She is sent to ultrasound for the answers.

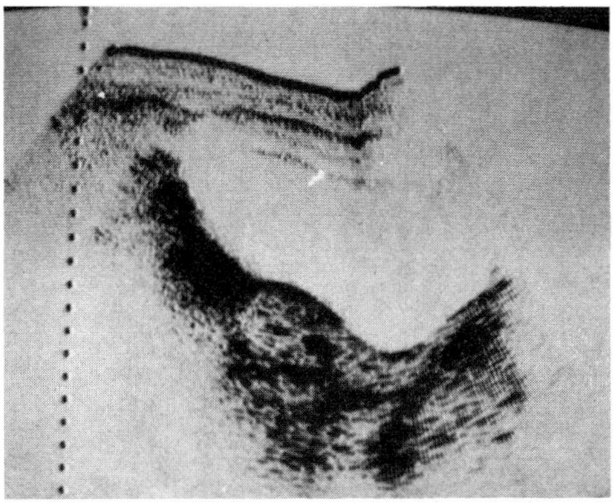

Figure 18–A. Longitudinal section 0L.

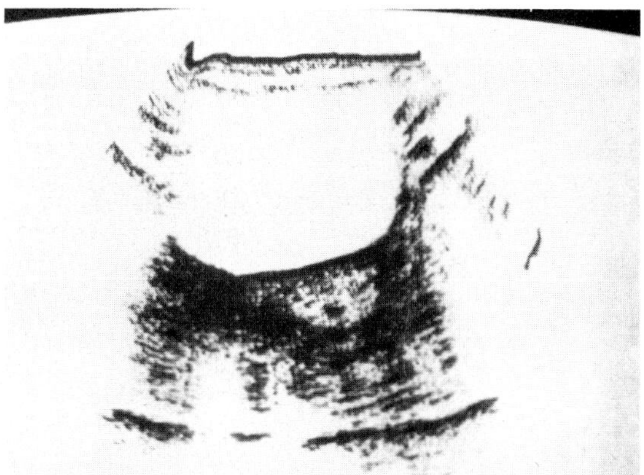

Figure 18–B. Transverse section –9.

DIAGNOSIS

Intrauterine IUD.

ULTRASOUND FINDINGS

There is a row of high-intensity echoes emanating from the center of the uterus. This is the characteristic appearance of an IUD, regardless of the type. The larger devices (such as the loops and Dalkon shield) give more prominent echoes than the smaller copper ones, but any intrauterine device will be detectable on the ultrasound scan. In the empty uterus it is usually not possible to identify the endometrial cavity and the entire organ is a uniform shade of gray. When an IUD is properly positioned, the high intensity echoes shown in these scans will appear.

If it is known that the patient is not pregnant, we find it more convenient to take a plain film of the pelvis first to see if the IUD can be located. An ultrasound scan is done only if there is a question of whether or not the device is intrauterine.

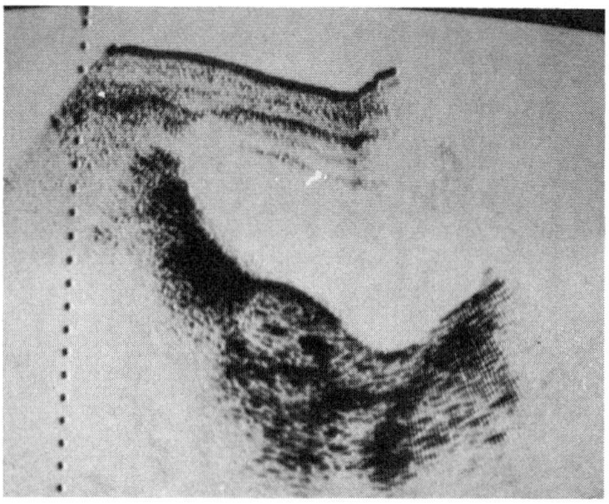

Figure 18–A. **Longitudinal section 0L.** IUD (normal).

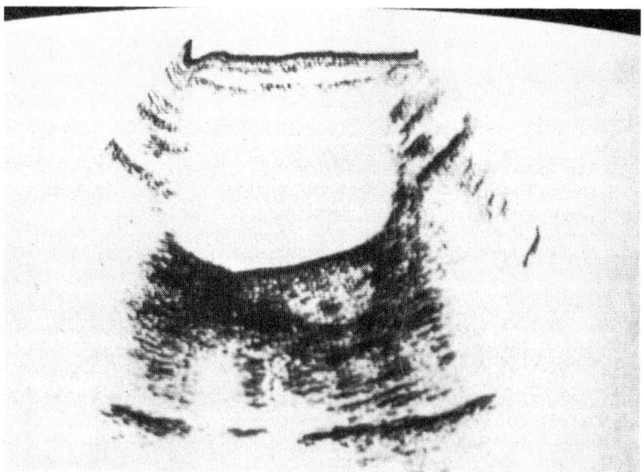

Figure 18–B. **Transverse section −9.** IUD (normal).

REFERENCES

Watt I, Watt E, Halliwell M, et al: Sonographic demonstration of intrauterine contraceptive devices. J Clin Ultra 5:378 (1977)

Wittman BK, Chow TT: Diagnostic ultrasound in the management of patients using intrauterine contraceptive devices. Br J Obst Gynecol 83:802 (1976)

Case 19

CLINICAL INFORMATION

The patient is a 43-year-old woman who is suffering from "indigestion and cramps." The pain is located in her midabdomen and is made worse by eating, particularly fatty food. On two occasions the cramps have caused her to vomit and this has brought relief. Her physical examination is unremarkable except for the presence of moderate obesity.

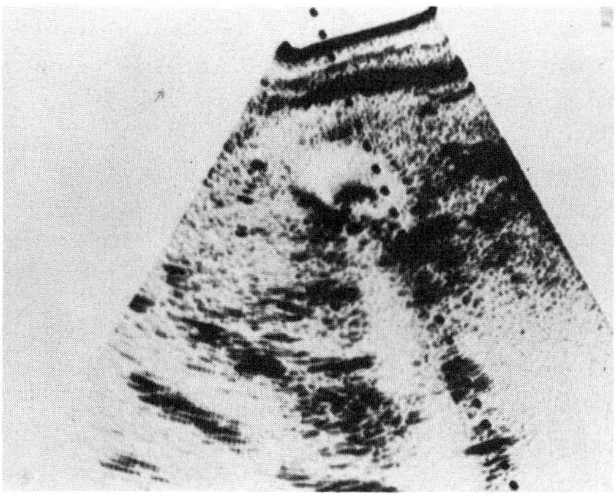

Figure 19–A. Transverse section +6.

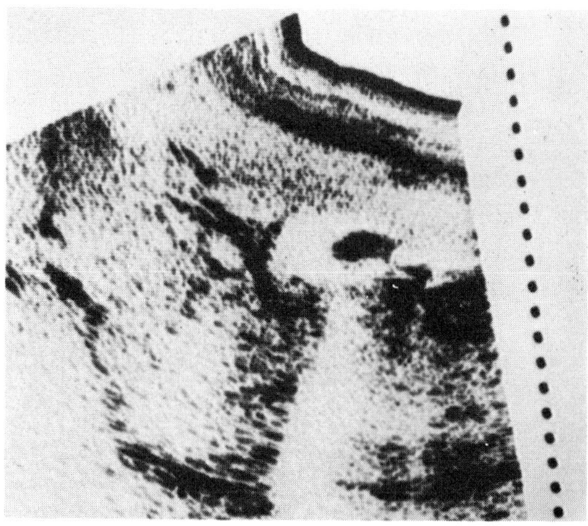

Figure 19–B. Longitudinal section 4R.

DIAGNOSIS

Gallstones.

ULTRASOUND FINDINGS

Both the cross-sectional and longitudinal scans show the "headlight" sign—the classic appearance of gallstones. When examining the gallbladder it is important to obtain scans at 5-mm. intervals parallel to the long axis of the organ. Usually this requires some oblique scans along the general line of the right subcostal margin; a few trial scans will quickly show the best scanning plane. The size of the gallbladder is not a particularly useful sign in deciding if it is diseased or normal. If the patient has just eaten, the gallbladder may be contracted and difficult to visualize, and some patients with no gallbladder disease have gallbladders nearly as large as the right kidney! The three important points to consider are: (1) Is the lumen echo-free or does it contain echo-producing structures? (2) Is the posterior wall a continuous, smooth line with no step-offs? (3) Is there uniform distribution of the sound beam in the tissues behind the gallbladder? Figures 19–C and 19–D illustrate the findings of a normal gallbladder. The lumen is echo-free, the back wall is a smooth, continuous line (arrow in Figure 19–C), and the echoes from the underlying tissue are uniform — there is no acoustic shadow (arrowheads in Figure 19–D).

As shown in Figures 19–A and 19–B, gallstones alter this appearance. The stones give rise to echoes within the lumen (arrow), the back wall is incomplete, and there is a large acoustic shadow in the underlying tissue *(s)*. The shadow arises because the stone reflects a considerable portion on the ultrasound and the weakened beam can produce only faint echoes from the underlying tissue. It is not necessary to have all three signs to make a diagnosis of stones; the presence of *any one of the three* is sufficient. Figure 19–E is from a patient with gallstones, in whom there is not a dramatic intraluminal echo. Notice, however, the very slight step-off in the posterior gallbladder wall (arrow); this is enough to make the diagnosis. This patient also has a prominent acoustic shadow (open arrows) to clinch the diagnosis. The gallbladder shown in Figure 19–F is contracted and it is difficult to say much about either the lumen or the back wall, yet the prominent acoustic shadow (straight arrow) allows us to make the diagnosis of stone (curved arrow) with confidence. Some small or low-density stones do not cast an acoustic shadow. Figures 19–G and 19–H are from two different patients with very small stones which do not cast an acoustic shadow; the diagnosis rests on identifying the small step-off on the posterior wall.

If meticulous technique is used, ultrasound can be as accurate as oral cholecystography in making the diagnosis of gallstones.

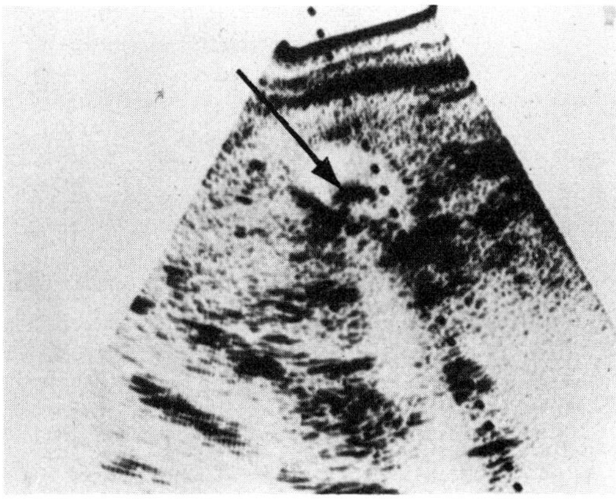

Figure 19–A. Transverse section +6.
Gallstone.

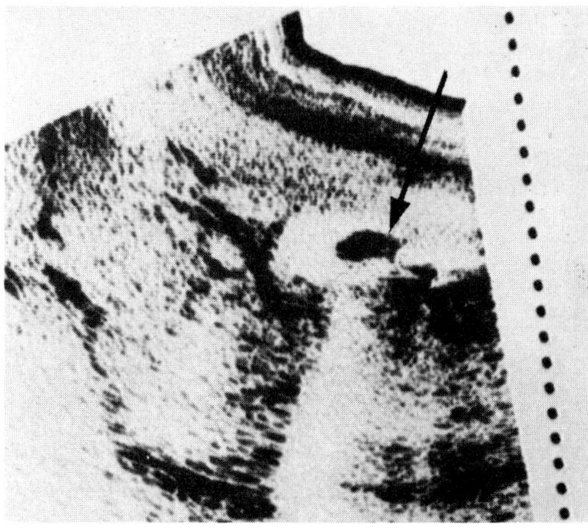

Figure 19–B. Longitudinal section 4R.
Gallstone.

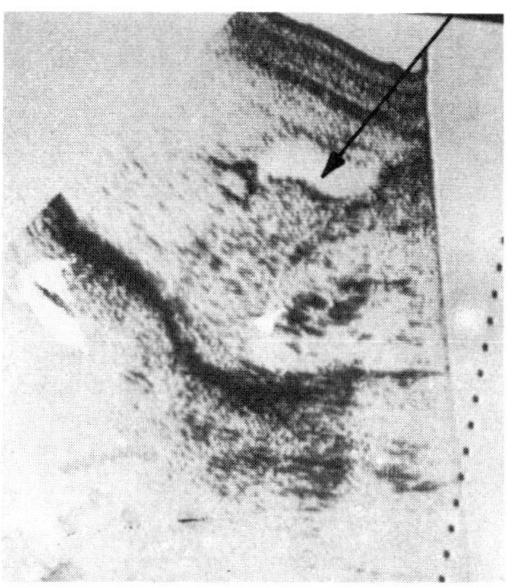

Figure 19–C. Longitudinal section 6R.
Normal gallbladder.

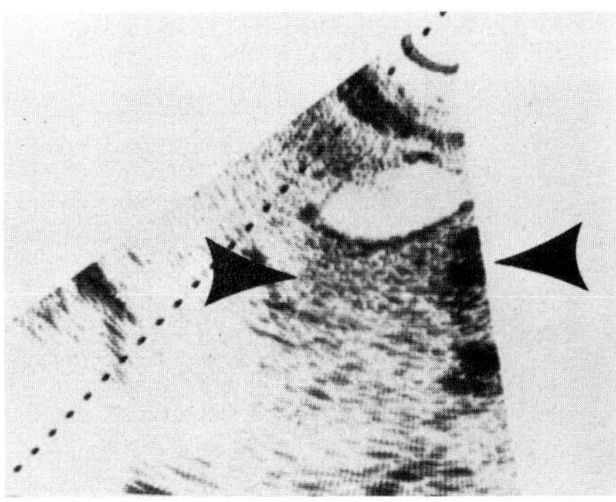

Figure 19–D. Longitudinal section 5R.
Normal gallbladder.

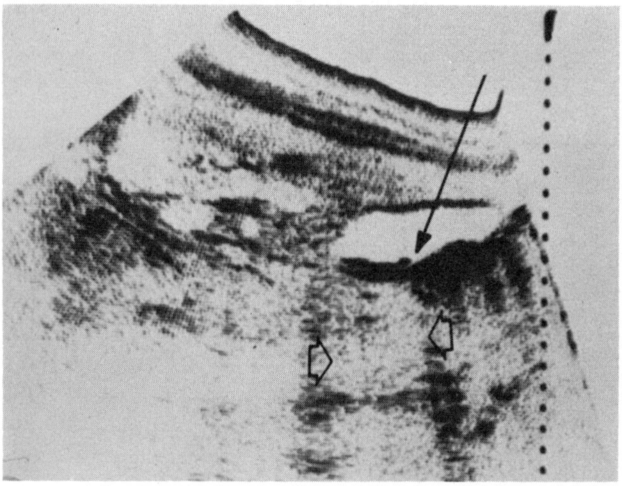

Figure 19–E. **Longitudinal section 4R.** Gallstones. Longitudinal section through the gallbladder.

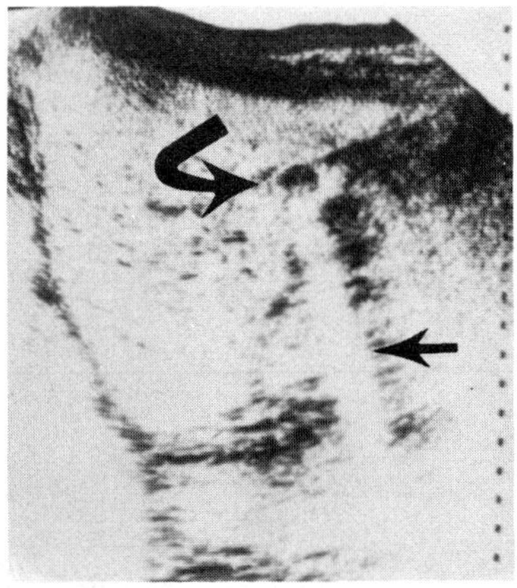

Figure 19–F. **Longitudinal section 5R.** Gallstones. Longitudinal section through the gallbladder.

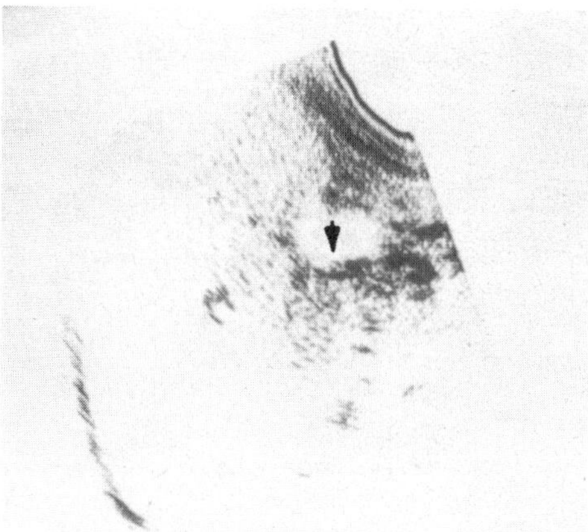

Figure 19–G. **Longitudinal section 4R.** Small gallstones. Longitudinal section through the gallbladder.

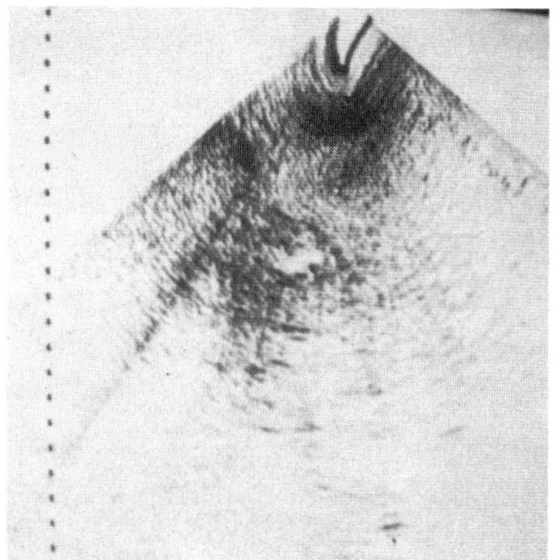

Figure 19–H. **Transverse section +8.** Small gallstones. Longitudinal section through the gallbladder.

REFERENCES

Crow HC, Bartrum RJ Jr, Foote SR: Expanded criteria for the ultrasonic diagnosis of gallstones. J Clin Ultra 4:289 (1976)

Arnon S, Rosenquist CJ: Gray scale cholecystosonography: an evaluation of accuracy. Am J Radiol 127:817 (1976)

Leopold GR, Amberg J, Gosink BB, et al: Gray scale ultrasonic cholecystography. Radiology 121:445 (1976)

Bartrum RJ Jr, Crow HC, Foote SR: Ultrasonic and radiographic cholecystography. N Engl J Med 296:538 (1977)

Lawson TL: Gray scale cholecystosonography: diagnostic criteria and accuracy. Radiology 122:247 (1977)

Case 20

CLINICAL INFORMATION

This 47-year-old man comes for evaluation of a mass in the right side of his thyroid gland, discovered during a routine physical examination. He has been in perfect health as far as he knows; the examination was the result of his having been promoted by the company he works for, and he is upset at having been told that there is something wrong. He is now being seen by an endocrinologist who has already determined that all of the parameters of thyroid function are normal. A technetium scan shows a "cold" region in the right lobe. Here are the ultrasound scans of this area; what shall we tell his physician?

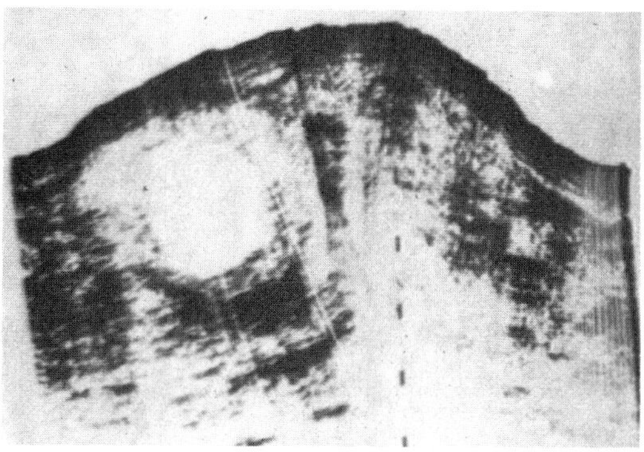

Figure 20–A. Transverse section +4.5.

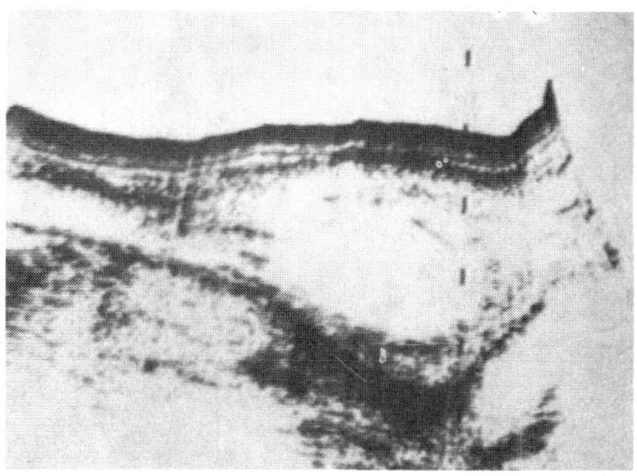

Figure 20–B. Longitudinal section 3R.

DIAGNOSIS

Cystic follicular adenoma.

ULTRASOUND FINDINGS

The lesion in the right lobe is obvious. There is a large area in which the echoes are relatively reduced in both number and intensity. In the center there are no echoes at all. On the transverse scan we can see the trachea (deviated toward the left), and, to its left, the normal left lobe of the thyroid is visible (arrow). Comparing the echo pattern of the left lobe with that of the right confirms that the right is indeed abnormal. Because of the echo-free nature of the center of this lesion, we can also say that it must be cystic; however, the cystic portion is surrounded by abnormal thyroid, so that this is not just an ordinary cyst but an area of cystic degeneration within a solid mass — a follicular adenoma. After we had made this scan our endocrinologist colleague put a small needle into the cystic area and aspirated some fluid.

Now what about the thyroid and ultrasound; what can we do for the clinician to answer questions about thyroid problems? Well, as you can see, the demonstration of cystic lesions is easy enough. We are usually called in to help when the examiner has found a solitary nodule which is "cold" on an isotope study. Our job is to decide whether or not the nodule is solid or cystic. Often the lesion is mixed, as in this case; and here again, aspiration can be of help. Malignant tumors usually do not form cysts large enough to aspirate. (Cytology is not as reliable here as in other areas.)

Figures 20–C and 20–D are scans from another patient who has a nearly identical history. The relatively echo-free areas, quite different from the normal thyroid, are easily seen on both scans. This also is a follicular adenoma, but notice in the longitudinal views, that the cystic areas are quite small compared with those in the first patient and could not be easily aspirated. Follicular adenoma is basically a solid lesion; the cystic change within it is variable, and this alters the ultrasound appearance from patient to patient.

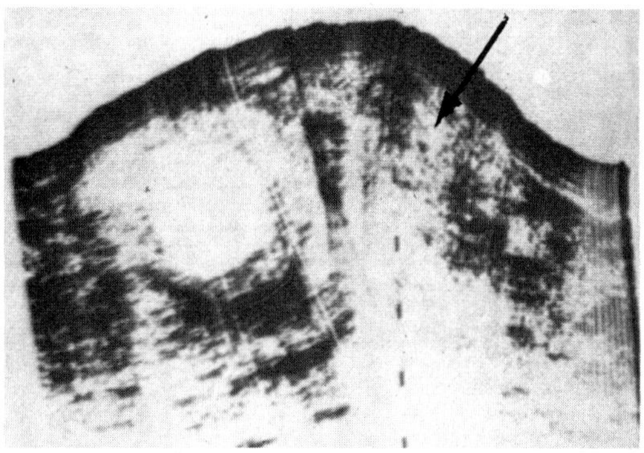

Figure 20–A. Transverse section +4.5. Cystic follicular adenoma.

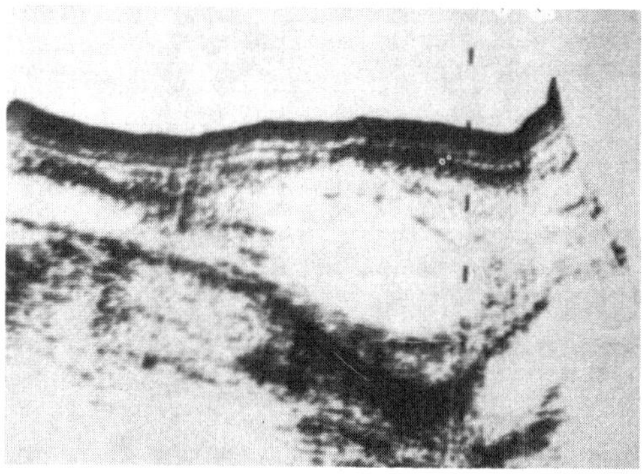

Figure 20–B. Longitudinal section 3R. Cystic follicular adenoma.

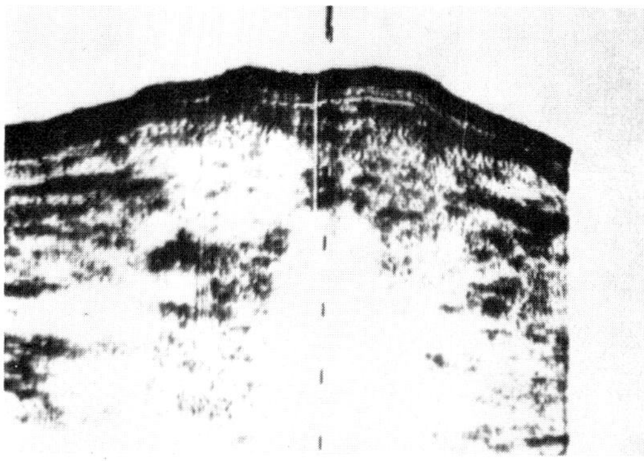

Figure 20–C. Transverse section +1. Follicular adenoma.

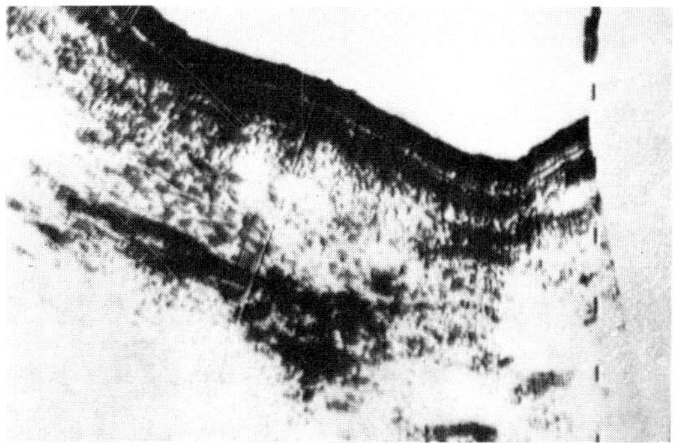

Figure 20–D. Longitudinal section 1R. Follicular adenoma.

REFERENCES

Sackler JP, Passalaqua AM, Blum M, et al: A spectrum of diseases of the thyroid gland as imaged by gray scale water bath sonography. Radiology 125:467 (1977)

Blum M, Goldman AB, Herskovic A, et al: Clinical applications of thyroid echography. N. Engl J Med 287:1164 (1972)

Case 21

CLINICAL INFORMATION

This 58-year-old man is a long-time alcoholic with biopsy-proven cirrhosis of the liver. He presents now with jaundice and a constant gnawing abdominal pain. When last seen by his doctor six months ago, he had an enlarged and firm liver. He was not jaundiced and his liver function studies were only mildly abnormal. There has been no change in his physical examination other than the obvious jaundice, but his laboratory examinations are now strikingly abnormal with a pattern suggesting both hepatocellular and obstructive jaundice. The serum amylase is not elevated.

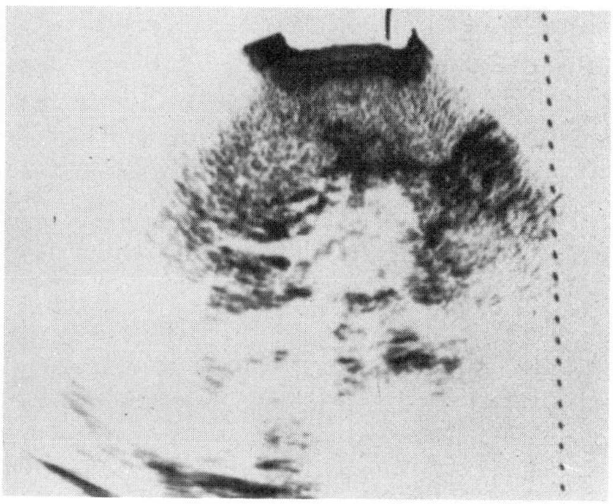

Figure 21–A. Transverse section +14.

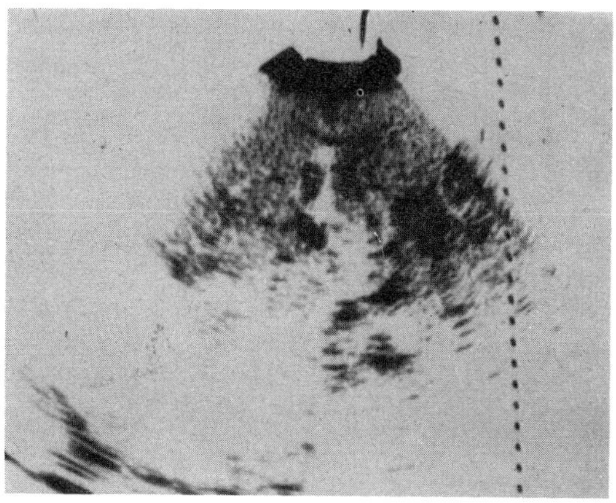

Figure 21–B. Transverse section +16.

DIAGNOSIS

Carcinoma of the head of the pancreas with obstruction of the common bile duct.

ULTRASOUND FINDINGS

The scans show the markedly increased liver parenchymal echoes that are typical of advanced cirrhosis and, in addition, reveal a large mass in the region of the head of the pancreas (Figure 21–A). This mass is relatively echo-free, but not completely echo-less and thus is not fluid-filled; it is a solid lesion. (Actually the mass contains more echoes than Figure 21–A suggests, because the unusually loud echoes from the liver parenchyma have caused us to scan at a lower gain than normal.) The more cranial sections (Figure 21–B) demonstrate a tortuous, tubular, echo-free structure, which is the dilated common bile duct (arrow). (We identified the portal vein as a separate structure in Figure 21–A.)

The presence of a solid mass localized to the region of the head of the pancreas means tumor; and by far the most common tumor in this location is carcinoma of the pancreas. This could also be a carcinoma of the ampulla of Vater or a tumor in the retroperitoneal lymph nodes; but when a mass is encountered in this area, one should generally put his money on pancreatic cancer. The normal serum amylase does not dissuade us from this diagnosis. In reviewing over 100 cases of proven pancreatic cancer, there were only two patients in whom the amylase was elevated, and both of these elevations were minor. In fact, it is the other way around: if the amylase is elevated, carcinoma of the pancreas is not a likely diagnosis. One further differential point: usually pancreatitis causes universal enlargement of the gland with diminished echo production within it. Tumors are more likely to be localized.

Figures 21–C through 21–E are from three other patients with carcinomas of the head of the pancreas. Notice that each has a slightly different appearance, ranging from weakly to strongly echogenic and from homogeneous to heterogeneous. The common denominator is the presence of the abnormal mass in the proper clinical setting.

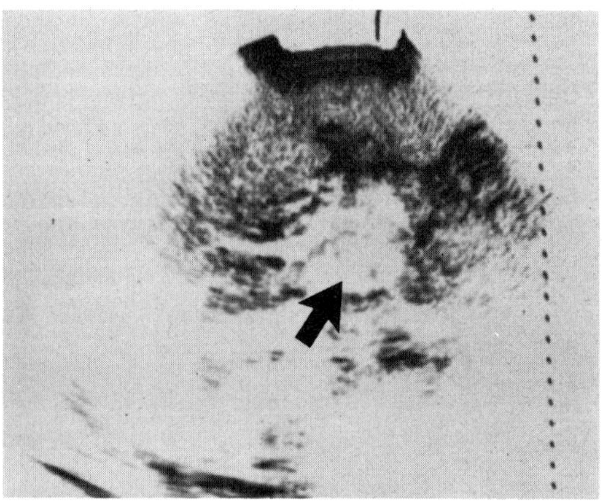

Figure 21–A. Transverse section +14. Carcinoma of the pancreas.

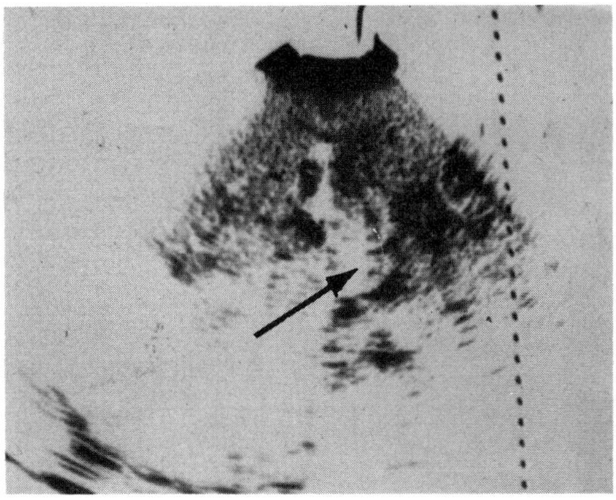

Figure 21–B. Transverse section +16. Carcinoma of the pancreas.

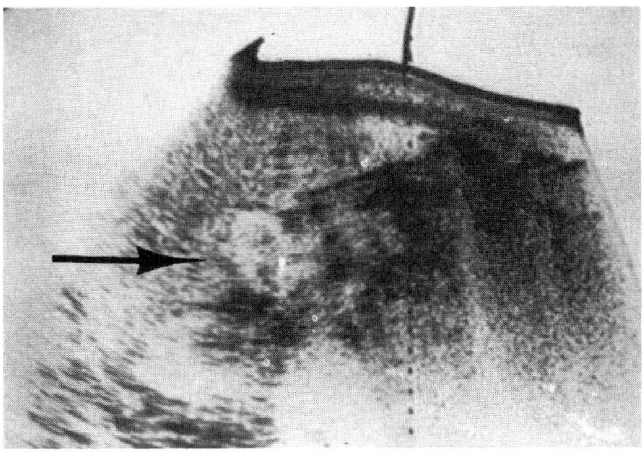

Figure 21–C. Transverse section +10. Carcinoma of the pancreas.

Figure 21–D. Transverse section +9. Carcinoma of the pancreas.

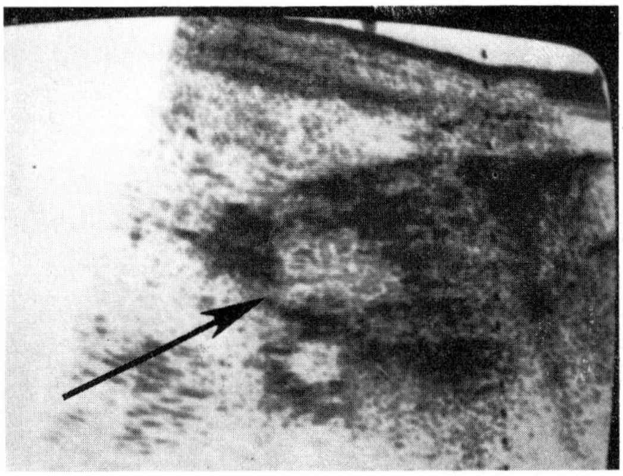

Figure 21–E. **Transverse section +10.** Carcinoma of the pancreas.

REFERENCES

Perlmutter GS: Chapter 7, Pancreas, in *Abdominal Gray Scale Ultrasonography.* BB Goldberg, editor. John Wiley & Sons. New York 1977

McCormack LR, Seat SG, Strum WB: Pancreatic carcinoma: Survival following detection by ultrasonic scanning. JAMA 238:240 (1977)

Doust BD: Ultrasonic examination of the pancreas. Rad Clin North Amer 13:467 (1975)

Case 22

CLINICAL INFORMATION

This 26-year-old woman is nine weeks pregnant but has experienced heavy vaginal bleeding for two days. The bleeding has tapered off but she has still been spotting for the past 24 hours.

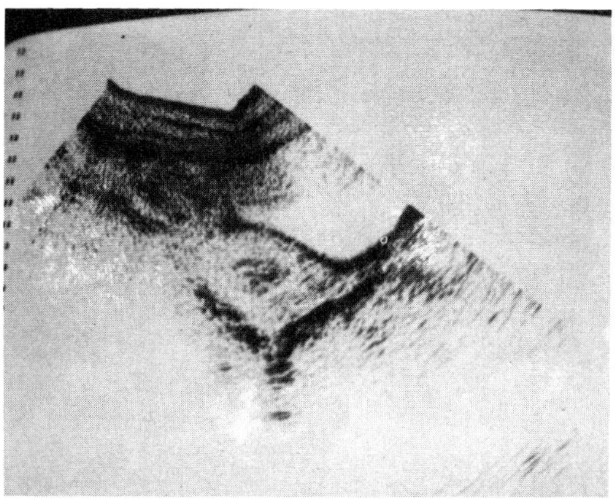

Figure 22–A. Longitudinal section 1.5R.

DIAGNOSIS

Incomplete abortion (retained products of conception).

ULTRASOUND FINDINGS

There is no gestational sac within the uterus, so we know that the patient has aborted. The endometrial cavity is not normal, however; it contains several loud echoes which represent retained products of conception. This patient needs a D & C to evacuate the uterus and prevent endometritis. Once a patient has had a complete abortion, the endometrial echoes become very faint or nonexistent and the ultrasonic picture of the uterus is very similar to the nonpregnant state. Figures 22–B is another example of an incomplete abortion.

We want to emphasize that the status of the pregnancy test is a very important factor in a case such as this. If it is positive, the possibility of gestational trophoblastic disease always has to be considered when a gestational sac cannot be found. Also, it is theoretically possible to have an intraperitoneal pregnancy not visible by ultrasound because of its position. This condition might produce such changes in the uterus. The obstetrician will know the results of the pregnancy test, and you will find this information invaluable in interpreting your findings.

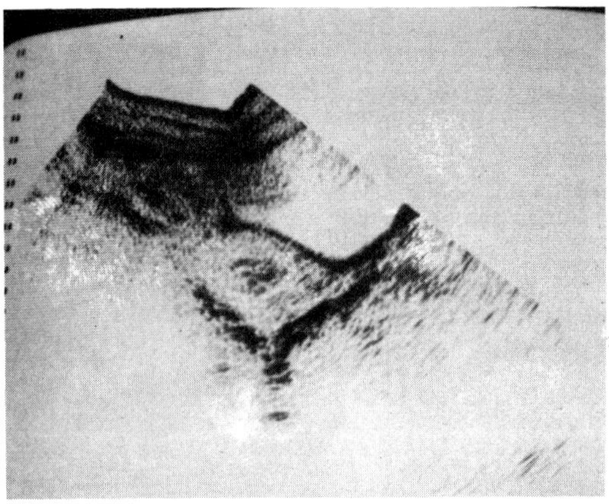

Figure 22–A. Longitudinal section 1.5R. Incomplete abortion.

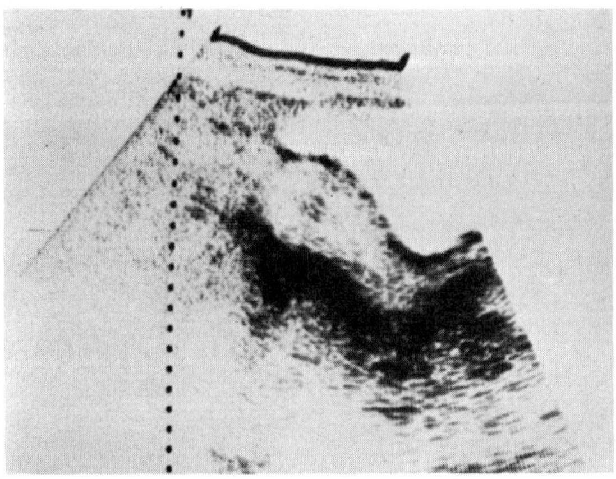

Figure 22–B. Longitudinal section 0L. Incomplete abortion.

Case 23

CLINICAL INFORMATION

This chronically ill patient is admitted for increasing dyspnea and weakness. She has been seen numerous times before for similar problems, but is distinctly worse today. The ultrasound technologist is having an awful day, and her troubles with this patient seem to be the culmination of everything: the patient is uncomfortable and, therefore, uncooperative; the scans do not look right; and she is uncertain of her findings. Here is a longitudinal scan from the right flank. What do you see? What do you think about the performance of the scanner?

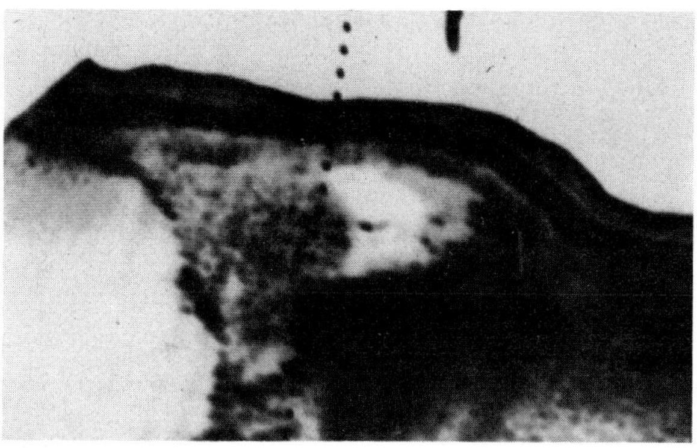

Figure 23–A. Longitudinal section 6R.

DIAGNOSIS

Large right pleural effusion and ascites.

ULTRASOUND FINDINGS

There is a very large echo-free area above the liver. The diaphragm is very clearly seen (Figure 23–A, arrow), and its course suggests that it is pushed down into the abdomen by a large effusion in the right hemithorax. The ultrasound scan is giving the same information which we have learned to seek on the chest radiograph (Figure 23–B). Here the entire right hemithorax is opacified, and the mediastinum is pushed toward the left; in other words the effusion must be under considerable pressure. This finding almost always indicates a malignant effusion or trauma. There is also some fluid in the belly, seen as the echo-free area below the tip of the liver. You have undoubtedly guessed that we have withheld some clinical information: the patient has metastatic ovarian carcinoma.

Now, what about the quality of the scan? Not very good, right? What would you do to improve it? The scanning technique is good, but the scanner is misbehaving (happens in even the best of families). Looking carefully, you will note that the "dots" are too large and fuzzy; and the gray scale is off. There are too few shades of gray; we can see only white, medium gray, and black. When this happens, call in the service man and ask him to adjust the scan converter. As the manufacturers have improved the quality of the products, problems with adjustment have proved to be less troublesome for the user; and, theoretically, at least, digital scan converters will eliminate most difficulties of this sort. We find it helpful to keep an example of a good scan handy as a means of checking the operation of the instrument. It is sometimes hard to recognize subtle changes in the performance of the scan converter until things have gotten this bad.

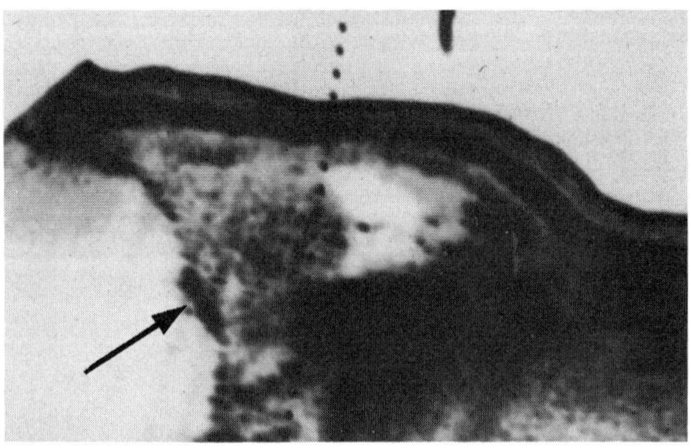

Figure 23–A. Longitudinal section 6R. Pleural effusion and ascites.

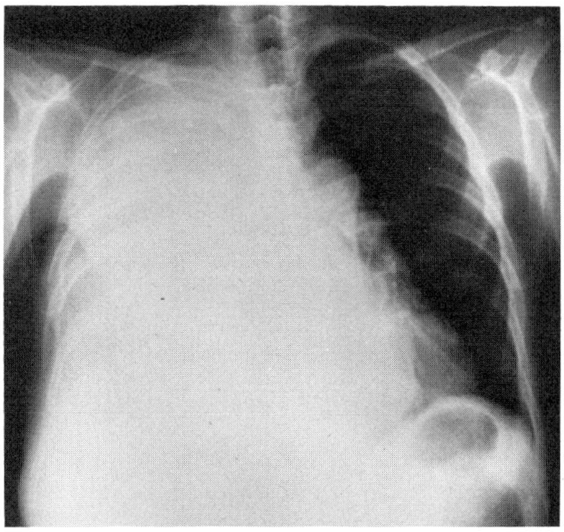

Figure 23–B.
Pleural effusion. Chest radiograph of patient in Figure 23–A.

REFERENCE

Doust BD, Baum JK, Maklad NF, et al: Ultrasonic evaluation of pleural opacities. Radiology 114:135 (1975)

Case 24

CLINICAL INFORMATION

This 23-year-old woman has had several episodes of right upper quadrant pain and nausea. She thinks that these pains are associated with eating fatty food. Her physical examination and basic laboratory studies are normal (she is only slightly overweight). As expected, her ultrasound scan shows gallstones; but on the longitudinal scans to the left of the gallbladder there is a round mass just off the tip of the liver (arrow). What do you think this is?

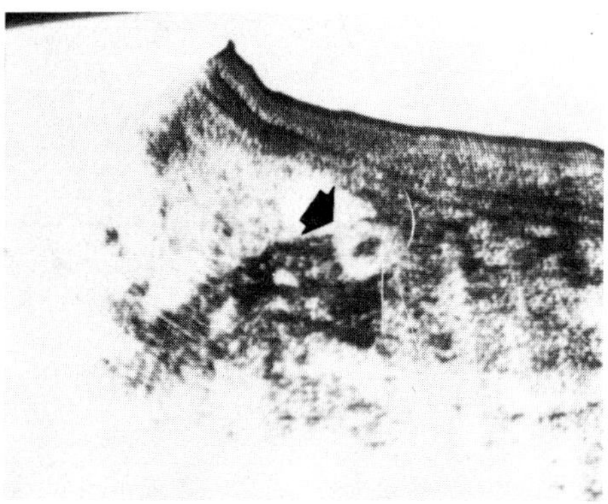

Figure 24–A. Longitudinal section 2L.

DIAGNOSIS

Normal transverse colon.

ULTRASOUND FINDINGS

The small "doughnut-like" mass could be seen on several different sections spanning about 5 cm. of the midabdomen. As always, our first consideration must be normal GI tract filled with fluid or solid material; in this case the location and shape are perfect for the transverse colon. The portions of the colon that are filled with air cannot be visualized on the scan and would generate only the common bowel gas artifacts, but the parts which are contracted or feces-filled will appear as a relatively echo-less mass with louder central echoes — the size of the central echoes depends on the volume of material in the lumen. Portions of the transverse colon can be seen in a high percentage of normal patients as well as in those with underlying pathology; and if we are familiar with its appearance, we will not be misled.

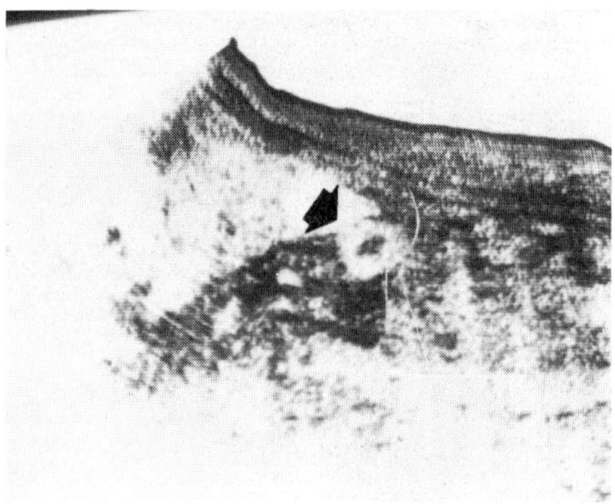

Figure 24–A. Longitudinal section 2L. Transverse colon (normal).

REFERENCE

Sample WF: Techniques for improved ultrasound delineation of normal anatomy . . . with gray scale ultrasound. Radiology 124:197 (1977)

Case 25

CLINICAL INFORMATION

This 22-year-old man had mononucleosis eight months ago and was thought to be fully recovered — he has been asymptomatic for over two months. At a recent follow-up physical examination there was a question of splenomegaly and he is sent to the ultrasound laboratory for evaluation of splenic size.

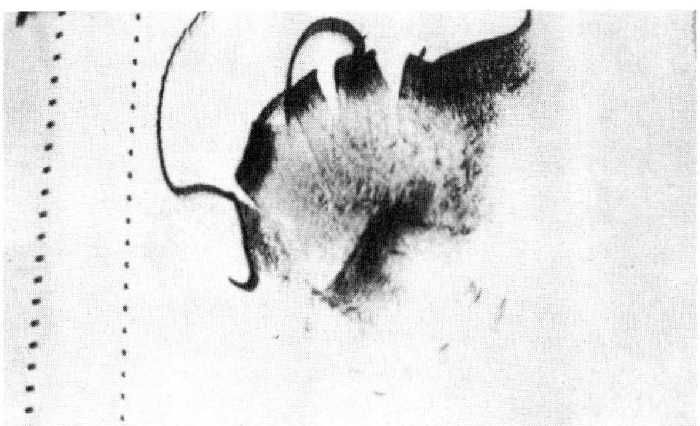

Figure 25–A. Transverse section +8. Right lateral decubitus scan.

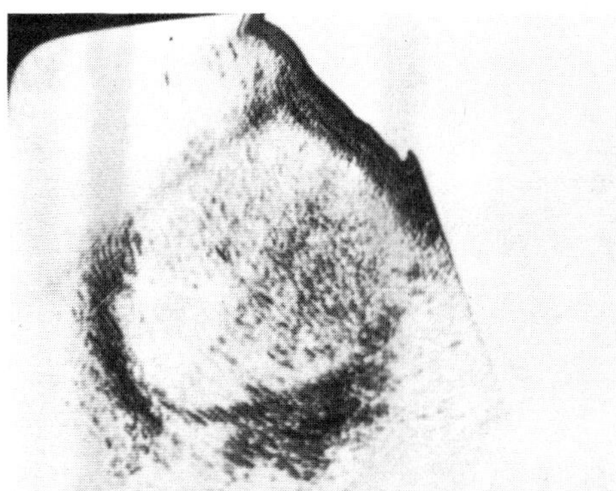

Figure 25–B.
Right lateral decubitus scan; section performed in an intercostal space.

DIAGNOSIS

Normal spleen, at upper limits of normal size.

ULTRASOUND FINDINGS

Making a scan of the normal spleen can be tricky business. It is located far posteriorly and tucked up under the left hemidiaphragm where there is no good acoustic window. The standard supine scan (Figure 25–C) usually shows nothing; the overlying stomach and colon with their inevitable gas shadow the area. The best approach is to have the patient roll onto his right side ("right lateral decubitus" position) or scan him prone. Sometimes varying degrees of obliquity are best. In this position only the ribs interfere with the scan, and these can be avoided by making a series of small sector scans in the interspaces, as has been done in Figure 25–A. It is also useful to move the scanning head so that the scanning plane is oriented along an intercostal space, as has been done in Figure 25–B.

The normal spleen produces only weak echoes; it is a very homogeneous organ and can be thought of as a "sac of blood." There are no rigorous criteria for size, but most radiologists seem to feel that as long as no single dimension is greater than 14 cm. the spleen is not enlarged. We have found the following rule-of-thumb useful: if the spleen cannot be visualized on a *supine* scan, it is not enlarged; if it can be *easily* seen on a *supine* scan, it is definitely enlarged.

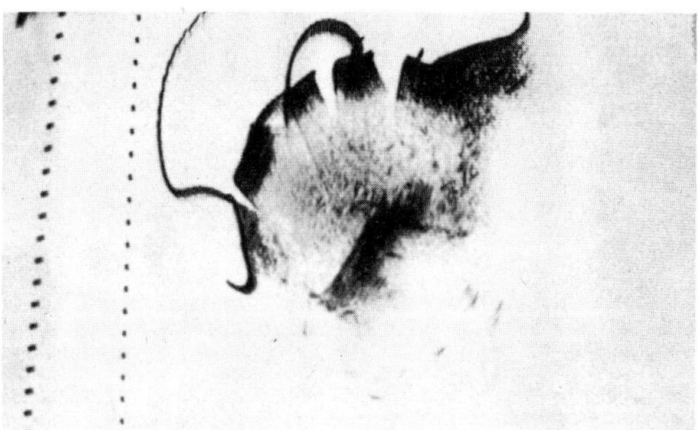

Figure 25–A. **Transverse section +8.** Normal spleen. Right lateral decubitus scan.

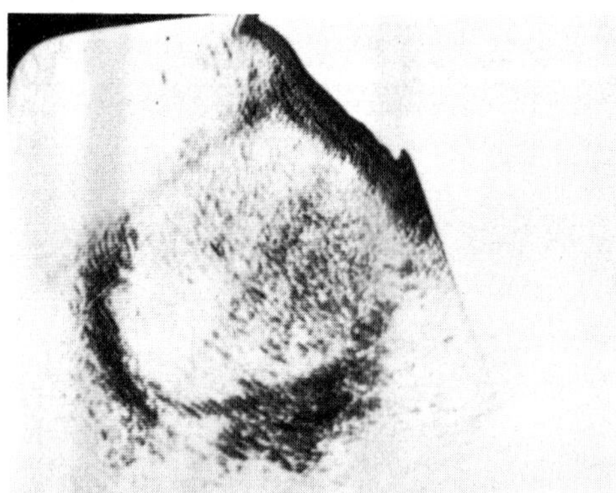

Figure 25–B.
Normal spleen. Right lateral decubitus scan; section performed in an intercostal space.

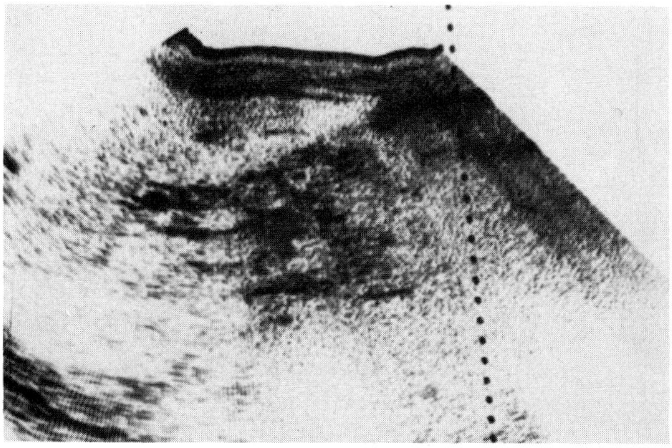

Figure 25–C. Transverse section +6. Normal abdomen (and spleen).

REFERENCES

Koga T, Morikawa Y: Ultrasonic determination of splenic size. Radiology 115:157 (1975)

Taylor KJW, Milan J: Differential diagnosis of chronic splenomegaly by gray scale ultrasonography. Br J Radiol 49:519 (1976)

Sample WF: Techniques for improved ultrasound delineation of normal anatomy . . . with gray scale ultrasound. Radiology 124:197 (1977)

Case 26

CLINICAL INFORMATION

This 84-year-old Grande Dame was sent to the ultrasound laboratory because of a firm, somewhat pulsatile right neck mass felt by the internist who tunes her up before her trip south each winter. This physician is a very careful fellow and even though the lady has no complaints referable to the mass, he is certain that it was not palpable last year. He does not think it is in the thyroid, but cannot be certain. Here is the scan; what is going on?

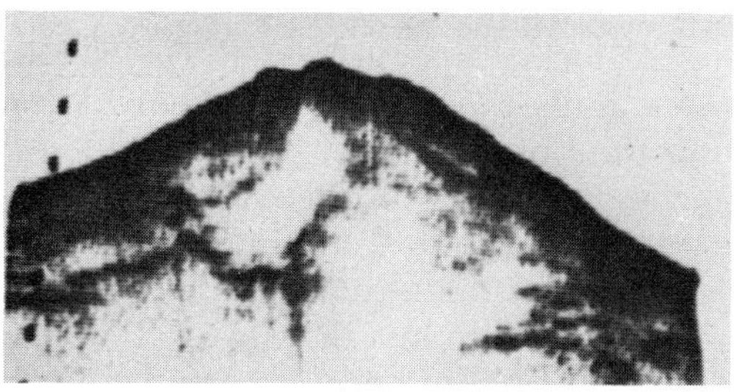

Figure 26–A. Transverse section +1.5.

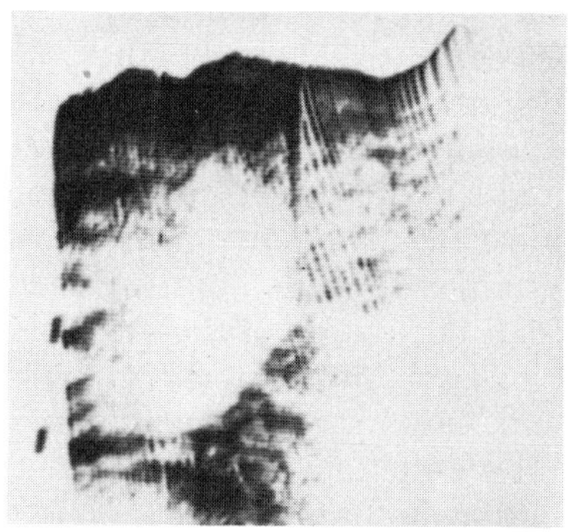

Figure 26–B. Transverse section +1.

DIAGNOSIS

Carotid artery aneurysm.

ULTRASOUND FINDINGS

Maybe it is time for a little anatomical review. Figure 26–C is a transverse scan through the thyroid showing both lobes as uniformly speckled areas adjacent to the trachea. (Remember, the trachea does not transmit sound because it is air-bearing, so there is always an acoustic shadow behind it. Normally, of course this will be in the midline.) Lateral to the lobes of the thyroid there are two small round echoless areas, the common carotid arteries (arrows). Behind the thyroid, particularly prominent on the right in this patient, are relatively echo-free zones, the sternocleidomastoid muscles. Sometimes the jugular veins can be seen, usually even farther laterally than the arteries. Because they vary in size much more than the arteries, they are not as constantly seen. Also, the carotid arteries are round, whereas the veins are elliptical.

Now returning to our patient, in Figure 26–B we are able to distinguish the right lobe of the thyroid by its homogeneous echo pattern. Lateral to it is a large echo-free mass, a "cyst." The question is whether or not the cyst is arising within the gland. On the basis of the scan alone this differentiation is not possible. Because he felt pulsation, the clinician thought there might be an aneurysm of the carotid. The findings of the ultrasound scan, coupled with the physical examination, enabled him to decide that this was a vascular rather than a thyroid problem.

Figure 26–D is a transverse scan from another patient with a somewhat similar picture, but here the cyst (arrow) lies within the gland; and the carotid is seen clearly lateral to the cystic area, obviously separate from it. How about Figure 26–E? Here the echo-free mass abuts the carotid, and there is no evidence of a wall dividing them. On the opposite side, both carotid and jugular vein can be seen. In this instance, just as in our original patient, it is necessary to resort to clinical information and physical findings. This mass was not pulsatile, and there was a large "cold" nodule on the radioisotope scan. Aspiration of the mass produced clear fluid from a thyroid cyst.

The point of all this is to remind you that extrathyroidal masses do occur, and they may simulate thyroid disease. Ultrasound can be of help in this differentiation, but it is most effective when used in combination with other clinical information.

Figure 26–A. Transverse section +1.5. Carotid artery aneurysm.

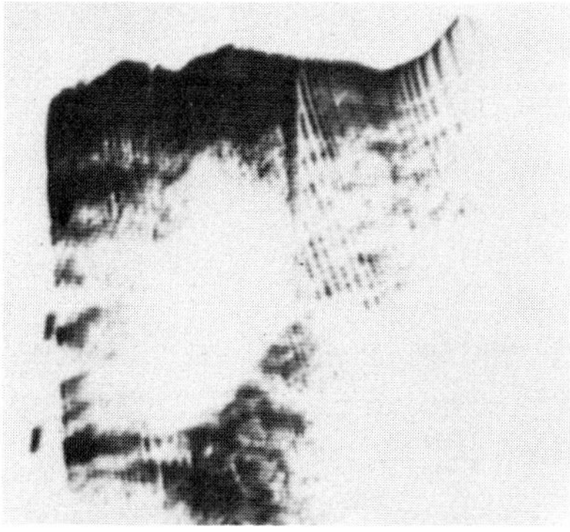

Figure 26–B. Transverse section +1. Carotid artery aneurysm.

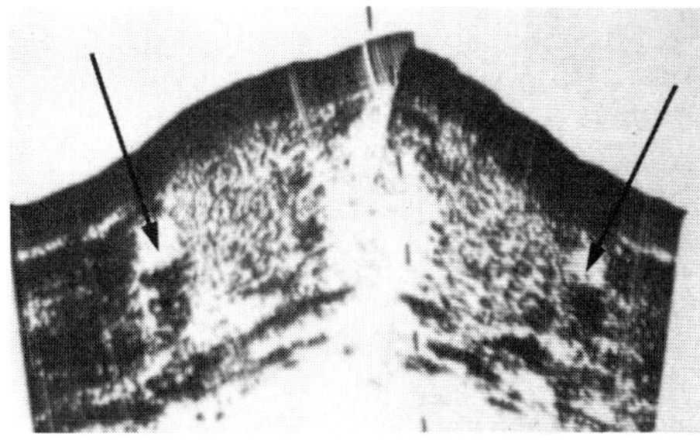

Figure 26–C. Transverse section +4. Normal thyroid.

Figure 26–D. Transverse section +3. Benign thyroid cyst.

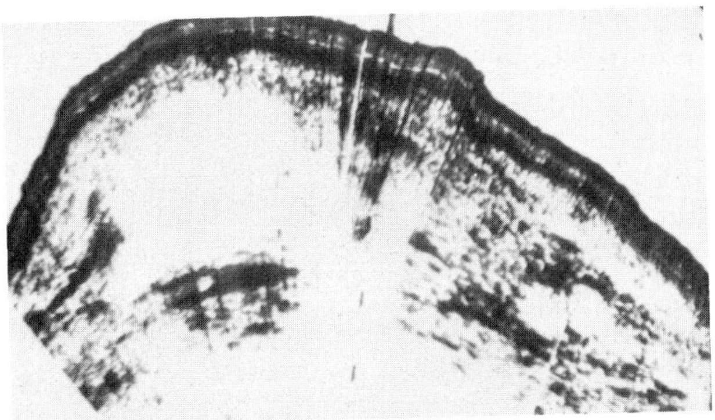

Figure 26–E. Transverse section +4. Benign thyroid cyst.

REFERENCES

Davis RP, Neiman HL, Yao JST, et al: Ultrasound scan in diagnosis of peripheral aneurysms. Arch Surg 112:55 (1977)

Gooding GAW, Herzog KA, Laing FC: Ultrasonographic assessment of neck masses. J Clin Ultra 5:248 (1977)

Case 27

CLINICAL INFORMATION

This 53-year-old alcoholic complains that he has been gaining a lot of weight recently and his pants no longer fit.

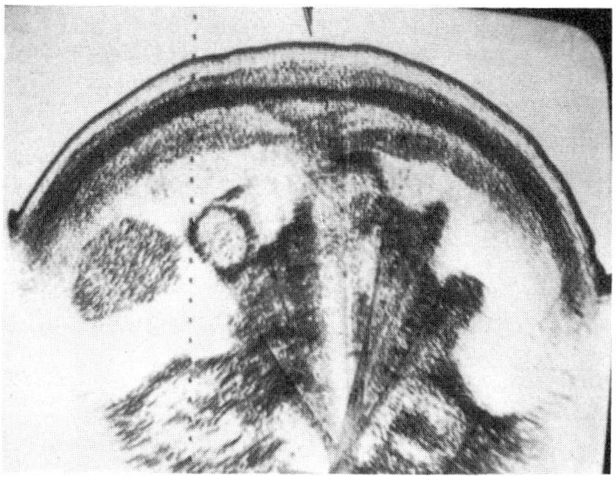

Figure 27–A. Transverse section +8.

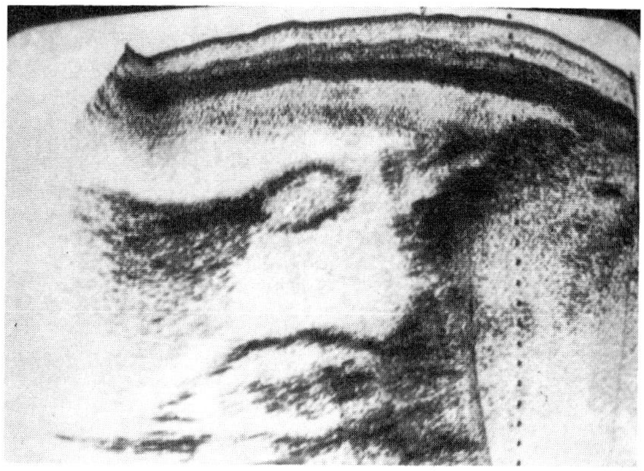

Figure 27–B. Longitudinal section 6R.

DIAGNOSIS

Massive ascites (from cirrhosis of the liver).

ULTRASOUND FINDINGS

The striking feature of these scans is the presence of large echo-free areas that surround and outline the intraperitoneal organs. This is probably the easiest diagnosis in abdominal ultrasound — massive ascites. The findings are straightforward: the flanks and portions of the central abdomen contain fluid; the tip of the right lobe of the liver and the gallbladder are floating in the fluid, and the normally amorphous loops of bowel appear as multiple echogenic "masses" because their margins are now clearly defined by the ascitic fluid. Loops of bowel that contain gas float to the central portions of the abdomen. Figures 27–A through 27–E are examples of these findings.

If you are really observant you will have noticed that the patient shown in Figures 27–A and 27–B has many low-level echoes in his gallbladder; this means that his bile is not normal in consistency. Two other questions to ponder: (1) Why is the left kidney seen so well in Figure 27–A? (2) What is producing the line of echoes shown by the arrow in Figure 27–E? (If you cannot answer the first question read Reference 3 below. If you need help with the second question, see Case 54.)

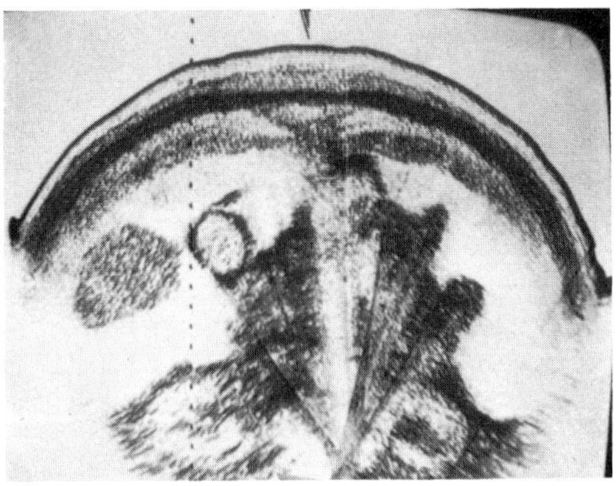

Figure 27–A. Transverse section +8. Ascites.

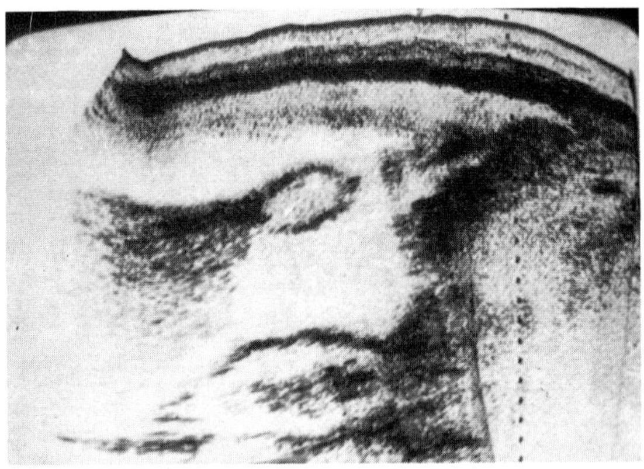

Figure 27–B. Longitudinal section 6R. Ascites.

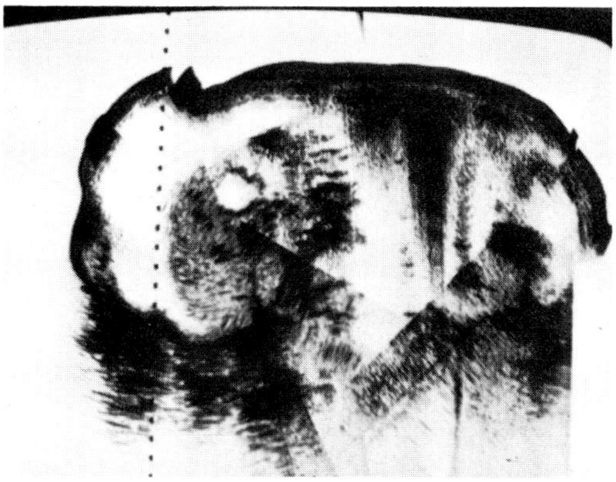

Figure 27–C. Transverse section +9. Ascites.

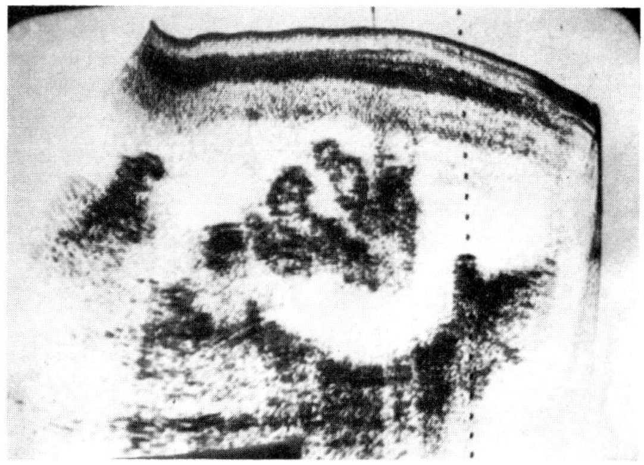

Figure 27–D. Longitudinal section 10L. Ascites.

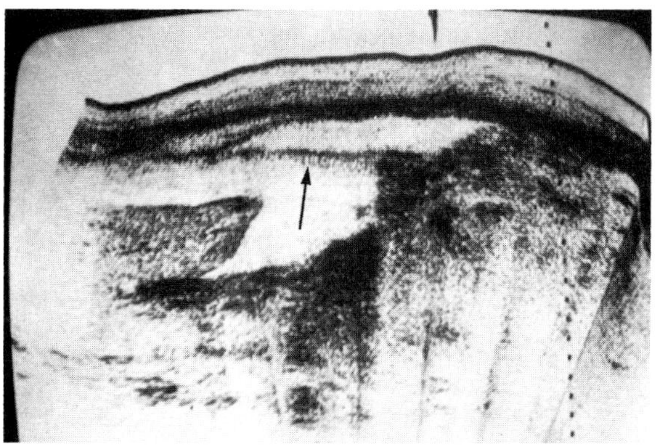

Figure 27–E. Longitudinal section 5R. Ascites.

REFERENCES

Yeh HC, Wolf BS: Ultrasonography in ascites. Radiology 124:783 (1977)

Goldberg BB, Goodman GA, Clearfield HR: Evaluation of ascites by ultrasound. Radiology 96:15 (1970)

Bartrum RJ Jr, Crow HC: *Gray-Scale Ultrasound: A Manual for Physicians and Technical Personnel.* Chapters 5 and 6. WB Saunders Co, Philadelphia (1977)

Case 28

CLINICAL HISTORY

This 59-year-old woman has pain in the right flank and hematuria. The flank pain is a dull ache, which has been present for about three months and actually seems to be improving slightly. The bloody urine has only recently appeared, and it is this which has brought her to the doctor. Aside from long-standing asthma, her health is good. She has lost about 10 pounds in the last six months and feels somewhat more tired than usual. Her physical examination is unremarkable; there is only minimal tenderness in the right flank and costovertebral angle. The hematocrit and white cell counts are normal, as are routine laboratory tests including creatinine. Urinalysis shows gross blood and a trace of sugar but no protein. The bacteria count is less than 100,000. Figures 28–A and 28–B are longitudinal and cross-sectional scans through the bladder, and Figure 28–C is a longitudinal scan of the right kidney.

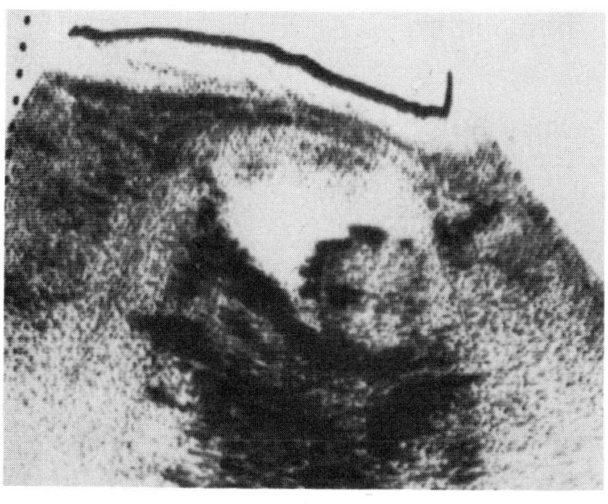

Figure 28–A. Longitudinal section 1R.

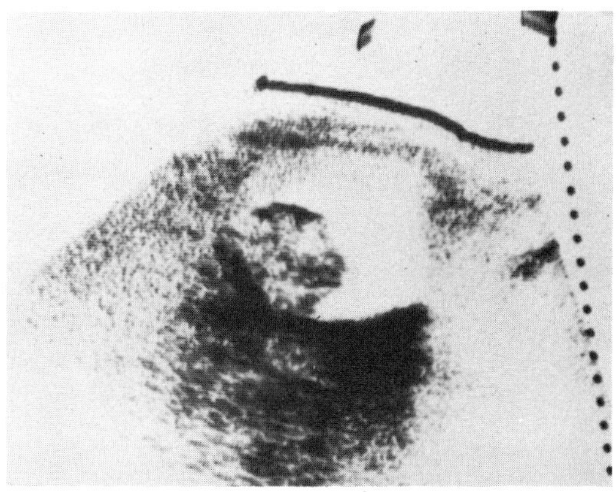

Figure 28–B. Transverse section –13.

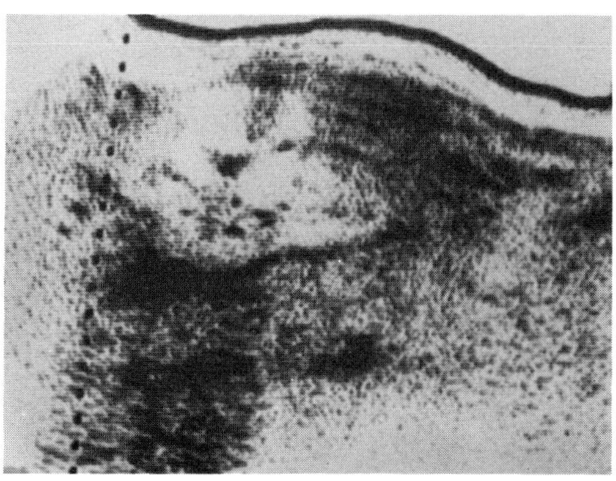

Figure 28–C. Longitudinal section 0L. Prone scan of right kidney.

DIAGNOSIS

Transitional cell carcinoma of the bladder obstructing the right ureter.

ULTRASOUND FINDINGS

On the posterior wall of the bladder there is a large echogenic mass that seems to be arising from within the bladder rather than invading from the outside. The scan of the kidney shows multiple fluid-filled areas characteristic of advanced hydronephrosis. How does this all fit together? There are only two sources for the mass in the bladder: stone and tumor. (A Foley catheter can also produce echoes but this patient was not catheterized.) We can eliminate stone because the mass seems fixed to the bladder wall, and a bladder stone should not cause obstruction of the right kidney. Thus we come down to bladder tumor. A cystogram (Figure 28–D) confirms our diagnosis.

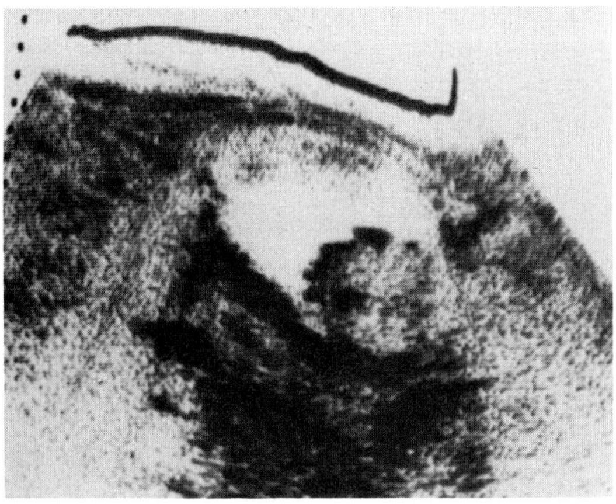

Figure 28–A. Longitudinal section 1R. Transitional cell carcinoma.

Figure 28–B. Transverse section −15. Transitional cell carcinoma.

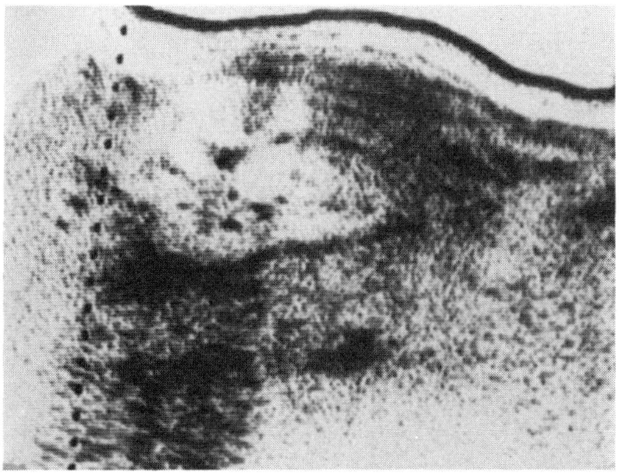

Figure 28–C. Longitudinal section 0L. Hydronephrosis. Prone scan of right kidney.

Figure 28–D. Transitional cell carcinoma. Cystogram from patient in Figures 28–A through 28–C.

REFERENCE

McLaughlin IS, Morley P, Deane RF, et al: Ultrasound in the staging of bladder tumors. Br J Radiol 47:51 (1975)

Case 29

CLINICAL INFORMATION

These two scans were made on different patients: One was normal and the other had metastatic cancer of the lung. They do not have much in common except for the presence of a "tornado" of echoes in the central abdomen. Although one of these is a simple "tornado" and the other is a tiger-striped "tornado," they are both caused by the same thing. What is it?

Figure 29–A. Longitudinal section 9R.

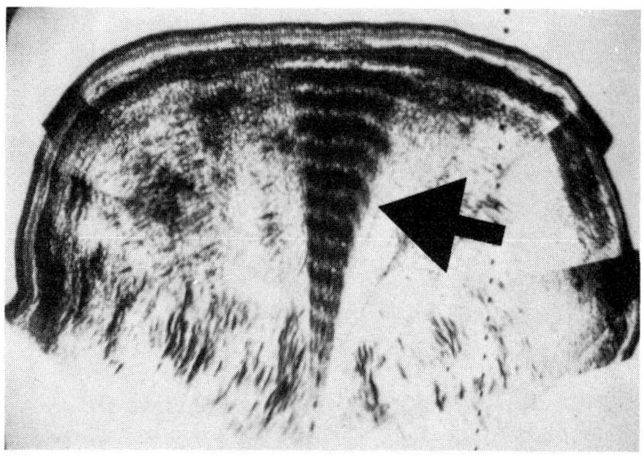

Figure 29–B. Transverse section +15.

DIAGNOSIS

Bowel gas artifacts.

ULTRASOUND FINDINGS

The "tornadoes" are not real echoes at all but rather reverberation artifacts caused by gas in the bowel. Whenever the sound beam crosses an air–soft tissue interface, it is totally reflected and begins to reverberate between the transducer and the interface. This has two effects: first, since the sound beam is totally reflected, the underlying tissues are "shadowed," and no echoes or information can be obtained about them. Second, the reverberating sound beams generate a shower of artifacts, which are projected into the underlying tissue and may be confused with real echoes by the unwary. Artifacts are inevitably produced whenever the sound beam passes through gas-containing bowel. Since we cannot tell which portions of the bowel contain gas, we try to avoid passing the sound beam through bowel and must always be skeptical of any "echoes" produced when the beam has passed through bowel. Figure 29–C shows extensive artifacts in another patient; despite the semblance of form, none of the "echoes" caudal to the gallbladder are real. Note in Figures 29–A and 29–C how the images of the structures deep in the abdomen are lost behind the gas. In Figure 29–A the acoustic shadow is relatively narrow, whereas in 29–C everything caudal to the gallbladder and the lower pole of the right kidney is obscured. Looking for these posterior structures is an easy way to check on the quality of the scan and avoid being misled by artifacts produced by gas. The reference cited discusses the problem of artifacts in greater detail. (Did you notice the liver metastases in Figure 29–B?)

Figure 29–A. Longitudinal section 9R.
Bowel gas artifacts.

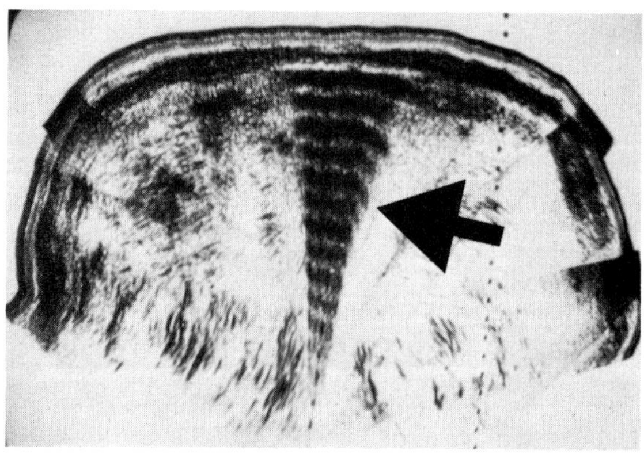

Figure 29–B. Transverse section +15.
Bowel gas artifacts.

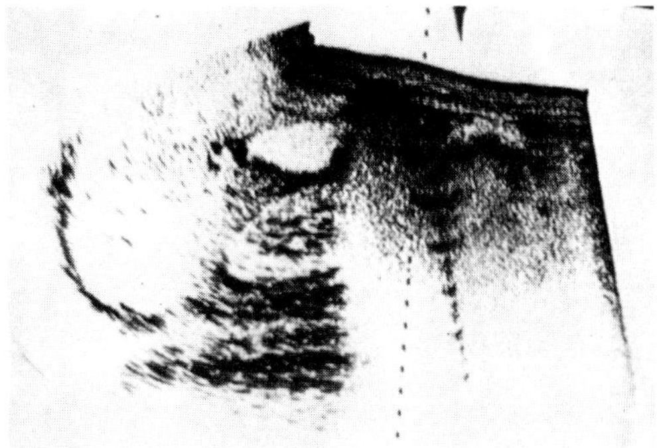

Figure 29–C. Longitudinal section 8R. Bowel gas artifacts.

REFERENCE

Bartrum RJ Jr, Crow HC: *Gray-Scale Ultrasound: A Manual for Physicians and Technical Personnel.* Chapters 5 and 6. WB Saunders Co, Philadelphia (1977)

Case 30

CLINICAL INFORMATION

This 34-year-old woman is two weeks late for her usually regular period. A pregnancy test is positive but there is still some question as to whether she is pregnant. There are two unusual collections of echoes in her uterus, indicated by the curved and straight arrows. What do they represent?

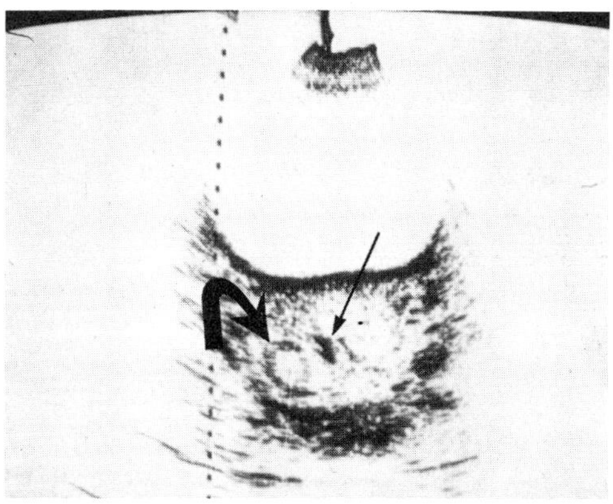

Figure 30–A. Transverse section −14.

DIAGNOSIS

Six-week gestational sac and IUD.

ULTRASOUND FINDINGS

The curved arrow points to a very typical gestational sac of about six weeks' age. The straight arrow indicates a collection of loud echoes adjacent to the sac and within the uterine cavity. If the pregnancy were not there, we would have no trouble recognizing these echoes as originating from an IUD. Well, nothing is perfect and even IUD's can fail. This woman conceived even though her device was in place and supposedly on the job. It is quite possible for this pregnancy to progress normally to term and there have been reports of babies delivered clutching an IUD triumphantly in the hand!

The obstetrician is always interested in knowing if the IUD is still intrauterine, since there is a higher incidence of perinatal maternal complications in patients with a retained IUD. If we are to have a chance of seeing it, we must examine the patient in the first trimester; after that there are too many echoes from the placenta and baby to identify an IUD reliably. Figure 30–B is another patient with the same problem.

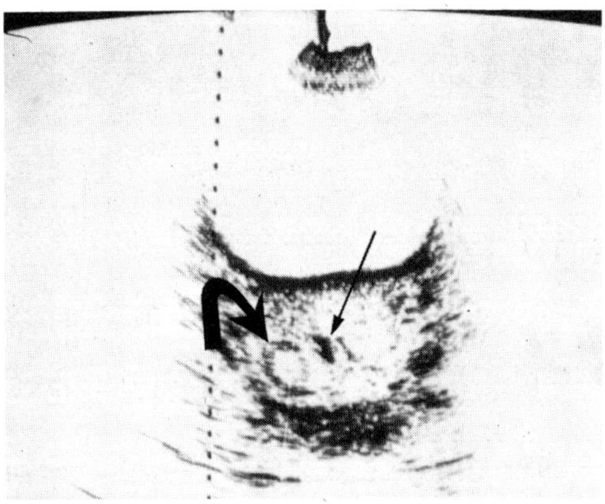

Figure 30–A. Transverse section −14. Gestational sac and IUD.

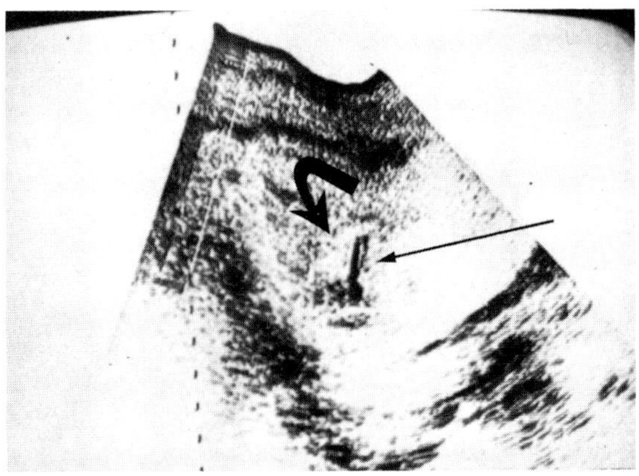

Figure 30–B. Longitudinal section 1L. Gestational sac and IUD.

REFERENCE

Wittman BK, Chow TT: Diagnostic ultrasound in the management of patients using intrauterine contraceptive devices. Br J Obst Gynecol 83:802 (1976)

Case 31

CLINICAL INFORMATION

This 53-year-old woman is really "sick." She feels poorly and weak, has had no appetite, and has lost 15 pounds in the last few months. She complains of a nagging and increasingly severe backache which prevents her from sleeping well. Her health has previously been quite good, and her only medical encounters have been a cholecystectomy 20 years ago and a cesarean section for her third child. Physical examination is not unusual; there are no palpable abdominal masses nor any area of undue tenderness. A full complement of laboratory examinations turns up only mild anemia (hematocrit 31), an elevated sedimentation rate, and borderline abnormal SGOT. An upper GI series shows a small hiatal hernia and some excess gastric secretions; the duodenal sweep is normal.

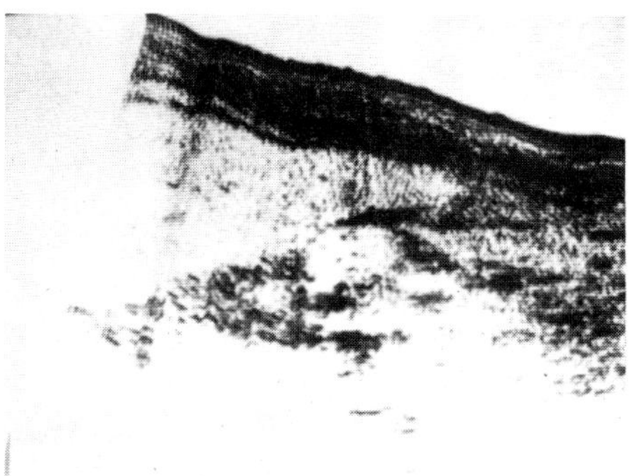

Figure 31–A. Longitudinal section 4L.

DIAGNOSIS

Carcinoma of the body of the pancreas.

ULTRASOUND FINDINGS

The longitudinal scan shows a round mass, moderately echo-producing, beneath the left lobe of the liver (arrow). We can see that the mass lies just caudal and anterior to the splenic vein and anterior to another vascular structure, the superior mesenteric artery (which could be followed to its origin from the aorta). By identifying the relation of the mass to the surrounding vessels, we can state with certainty that it is arising in the pancreas. We have found that longitudinal scans are often more helpful than transverse ones in defining and localizing masses in the body of the pancreas. This is probably because it is easier to stay over the small acoustic window provided by the left lobe of the liver when scanning longitudinally. The splenic vein is also very easy to identify in longitudinal scans, because it is cut in cross section and appears as a characteristic echo-free "hole" just underneath the liver. Once the splenic vein has been found, we can locate the pancreas. The diagnosis in this case is easy; pancreatic masses are either tumors or pancreatitis. All the clinical data and the localized nature of this mass point to cancer. The diagnosis was verified by percutaneous aspiration biopsy.

Tumors of the body of the pancreas can be seen also in transverse scans, as Figures 31–B and 31–C illustrate.

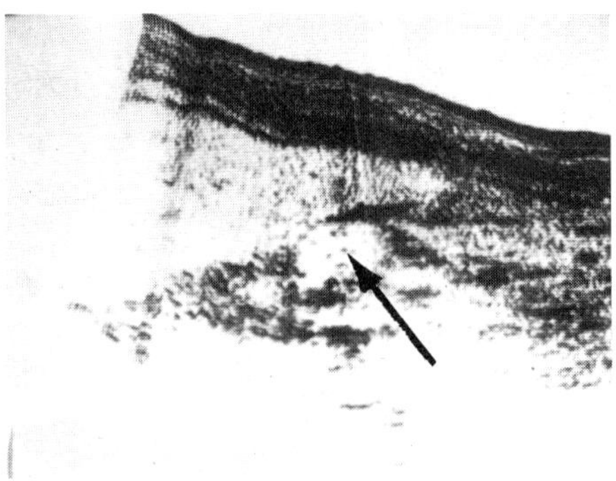

Figure 31–A. **Longitudinal section 4L.** Carcinoma of the pancreas.

Figure 31–B. **Transverse section +8.** Carcinoma of the pancreas.

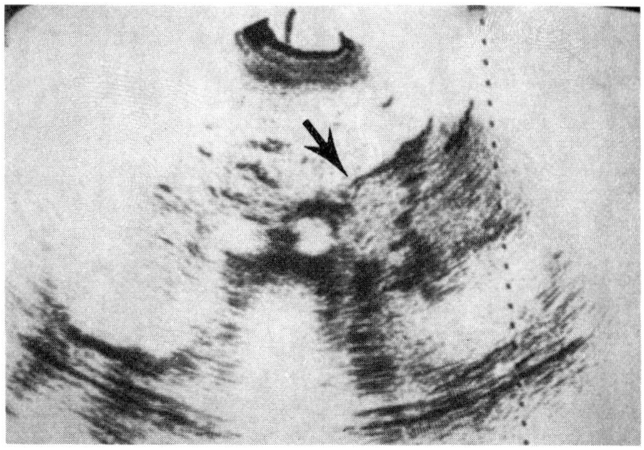

Figure 31–C. Transverse section +12. Carcinoma of the pancreas.

REFERENCES

Perlmutter GS: Chapter 7, Pancreas in *Abdominal Gray Scale Ultrasonography.* BB Goldberg, editor. John Wiley & Sons. New York 1977

Sample WF, Po JB, Gray RK, et al: Gray scale ultrasonography: Techniques in pancreatic scanning. Appl Radiol 4:63 (1975)

Doust BD: Ultrasonic examination of the pancreas. Rad Clin North Amer 13:467 (1975)

Case 32

CLINICAL INFORMATION

This 22-year-old pregnant woman's uterus is considerably larger than the calculated gestational age of 24 weeks. How many babies are there? If more than one, are they monozygotic or multizygotic?

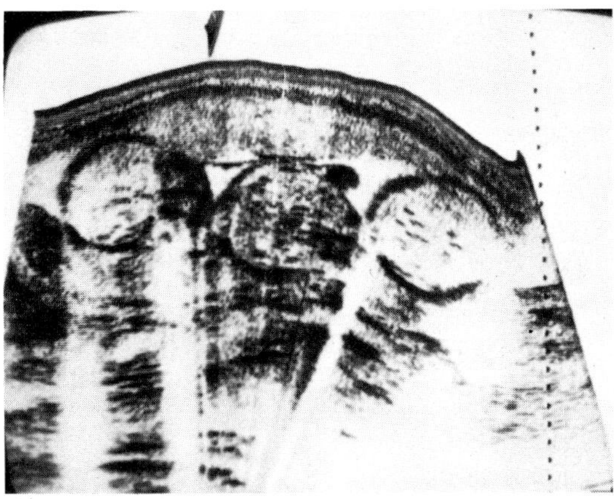

Figure 32–A.
Oblique scan of pregnant uterus.

DIAGNOSIS

Twins (dizygotic).

ULTRASOUND FINDINGS

There are two heads flanking an abdomen (we recognize the abdomen by its multiple internal echoes arising from bowel and liver). Identifying twins can be a tougher problem than meets the eye, particularly without a real-time scanner. In the second trimester a single active baby can change position rapidly, and its head may be found in many different parts of the uterus during the course of an examination. When two babies are jostling around, things can get confusing in a hurry! In order to be certain of our diagnosis we must be able to make a single scan which includes both fetal heads, as in Figure 32–A.

There are two placentas in this case, one on the anterior and the other on the posterior wall of the uterus. The presence of two placentas tells us that we are dealing with dizygotic twins. If dizygotic twins are scanned in the first trimester, we will see two separate gestational sacs (Figure 32–B). Monozygotic twins have only a single placenta (Figure 32–C). (The presence of a multiple gestation does not affect the rate of growth of the fetal skull; the same BPD charts can be used for twin and single gestations.)

When there are more than two babies it takes a skilled scanner to include all the heads on a single picture (Figure 32–D).

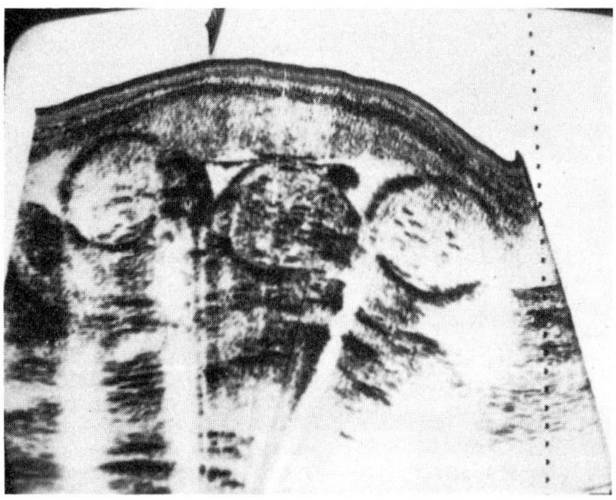

Figure 32–A.
Twins. Oblique scan of pregnant uterus.

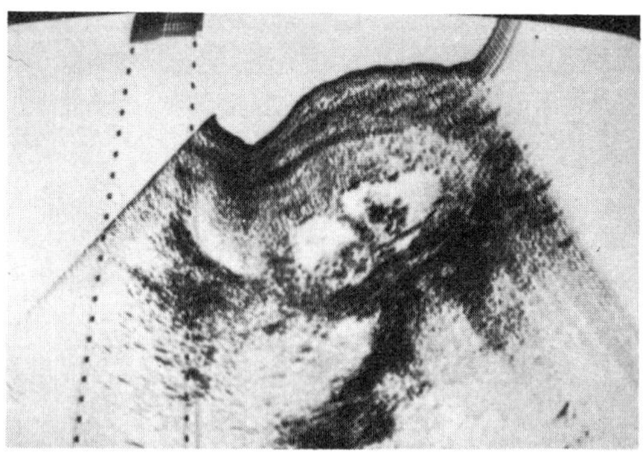

Figure 32–B. Longitudinal section 2R.
Twin gestational sacs.

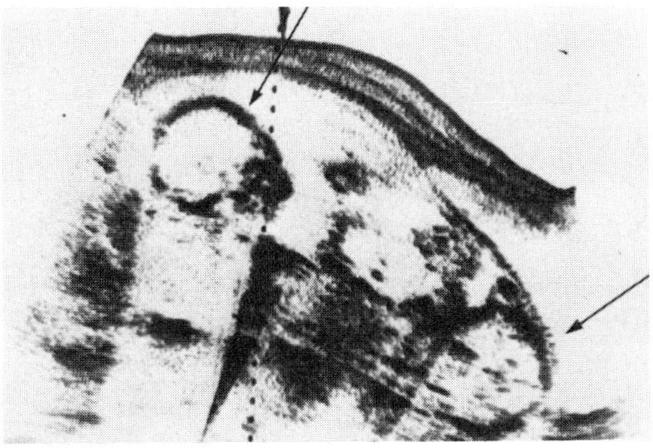

Figure 32–C. Longitudinal section 3R. Twins (arrows point to heads).

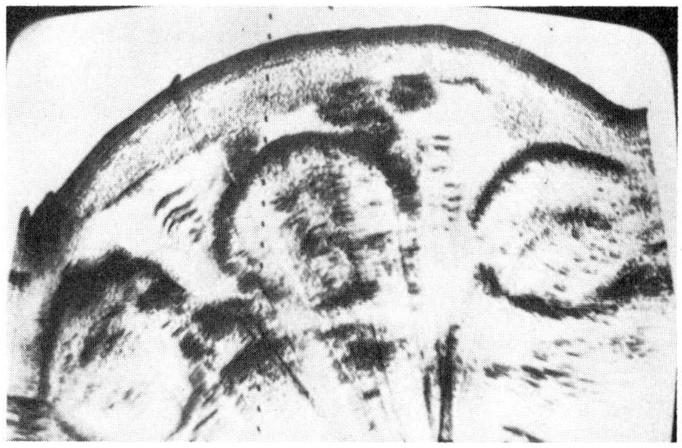

Figure 32–D.
Triplets. Oblique scan of pregnant uterus.

REFERENCE

Scheer K: The role of diagnostic ultrasound in the management of multiple pregnancy. Chapter in, *Ultrasonography in Obstetrics and Gynecology*, Sanders & James, editors. Appleton-Century-Crofts. New York, 1977

Case 33

CLINICAL INFORMATION

This 52-year-old man is bothered by episodic upper abdominal pain. The pain is neither severe nor crampy but occurs fairly regularly about two hours after eating. There is no particular type of food that precipitates the attacks, nor is there anything he can do to abort them (antacids have been ineffective). His general health is good, but upon close questioning he admits to some symptoms that suggest coronary artery disease. Ultrasound is requested as the first procedure in a planned evaluation of nonspecific upper abdominal and chest pain. Figure 33–A is a scan of the gallbladder made with the patient supine; Figure 33–B is a similar scan obtained with the patient upright.

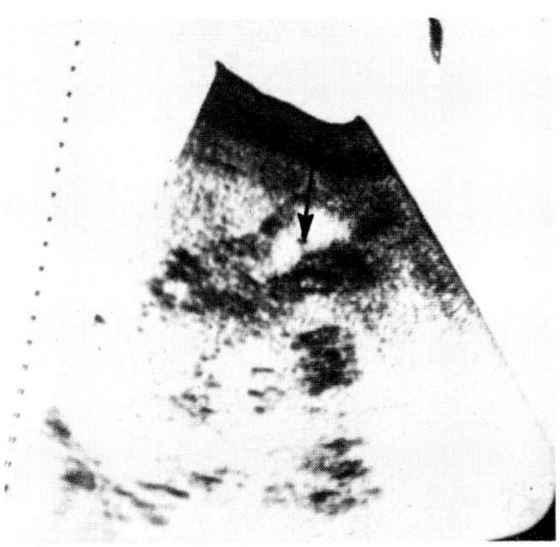

Figure 33–A.
Supine scan; longitudinal section through gallbladder.

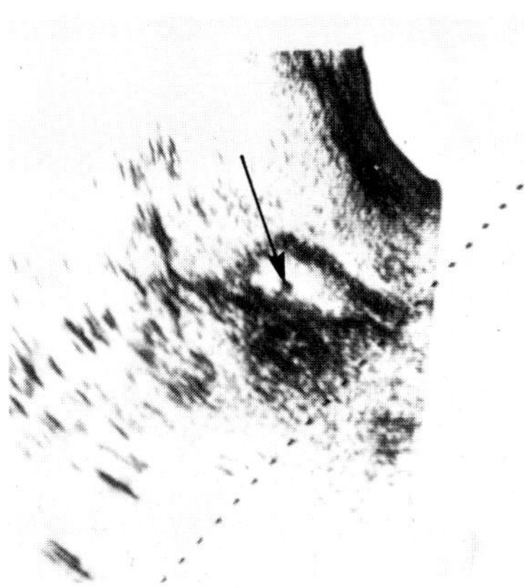

Figure 33–B.
Upright scan; longitudinal section through gallbladder.

DIAGNOSIS

Polyp (cholesterol) of the gallbladder.

ULTRASOUND FINDINGS

There is a small intraluminal echo in the midportion of the gallbladder (arrow). This echo is somewhat unusual for a gallstone since it seems to be attached to the posterior gallbladder wall and does not cast a shadow (a circumstance which implies that it is not very dense). We suspected that it might be a cholesterol polyp and performed an upright scan, which confirmed our suspicion — the echo does not fall to the dependent portion of the gallbladder but remains attached to the wall.

Upright scans are very useful in sorting out problems in the gallbladder, particularly the existence of a small stone, as shown in Figure 33–C. In the supine scan there is a small echo in the neck of the gallbladder that seems to be a stone (wavy arrow); the upright scan shows that the echo has shifted to the dependent portion of the gallbladder (arrowhead) and must indeed be a stone. Contrast this with the patient illustrated in Figure 33–D. The supine scan again shows the suspicious echo in the neck (arrow), but the upright scan is normal. This patient did not have gallstones. We do not perform upright scans in every patient since many times the diagnosis is obvious on supine scans; however, when there is a question in our minds as to what is going on, we do examine the patient upright, and this almost always clears up the confusion.

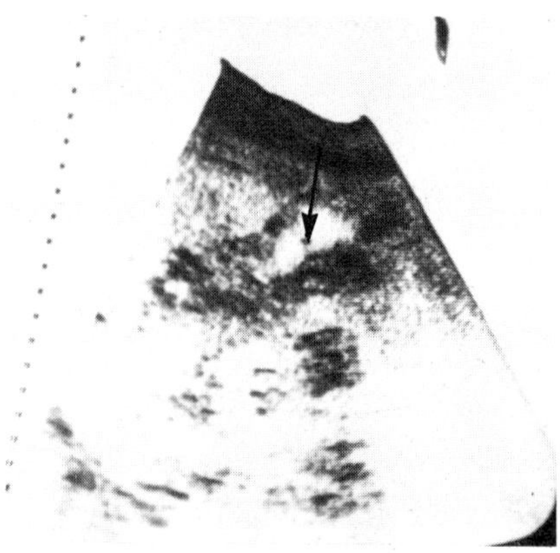

Figure 33–A.
Cholesterol polyp of gallbladder. Supine scan; longitudinal section through gallbladder.

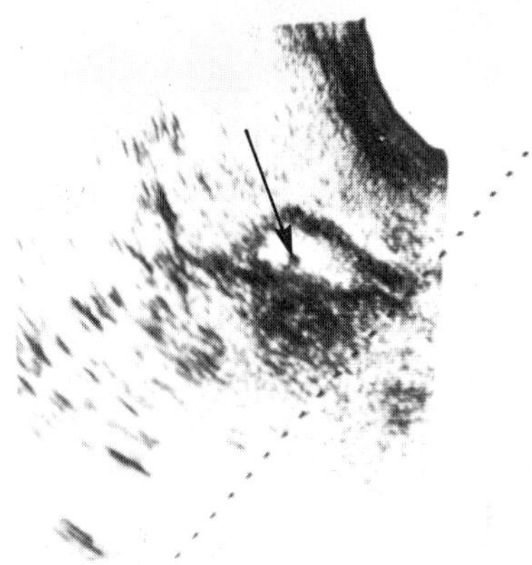

Figure 33–B.
Cholesterol polyp of gallbladder. Upright scan; longitudinal section through gallbladder.

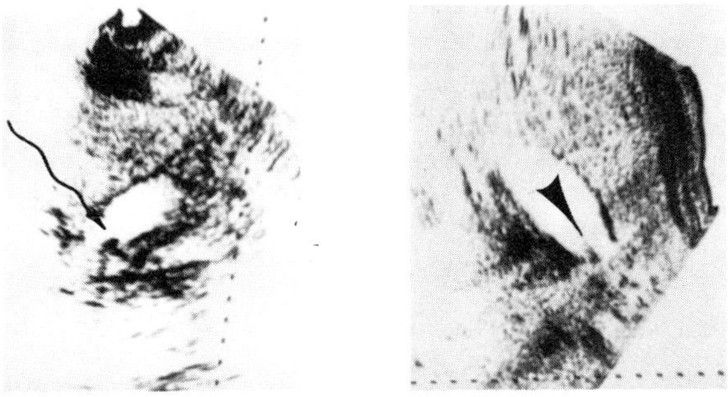

Figure 33–C.
Small gallstone. Supine and upright scans through long axis of gallbladder.

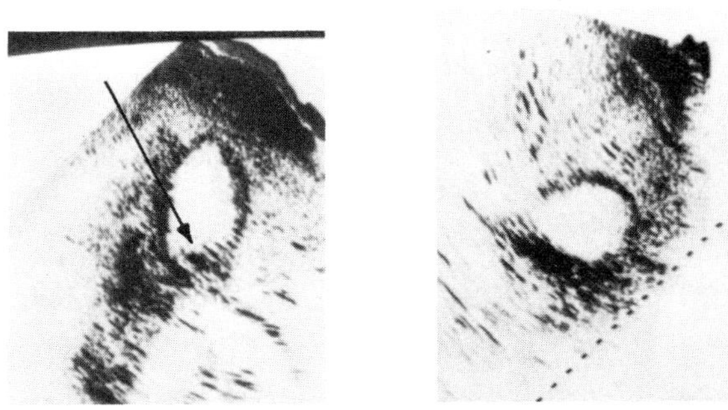

Figure 33–D.
Normal gallbladder. Supine and upright scans through long axis of gallbladder.

REFERENCES

Crow HC, Bartrum RJ Jr, Foote SR: Expanded criteria for the ultrasonic diagnosis of gallstones. J Clin Ultra 4:289 (1976)

Albarelli JN: Erect cholecystosonography. J Clin Ultra 3:309 (1975)

Foster SC, McLaughlin SM: Improvement in the ultrasonic visualization of the gallbladder by using the left lateral decubitus position. J Clin Ultra 5:253 (1977)

Case 34

CLINICAL INFORMATION

This 26-year-old man had severe poststreptococcal glomerulonephritis, which led to total renal failure 18 months ago. He was maintained on hemodialysis for 15 months, and 10 weeks ago had a cadaver renal transplant. The initial post-transplant period went well, but over the past three weeks he has noticed increasing signs of renal failure — weight gain, headaches, and some pain in the transplant site. His serum creatinine, which had been running between 2.0 and 3.0, is now 4.6. An IVP was performed but showed only minimal excretion of contrast. The calyces cannot be definitely identified, but the radiologist is suspicious that there may be some hydronephrosis. Unfortunately, there is no previous IVP to use as a baseline. The patient is sent for ultrasound examination to help in determining if his increasing renal failure is due to ureteral obstruction or rejection. The technician has obtained scans through the long axis of the transplant (Figure 34–A) and through the short axis (Figure 34–B).

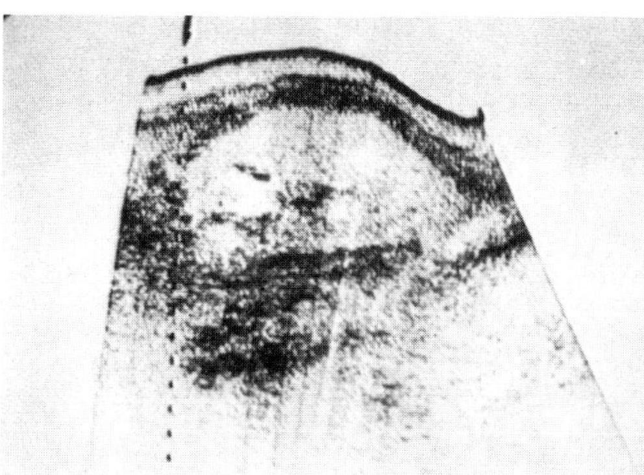

Figure 34–A.
Longitudinal section of transplanted kidney.

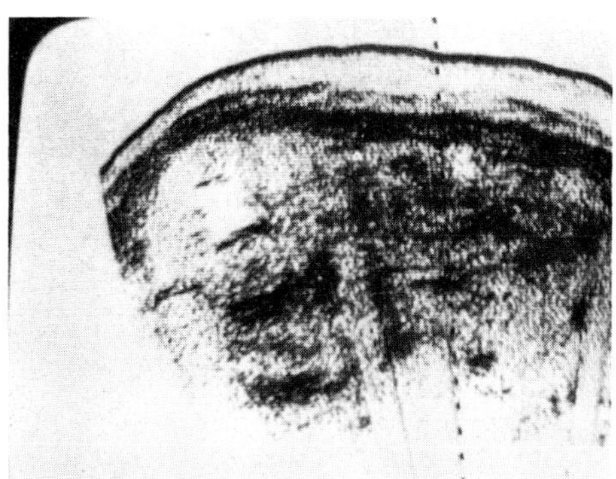

Figure 34–B.
Transverse section of transplanted kidney.

DIAGNOSIS

Transplant rejection, no obstructive uropathy.

ULTRASOUND FINDINGS

As we can see, the transplanted kidney can be very nicely visualized with ultrasound; it is easily accessible because bowel and bone do not interfere with the picture. Interpreting the scans, however, has to be somewhat subjective. Following transplantation the kidney undergoes hypertrophic growth, and there may normally be a mild degree of hydronephrosis as well. This dilatation of the collecting system is not well understood, but it is probably related to partial devascularization of the ureter during surgery or to partial stenosis at the anastomosis of the ureter to the bladder. Whatever the cause, many transplanted kidneys function well in spite of this mild degree of hydronephrosis; and it is, therefore, difficult to say that transplant failure is the basis of obstruction if mild hydronephrosis is the only finding. This case illustrates the problem. The renal pelvis is slightly dilated on these scans (the relatively echo-free area seen in both figures), but because this degree of enlargement can be normal we are uncertain as to what significance to attach to the finding. If this patient had had a baseline scan shortly after transplantation, say within a week, we could then tell whether the hydronephrosis was worsening and could make a more definite statement. Furthermore, if there was marked hydronephrosis, we could confidently say that there was a ureteral obstruction because the physiologic post-transplant hydronephrosis is never massive.

On this scan the only certain ultrasonic sign of renal transplant rejection is the overall renal enlargement. Once again, it would have been helpful to have a baseline study, as this would have allowed us to determine whether or not the kidney is actively enlarging. When rejection occurs, the kidney gets larger because of edema. Differentiating this from the expected renal hypertrophy can be impossible unless one has had the opportunity to observe the kidney in its "normal" post-transplant functioning state.

In reporting this case the ultrasonographer must hedge. He can say that there is mild hydronephrosis, which is consistent with normal post-transplantation dilatation; and although obstructive uropathy cannot be completely excluded, it is unlikely. The increase in renal size is worrisome and might indicate edema due to rejection.

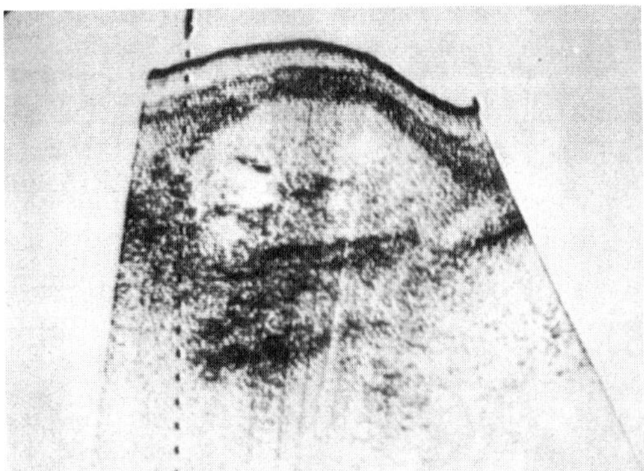

Figure 34–A.
Transplant rejection. Longitudinal section of transplanted kidney.

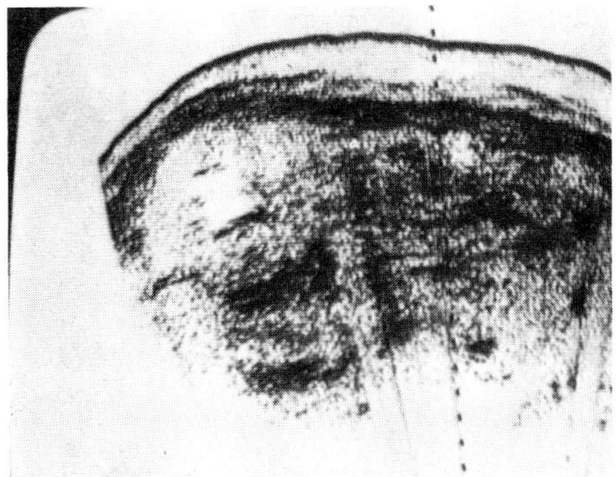

Figure 34–B.
Transplant rejection. Transverse section of transplanted kidney.

REFERENCES

Rosenfield AT, Taylor KJW: Obstructive uropathy in the transplanted kidney: evaluation by gray scale sonography. J Urol 116:49 (1976)

Bartrum RJ Jr, Smith EM, D'Orsi CJ, et al: Evaluation of renal transplants with ultrasound. Radiology 118:405 (1975)

Case 35

CLINICAL INFORMATION

This 44-year-old man feels weak, is frequently nauseated, and his family says he has the "yellow jaundice." His health has not been good in the past few years: he used to have a "drinking problem" which caused chronic stomach ulcers and several episodes of pancreatitis, which have evolved into a clinical picture suggesting chronic relapsing pancreatitis. He claims that he has stopped drinking. His physical examination is marked by obvious clinical jaundice, but the liver does not seem to be enlarged and there is no evidence of ascites or an intra-abdominal mass. The total bilirubin is 5.7 with a direct component of 4.2. The SGOT is 85 (normal less than 40), the alkaline phosphatase 370 (normal less than 120), and the amylase 65. Total protein is 6.2 with an albumin of 3.9. The prothrombin time is 140 per cent, the sedimentation rate 37, the hematocrit 40, and the white cell count 8,300. This limited sector scan was made in the right lobe of the liver.

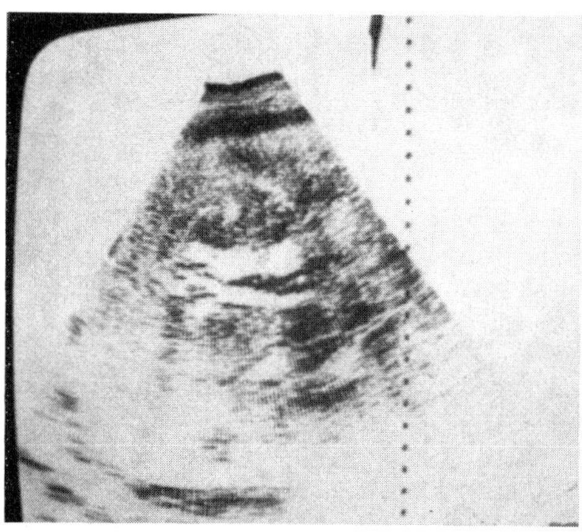

Figure 35–A. Transverse section +13.

DIAGNOSIS

Dilated intrahepatic bile ducts (obstructive jaundice).

ULTRASOUND FINDINGS

There are two parallel tubular structures in the liver: one of these is a portal venous radicle and the other an intrahepatic bile duct. The more posterior is the portal vein; the other is the greatly enlarged right common hepatic duct. Under normal conditions the intrahepatic bile ducts are rarely seen; they are quite small and usually not clearly separated from the adjacent portal venous radicle. On those occasions when normal ducts are seen they do not measure more than 2 mm. in diameter (Figure 35–B). When the ducts are dilated, however, they are easy to see and appear as a second tubular structure running parallel to the vein. This has been termed the "double lumen" or "double-barreled shotgun" sign and is pathognomonic for an enlarged biliary ductal system. A single tubular structure in the liver must be considered a vein, and only when we see the double lumen do we feel confident we are dealing with obstructive jaundice.

Some other characteristics of dilated bile ducts have been described: the ducts are said to taper sharply while veins remain relatively constant in size (Figure 35–C) and massively dilated ducts fill the region of the porta hepatis with multiple circular structures resembling a string of poppet beads (Figure 35–D). While these ancillary signs are often seen and can be useful, we feel that the "double lumen sign" is the *sine qua non* and hesitate to make a diagnosis of dilated ducts without it.

(This patient's distal common bile duct was narrowed by chronic pancreatitis; there was no tumor.)

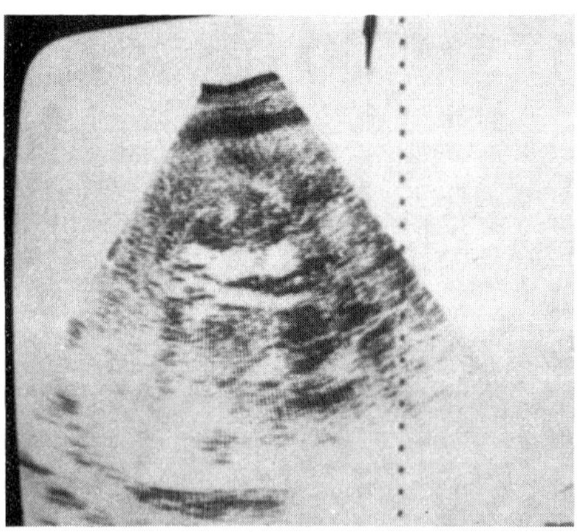

Figure 35–A. Transverse section +13. Dilated bile duct.

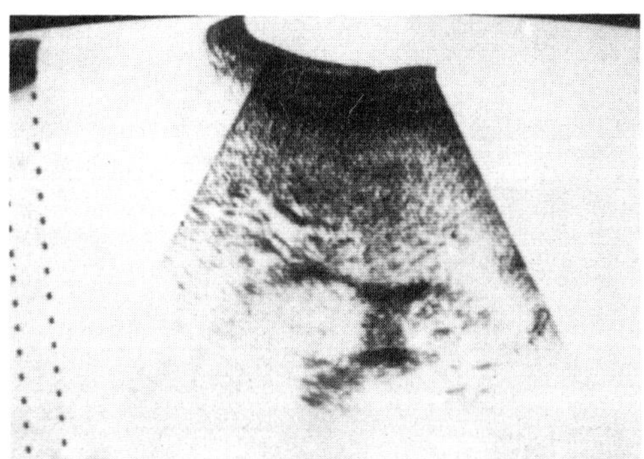

Figure 35–B. Transverse section +16. Normal bile duct.

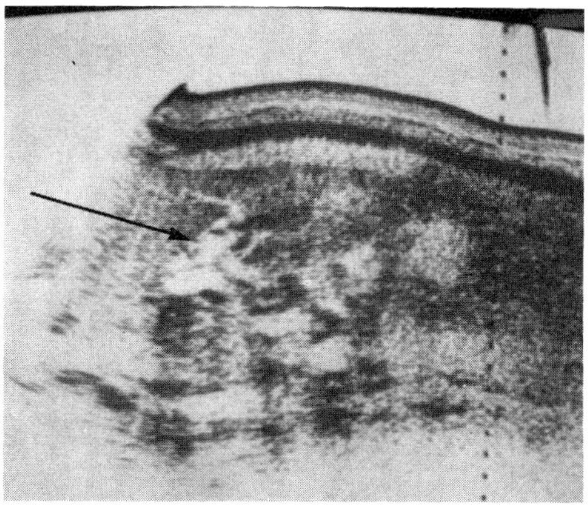

Figure 35–C. Longitudinal section 2R. Dilated bile duct.

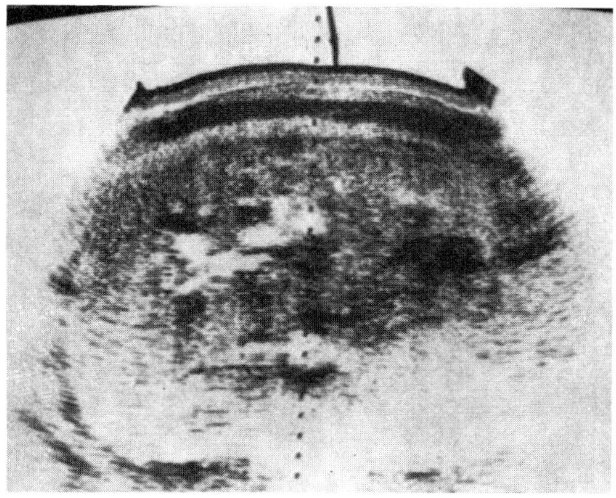

Figure 35–D. Transverse section +12. Dilated bile ducts.

REFERENCES

Weill F, Eisencher A, Zeltner F: Ultrasonic study of the normal and dilated biliary tree. Radiology 127:221 (1978)

Conrad MR, Landay MJ, James JO: Sonographic "parallel channel" sign of biliary tree enlargement in mild to moderate obstructive jaundice. Am J Radiol 130:279 (1978)

Neiman HL, Mintzner RA: Accuracy of biliary duct ultrasound: comparison with cholangiography. Am J Radiol 129:979 (1977)

Case 36

CLINICAL INFORMATION

Figures 36–A, 36–B and 36–C are from three different patients, all of whom are being examined to "rule out gallstones." Which, if any of them, have stones? How certain are you of your diagnosis in each case?

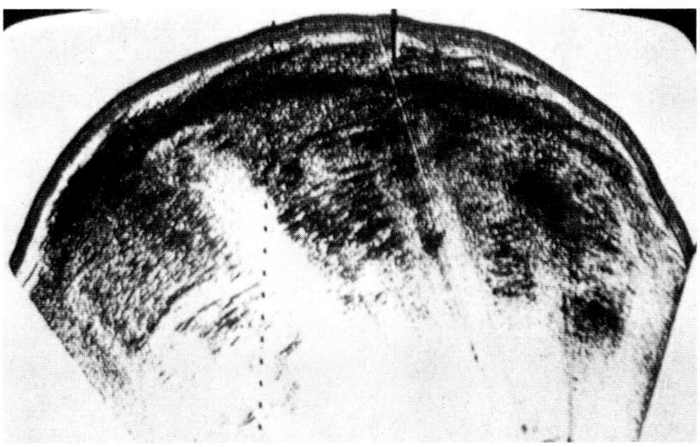

Figure 36–A. Transverse section +7.

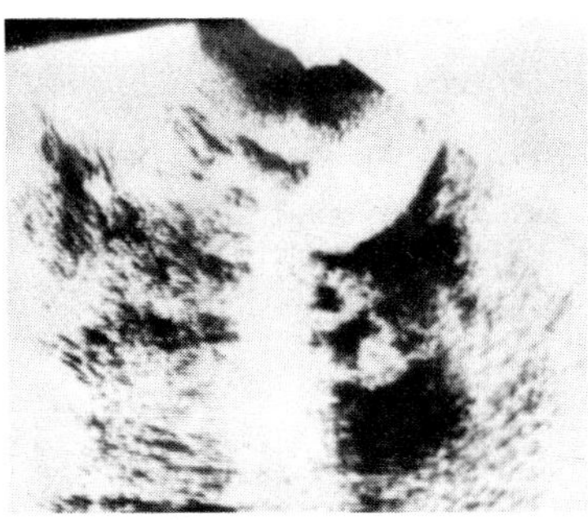

Figure 36–B. Longitudinal section 5R.

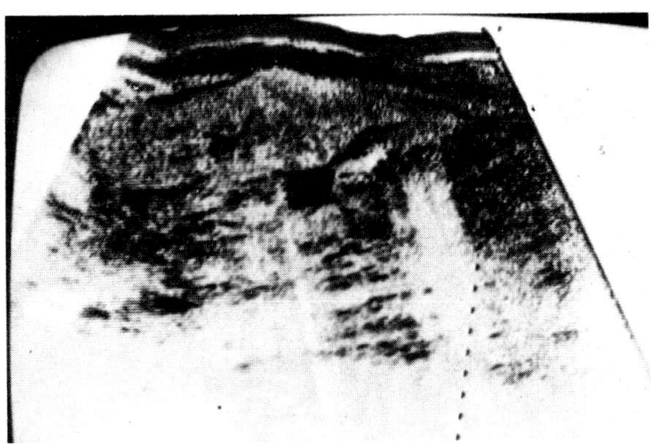

Figure 36–C. Longitudinal section through gallbladder.

DIAGNOSIS

All three patients have gallstones.

ULTRASOUND FINDINGS

We can say with a high degree of confidence that all of these patients have gallstones. We are basing this diagnosis on the presence of an ultrasonic shadow arising from the region of the gallbladder (arrows). In the discussion of Case 19 we mentioned that there are three ultrasonic signs of gallstones: the presence of an intraluminal echo, an irregularity in the posterior wall of the gallbladder, and the presence of an acoustic shadow in the tissues underlying the gallbladder. Any one of these signs is sufficient to make the diagnosis of stones — it is not necessary for all three to be present.

The ultrasonic "shadow sign" arises because dense gallstones reflect a large percentage of the sound and leave a greatly weakened beam to traverse the underlying tissues. This "anemic" sound beam can generate only very weak (or no) echoes, and thus tissues behind a gallstone are said to be in an ultrasonic "shadow." When the gallbladder is completely filled with stones or a stone is impacted in the neck of the gallbladder or cystic duct, there is no surrounding bile and it can be very difficult to identify the echoes from the stones themselves. They still cast an ultrasonic shadow, however, and the presence of the shadow is a dead giveaway to their existence. Figures 36–A through 36–E are examples of this phenomenon. In Figures 36–A and 36–D the gallbladder is completely filled with stones and the only thing we see is the shadow (arrows). In Figures 36–C and 36–E there are still small pockets of bile but the prominent feature is the shadow (arrows). In Figure 36–B we clearly see the distal portions of the gallbladder, but there is a dramatic shadow obscuring the neck (arrow); this shadow arises from a large stone impacted in the infundibulum of the gallbladder.

We must be careful in using the "shadow sign," since air in the bowel also casts an ultrasonic shadow, which can be indistinguishable from that caused by gallstones. The trick is to locate the origin of the shadow; if it arises from the gallbladder fossa and *if the gallbladder is not seen,* then we can be confident that we are dealing with gallstones. We should be able to identify the gallbladder in every patient we scan (assuming, of course, it has not been removed); if it can't be seen, we must be very suspicious that we are dealing with a gallbladder full of stones and hunt carefully for a "shadow sign."

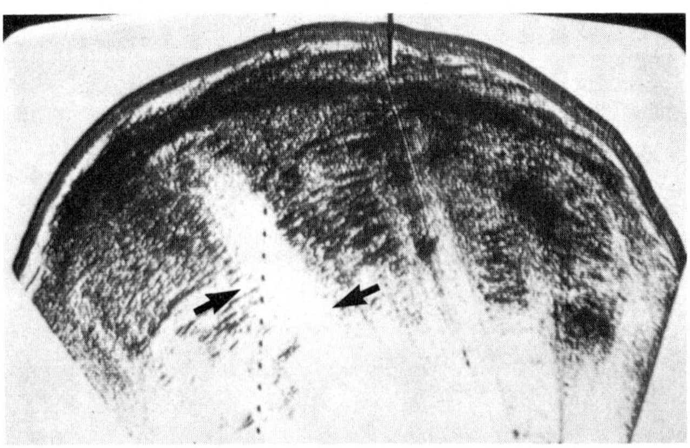

Figure 36–A. Transverse section +7. Gallstones (shadow sign).

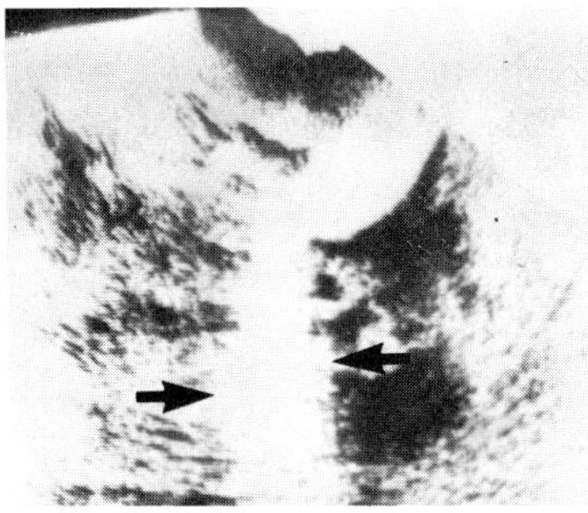

Figure 36–B. **Longitudinal section 5R.** Gallstone in gallbladder neck.

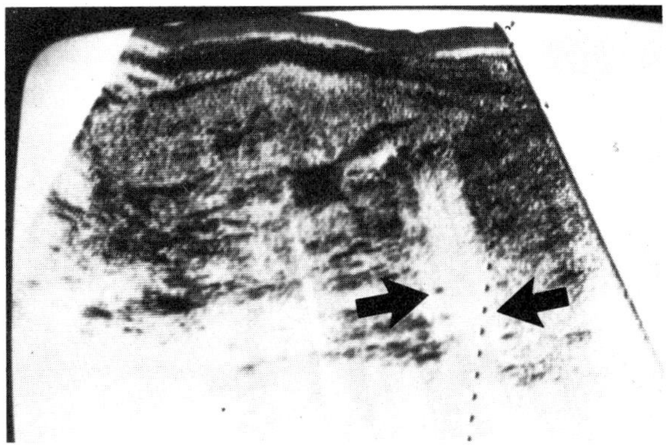

Figure 36–C. Gallstones. Longitudinal section through gallbladder.

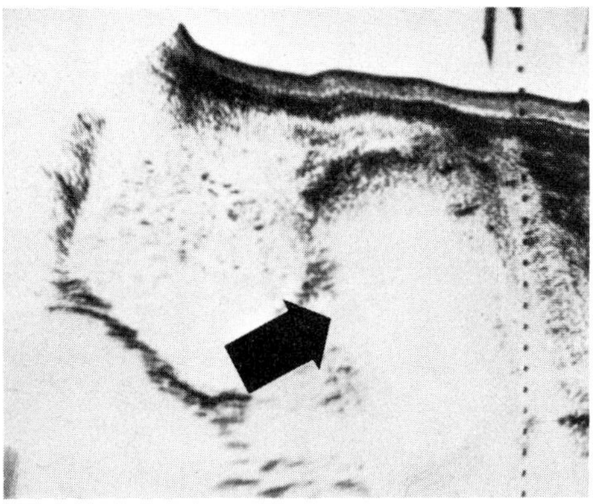

Figure 36–D. **Longitudinal section 10R.** Gallstones.

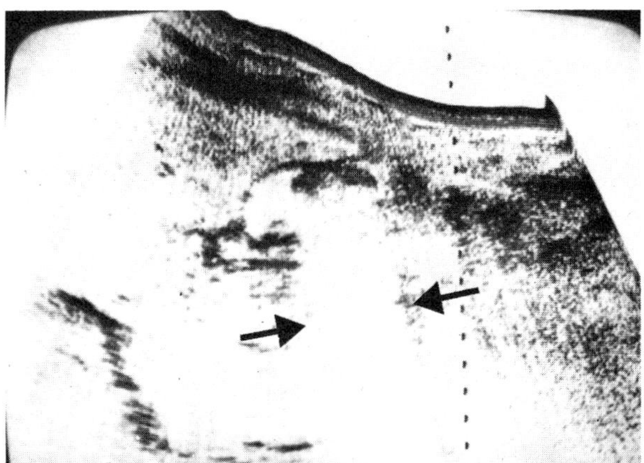

Figure 36–E. **Longitudinal section 8R.** Gallstones.

REFERENCE

Crow HC, Bartrum RJ Jr, Foote SR: Expanded criteria for the ultrasonic diagnosis of gallstones. J Clin Ultra 4:289 (1976)

Case 37

CLINICAL INFORMATION

This 29-year-old woman is eight weeks pregnant but has had intermittent vaginal bleeding for the past three days. She has not passed any tissue or blood clots. Is the pregnancy still viable?

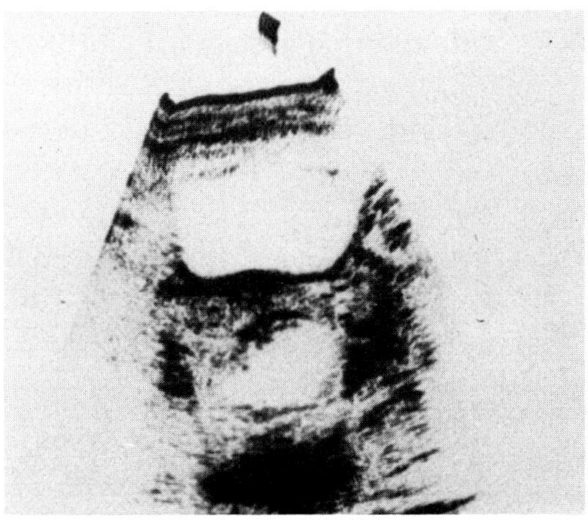

Figure 37–A. Transverse section –14.

DIAGNOSIS

Nonviable pregnancy (missed abortion).

ULTRASOUND FINDINGS

You should have missed this case: if not, then you are over-reading the scan. This uterus contains a gestational sac which is appropriate in size for the eight-week gestational age. There is nothing about the appearance of the sac that allows us to decide if the pregnancy is viable. In these situations we must always take the conservative position and assume that the pregnancy is viable until it has definitely proved itself otherwise. If the appearance of the sac or early embryo is compatible with the gestational age (see Case 8), we must wait for four weeks and look for signs of continued growth. During this period the patient will usually progress to clinical abortion if the pregnancy is not viable. If she has still not passed tissue and there has been no ultrasonic evidence of growth over a four-week period, we can conclude that the pregnancy is dead. The four-week observation period is valid *only in the first trimester*; fetal growth is too variable in the later stages of pregnancy to use this criterion. In the second and third trimester the diagnosis of fetal demise is made by observing the fetal heart.

Sometimes the gestational sac will be obviously disrupted and inappropriate for the predicted age. In these situations we can suggest that the pregnancy is not viable, but we still recommend waiting for the passage of tissue before proceeding to a D & C. Figures 37–B and 37–C are examples of nonviable pregnancies with distorted gestational sacs.

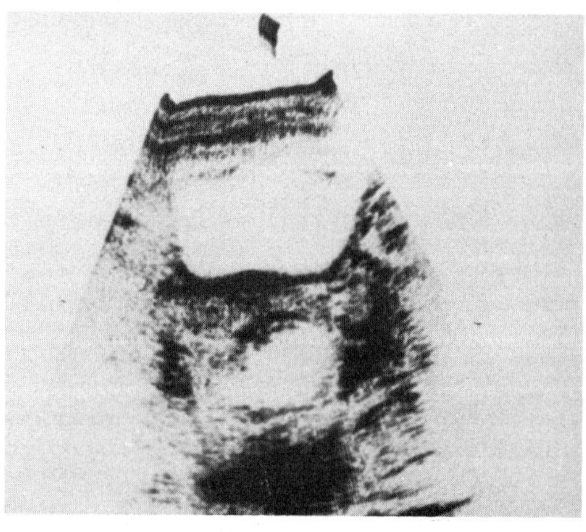

Figure 37–A. Missed abortion. **Transverse section −14.**

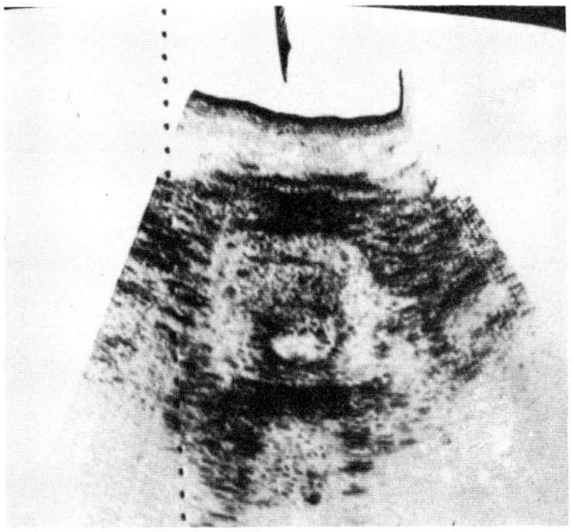

Figure 37–B. Transverse section −12. Missed abortion.

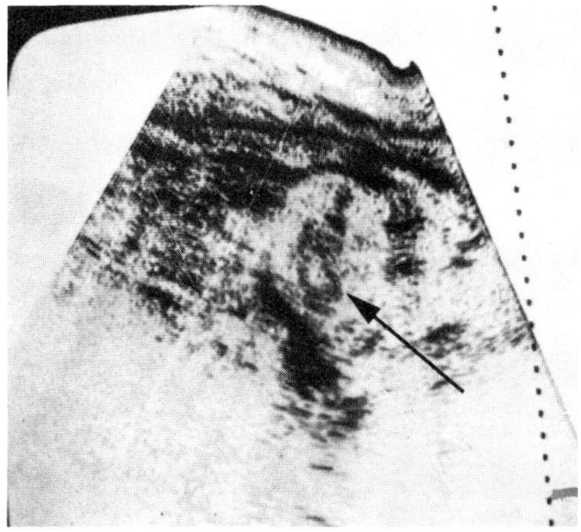

Figure 37–C. Longitudinal section 1L. Missed abortion.

REFERENCES

Hobbins JC, Winsberg F: *Ultrasonography in Obstetrics and Gynecology.* Chapter 2. Williams & Wilkins Co., Baltimore (1977)

Drumm JE, Clinch F: Ultrasound in the management of clinically diagnosed threatened abortion. Br Med J 2:424 (1975)

Case 38

CLINICAL INFORMATION

This 33-year-old woman has episodic midepigastric pain that is unrelated to eating and is worse at night. Physical examination and routine laboratory tests are normal, as is an oral cholecystogram. An upper gastrointestinal series shows only a small hiatal hernia without gastroesophageal reflux. The abdominal ultrasound scan is unremarkable except for the presence of a band of loud echoes in the right lobe of the liver (arrow). How do you explain this finding?

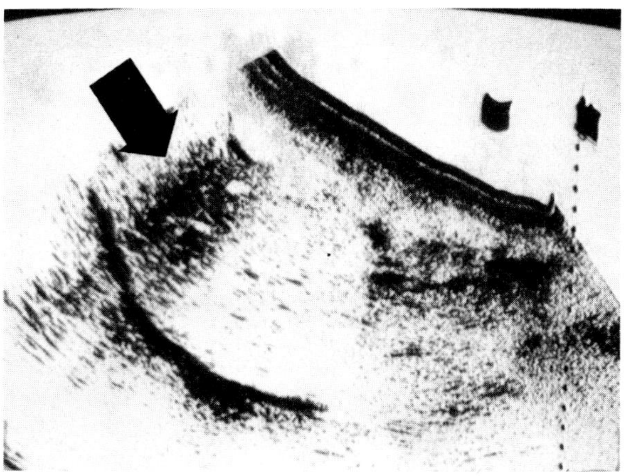

Figure 38–A. Longitudinal section 10R.

DIAGNOSIS

Normal liver showing effects of increased sound transmission through the gallbladder.

ULTRASOUND FINDINGS

The band of loud echoes in this liver are caused by "increased through transmission" of the sound beam by the echo-free gallbladder. This phenomenon occurs when the sound beam passes through any echo-free (i.e., fluid-filled) area and is the ultrasonic opposite of a "shadow sign." The portion of the sound beam that passes through an echoless area is not being attenuated; and when it emerges from the area, it is relatively stronger than the other portions of the beam, which do not have the benefit of the echo-free journey. The subsequent echoes that are produced by this unattenuated beam will be louder than those produced by its attenuated neighbors. This phenomenon of "increased through transmission" is actually one of our diagnostic criteria for distinguishing a fluid-filled or "cystic" area. In this case the "cystic" area is the patient's gallbladder; the gallbladder itself is not well seen because it is located very superficially in the region where the transducer focus is poor (we switched to a transducer with a near field focus to evaluate this gallbladder). "Increased through transmission" is easy to recognize if the ultrasonographer is aware of it; however, the uninitiated might be misled into confusing it with a focal liver lesion.

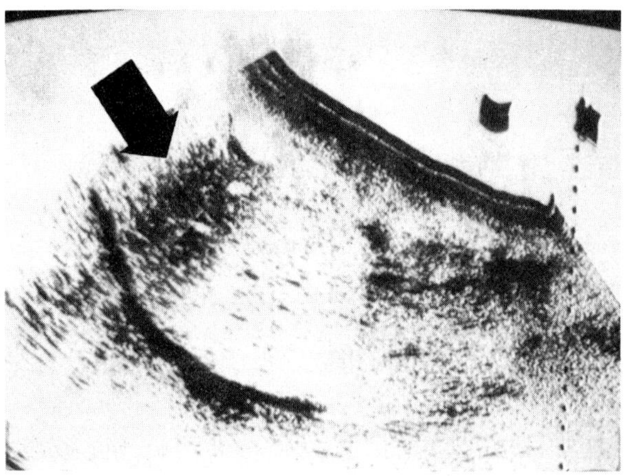

Figure 38–A. **Longitudinal section 10R.** Increased sound transmission through gallbladder (normal liver).

Case 39

CLINICAL INFORMATION

This 34-year-old woman has had three episodes of crampy pelvic pain. The most recent was quite severe and associated with nausea and vomiting. Her health is otherwise good. She does not think she is pregnant and has never had any previous gynecologic problems. Her only previous hospitalizations were for peritonitis from a ruptured appendix 15 months ago and normal uncomplicated childbirth three years ago. Physical examination reveals a normal uterus and an irregular, soft, nontender mass occupying most of the right half of the pelvis. The right ovary cannot be felt. She is afebrile and routine laboratory data, including pregnancy test and cervical culture, are normal.

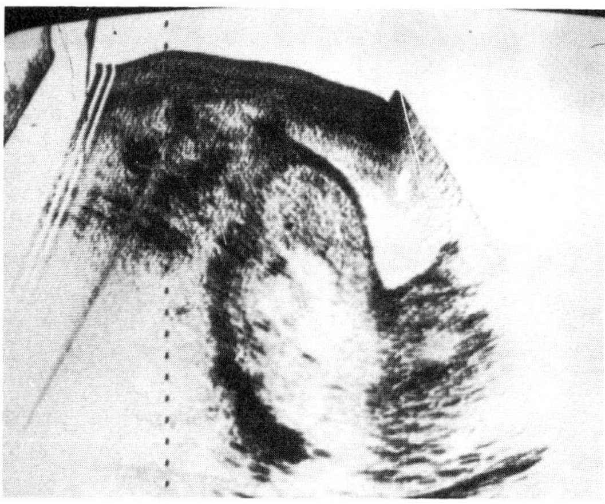

Figure 39–A. Longitudinal section 1R.

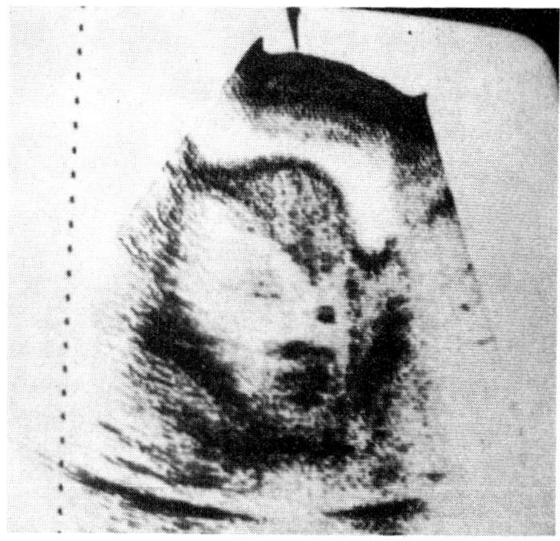

Figure 39–B. Transverse section –15.

DIAGNOSIS

Small bowel adhesions secondary to prior peritonitis.

ULTRASOUND FINDINGS

Although the findings are rather straightforward in this case, the diagnosis is not. There is an obvious mass in the right pelvis which elevates the uterus and displaces it toward the left. The mass is mostly echo-free but does have occasional internal echoes, suggesting septation. It appears we are dealing with a multilocular cystic structure. Could this be arising from the ovary? A corpus luteum cyst should not be this large in the presence of a normal pregnancy test. An endometrioma is unlikely because of the prior pregnancy. Simple ovarian cysts are not usually multilocular. About the only ovarian mass which could be considered is a cystadenoma or cystadenocarcinoma. What else could this be? Ectopic pregnancy is excluded by the normal pregnancy test and pelvic abscess seems unlikely because of the normal cervical culture and lack of tenderness, fever, or elevated white count. We have no really satisfactory explanation — it seems that cystadenoma or cystadenocarcinoma are the best diagnoses — by default!

At exploratory surgery there were multiple matted loops of small bowel entangled with and adhering to the uterus and right adnexal structures. The right ovary was normal. The adhesions were presumably secondary to the previous peritonitis and it is probably fluid in the loops of small bowel which produce the appearance of a multilocular cyst. Sometimes we cannot make the correct diagnosis even with a good scan and thorough clinical history — a worthwhile lesson in humility for those who claim ultrasound has all the answers.

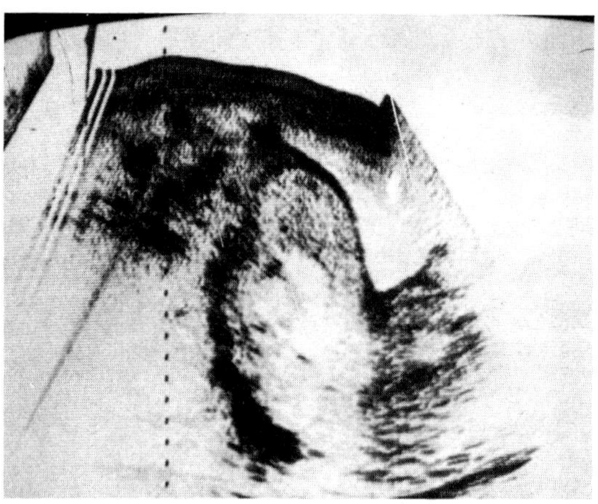

Figure 39–A. Longitudinal section 1R. Small bowel adhesions.

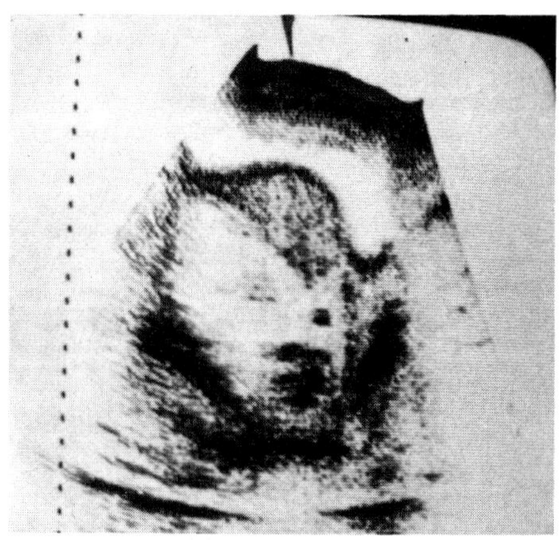

Figure 39–B. Transverse section −15. Small bowel adhesions.

Case 40

CLINICAL INFORMATION

This 78-year-old man has abdominal and back pain, which has been bothering him for several months. He says that he had "the yellow jaundice" several years ago, but otherwise considers himself to be healthy. Physical examination is not remarkable except for signs of chronic pulmonary disease. The technician comes to you with this scan saying that the man has a tumor in the left upper quadrant; what do you think?

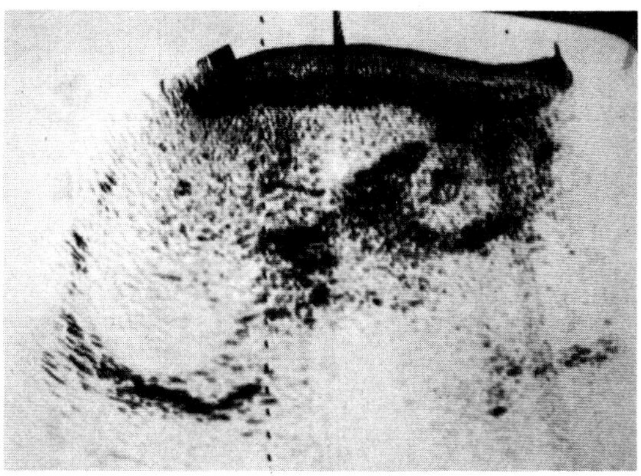

Figure 40–A. Transverse section +9.

DIAGNOSIS

Normal stomach containing food.

ULTRASOUND FINDINGS

There is a well-defined mass in the left upper quadrant just beneath the left lobe of the liver. The presence of a back wall and the fact that it can be easily reproduced on scans performed in different planes make us confident that we are dealing with a real mass and not an artifact. Could this be some normal structure? The stomach is the organ we expect to find in this region, but it should not appear as a mass because the patient had been fasting and the stomach should be empty. Just to be sure, we questioned the patient and discovered that he had eaten two eggs, six strips of bacon, a bowl of oatmeal, and a stack of pancakes for breakfast an hour before. Nobody told him ". . . anything about skipping breakfast." This is the reason for having patients fast before abdominal ultrasound examinations; we do not want any extra "masses" around. (Fasting is *not* done to decrease bowel gas nor to torment the patient.) Although we suspect that we are seeing only stomach contents, we cannot exclude a stomach tumor or other pathologic mass. We explained to our friend, therefore, that it is important that he have an empty stomach when the examination is performed and we rescheduled him for the next day. That scan (Figure 40–B) is more normal in appearance; the left upper quadrant is obscured by gas, and the previous day's mass is gone.

Whenever we suspect that a mass may actually be a part of the normal GI tract containing food or feces, we request a repeat scan on a different day. In most cases the mass will have disappeared. In the rare situation where the appearance is unchanged, we recommend either a barium study or percutaneous aspiration biopsy to confirm the diagnosis. (There is no danger in biopsy of either the stomach or feces-filled colon if a thin needle is used.)

Figure 40–C shows another mass in the left upper quadrant that turned out to be normal GI tract — a feces-filled splenic flexure of the colon.

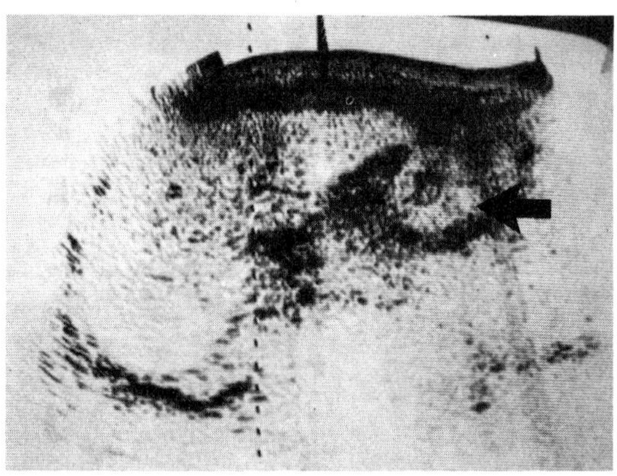

Figure 40–A. Transverse section +9. Food in stomach (normal).

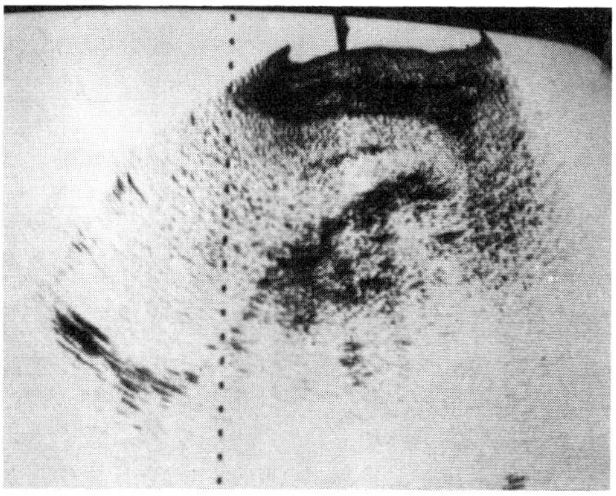

Figure 40–B. Transverse section +9. Normal stomach (same patient as in Figure 40–A, scanned after fasting).

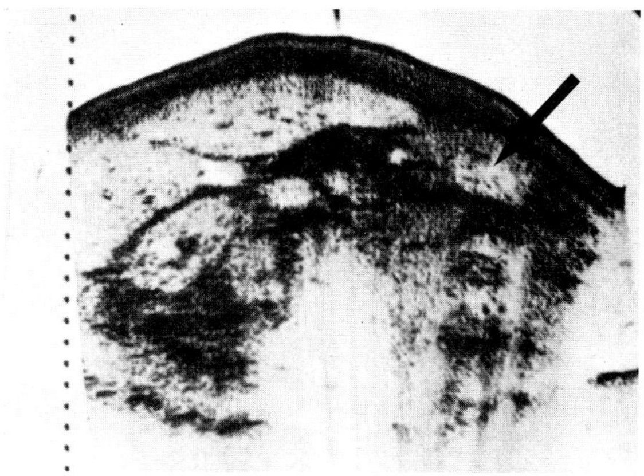

Figure 40–C. Transverse section +5. Feces-filled colon (normal).

REFERENCE

Teele RL, Rosenfield AT, Freedman GS: The anatomic splenic flexure: an ultrasonic renal imposter. Am J Radiol 128:115 (1977)

Case 41

CLINICAL INFORMATION

This 52-year-old business executive is well and has no medical complaints, but his company requires a routine physical examination every 18 months for all its executives. Although in the past these have always been normal, his most recent examination revealed a blood pressure of 150/93. All the basic laboratory studies, including a urinalysis, are normal; because of the hypertension an IVP was performed. The films demonstrate a large mass in the lower pole of the right kidney; nephrotomograms suggest that this is a solid lesion rather than a renal cyst (Figure 41–A). We have obtained a longitudinal ultrasound scan through the kidney and the mass (Figure 41–B).

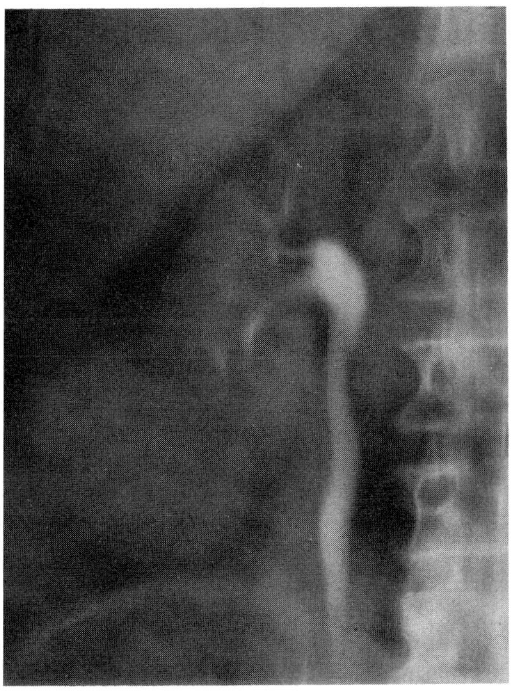

Figure 41–A. Nephrotomogram of right kidney.

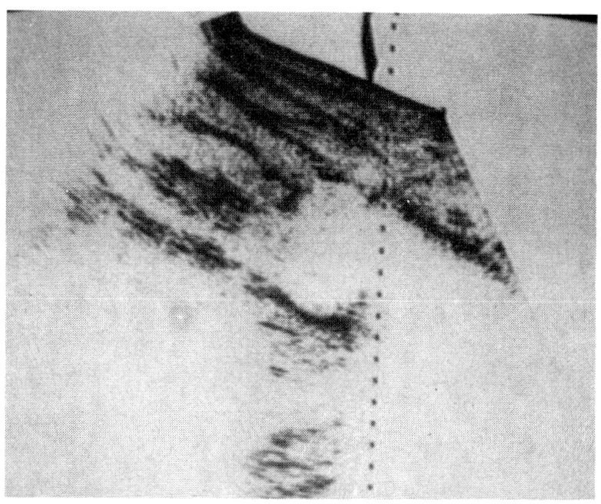

Figure 41–B. Longitudinal section 0L. Prone scan of right kidney.

DIAGNOSIS

Renal cyst.

ULTRASOUND FINDINGS

Probably the most frequent indication for an ultrasound scan of the kidney is the evaluation of a mass found on an IVP. The majority of such cases are simple renal cysts, and the ultrasonographer should be able to identify these benign lesions quite accurately. Not only should he be able to find these lesions, but he must also have the courage to diagnose them definitively as a "benign cyst" for which no further diagnostic or therapeutic procedures are needed. A wishy-washy ultrasound report serves no one well; if the ultrasonographer equivocates on a case and defers to angiography, the ultrasound examination will have been a waste of time and money. Of course common sense is also necessary; there are some cases in which it simply is not possible to classify the renal mass accurately (just as some gastric ulcers defy categorization as benign or malignant), and in these situations the ultrasonographer must have the wisdom to recommend further studies, such as cyst puncture (or biopsy) or arteriography. We have found that these equivocal cases are infrequent.

This scan (Figure 41–B) shows all the classic findings of a benign renal cyst; let us go over them carefully and also examine some less typical scans of other renal cysts (Figures 41–C through 41–G) so that we can become confident in making this diagnosis.

(1) A cyst should have no internal echoes. There are no interfaces in a cyst and consequently no echoes should be generated. The absence of internal echoes is relatively easy to demonstrate with gray-scale scanners, *providing that the gain and the TCG controls are properly set to minimize artifacts.* Initial scans should be made at a very low-gain setting so that both the mass and the renal parenchyma are echo-free. The gain should then be slowly increased until the renal parenchyma fills in with low-level echoes; if at this gain setting there are no echoes within the mass, we can say that the lesion is echo-free. On the other hand, if echoes are present at this level, the mass cannot be considered a cyst, and further work-up is indicated. Sometimes artifactual echoes will appear within a cyst, even at low-gain settings; these artifacts can usually be recognized because they do not appear uniformly in the mass but rather are seen near the anterior wall (wiggly arrows in Figures 41–C, 41–D, and 41–G).

(2) A cyst should show "increased through transmission." This sign is just the opposite of an ultrasonic "shadow sign"; because a cyst does not generate internal echoes the sound beam passes through it intact and is able to generate louder echoes in the tissue behind the cyst. (You recall that the "shadow sign," most commonly seen with gallstones, occurs because most of the sound beam is reflected by strong echo interfaces and is not available to produce echoes in underlying tissues.) Increased through transmission is clearly seen in Figures 41–B, 41–C, and 41–E (between stubby arrows).

(3) A cyst must have a *smooth, continuous back wall.* The entire wall of the cyst is smooth and continuous, of course, but ultrasound can demonstrate this only on the back wall; the front and side walls are usually not clearly defined and smooth. We feel that this is the most sensitive and important

finding in diagnosing a benign cyst. We have never seen (nor been shown) a tumor, cyst wall tumor, necrotic tumor, or abscess which has a smooth, continuous back wall. Consequently, if such a back wall can be seen on the ultrasound scan, we call the mass a cyst unequivocally and recommend no further procedures. Conversely, if despite careful scanning we cannot get the back wall to be smooth and continuous, we always give an equivocal reading and recommend cyst puncture. We do this even though the other two signs of a cyst are present; a smooth back wall is the *sine qua non* of a renal cyst. (Many of these equivocal cases ultimately prove to be renal cysts, but some turn out to be tumors or infections.) All the cysts on these pages have the smooth back wall (small arrows on Figures 41–C through 41–F).

Usually we require all three ultrasonic signs before diagnosing a cyst; however, on occasion it is not possible to demonstrate increased through transmission or clear the mass of artifactual echoes. If there is a strong back wall echo and the IVP and clinical findings are compatible with a benign cyst, we will also diagnose these cases as definitely cystic.

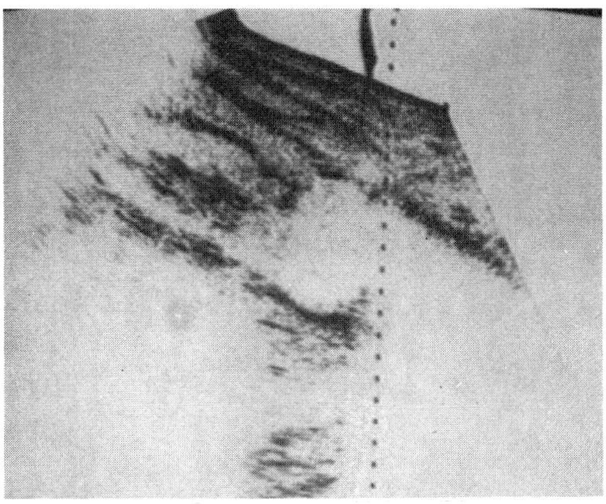

Figure 41–B. Longitudinal section 0L. Renal cyst. Prone scan of right kidney.

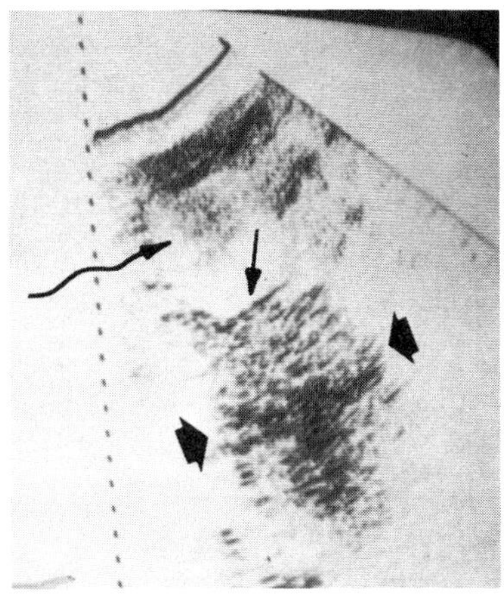

Figure 41–C. Renal cyst.

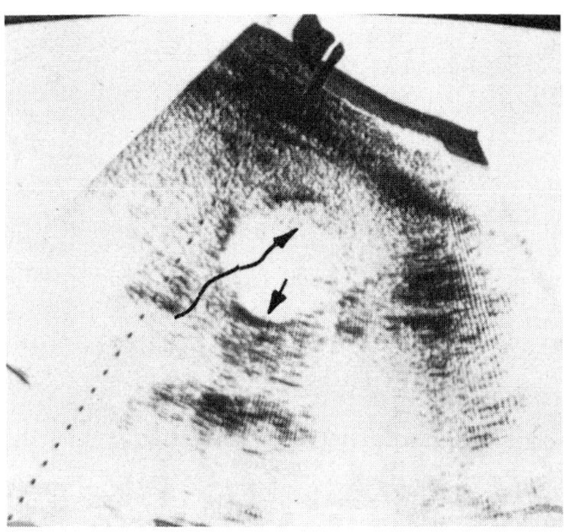

Figure 41–D. Renal cyst.

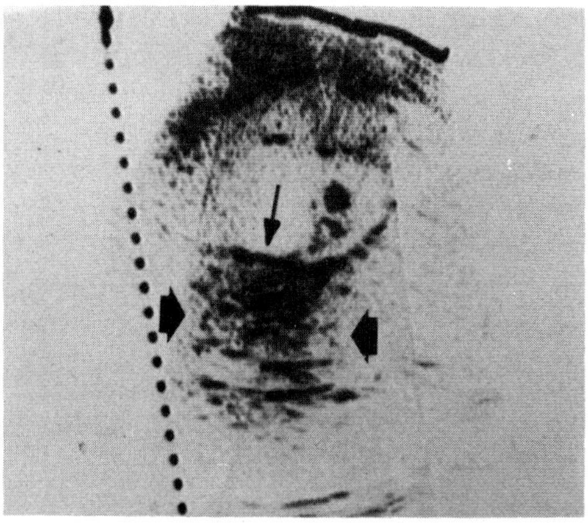

Figure 41–E.
Renal cyst.

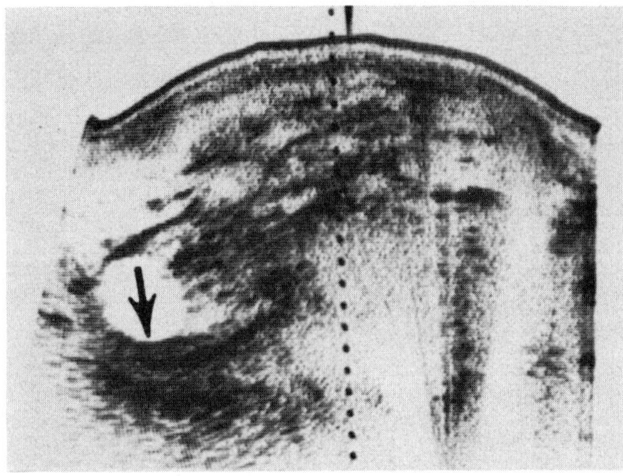

Figure 41–F.
Renal cyst.

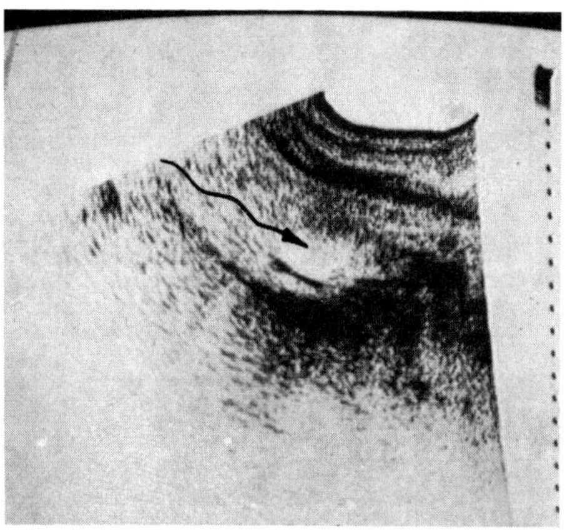

Figure 41–G.
Renal cyst.

REFERENCES

Green WM, King DL, Cassarella WJ: A reappraisal of sonolucent renal masses. Radiology 121:163 (1976)

Pollack HM, Goldberg BB, Morales JO, et al: A systemized approach to the diagnosis of renal masses. Radiology 113:653 (1974)

Pollack MM, Goldberg BB: Chapter 10, Kidney, in *Abdominal Gray Scale Ultrasonography*. Goldberg BB, ed. John Wiley & Sons, New York (1977)

Sanders RC: Renal ultrasound. Rad Clin N Am 13:417 (1975)

Case 42

CLINICAL INFORMATION

This 47-year-old lawyer has been under considerable stress recently, and his ulcer is acting up again. He has lived with recurrent duodenal ulcer for seven years and has usually been able to manage attacks with Librax and antacids. The present episode has lasted over a month; and because surgery has been discussed in the past, he wonders if it is now the time to have something definitive done. His doctor gets another upper GI series (the patient's fifth), which is not much different from one performed two years ago. The duodenal bulb is spastic and scarred, but there are no definite ulcer craters. A subsequent oral cholecystogram shows only faint opacification of the gallbladder; therefore, an ultrasound examination is ordered.

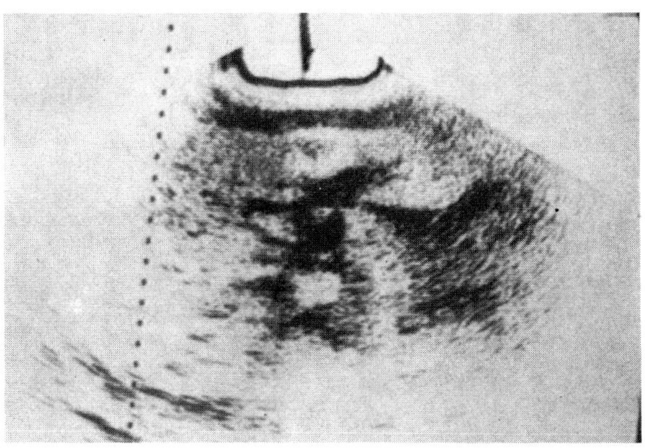

Figure 42–A. Transverse section +10.5.

DIAGNOSIS

Carcinoma of the tail of the pancreas.

ULTRASOUND FINDINGS

The tail of the pancreas is quite prominent and easily seen in this case. This always raises a red flag; because of air in the stomach it is unusual to be able to see the tail of the pancreas at all, let alone well. The thickness of the tail of our patient's pancreas is just about 3 cm., which is in the indeterminate range. We do not like pancreatic measurements; invariably they involve a good deal of overlap between normal and abnormal, and this simply is not good enough to provide a definitive answer for the individual patient. This is where clinical judgment and experience come into play; the ultrasonographer must serve as a consultant and give an informed opinion as to whether the findings are normal or abnormal. We tend to lean toward overcalling these cases, preferring to have more false positive than false negative readings. Now that percutaneous aspiration biopsy is available, it is a simple matter with biopsy of *all* suspicious areas to get a more precise, cytologic diagnosis.

Our concerns for this patient proved justified; the biopsy yielded "adenocarcinoma, probably of pancreatic origin." The ultrasonographer must maintain a high index of suspicion for pancreatic neoplasm, because it can present as only a slight focal irregularity of the gland. Figure 42–B shows another small tumor of the tail of the pancreas, which was subsequently verified by CT scan and arteriography.

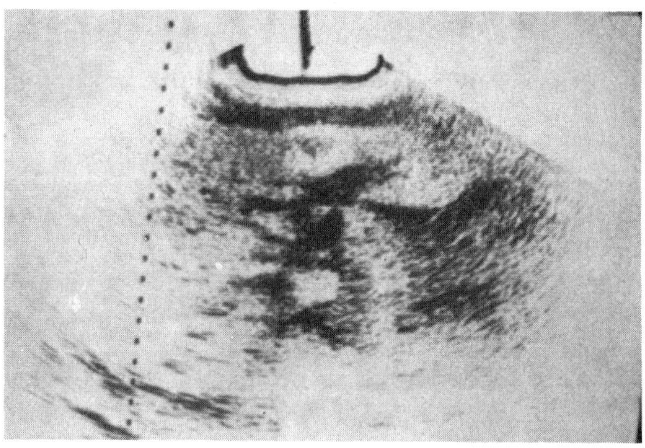

Figure 42–A. Transverse section +10.5. Carcinoma of tail of pancreas.

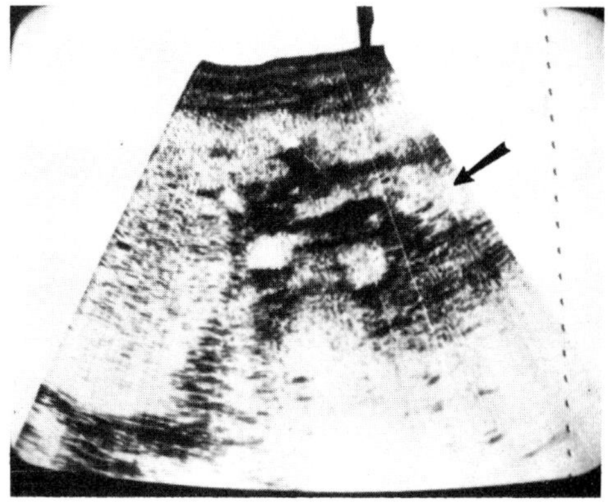

Figure 42–B. Transverse section +8. Carcinoma of tail of pancreas.

REFERENCES

Weill F, Schraub A, Eisencher A, et al: Ultrasonography of the normal pancreas: success rate and criteria for normality. Radiology 123:417 (1977)

Haber K, Freimanis AK, Asher WM: Demonstration and dimensional analysis of the normal pancreas with gray scale echography. Am J Radiol 126:624 (1976)

Smith EH, Bartrum RJ Jr, Chang YC, et al: Percutaneous aspiration biopsy of the pancreas under ultrasonic guidance. N Engl J Med 292:825 (1975)

Case 43

CLINICAL INFORMATION

This 24-year-old woman is 34 weeks pregnant by dates and has had a little vaginal bleeding. Ultrasound is requested to localize the placenta. Where is it?

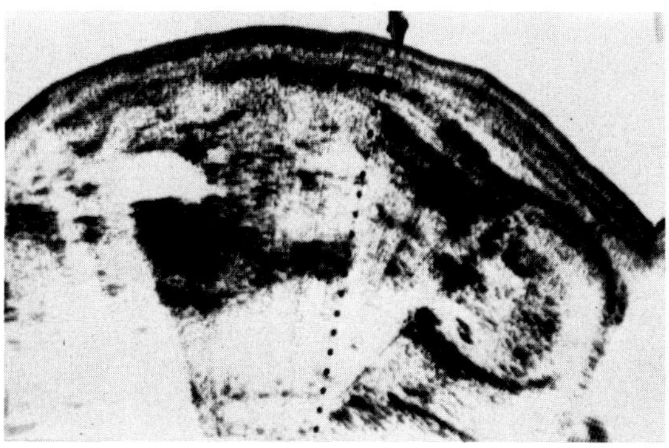

Figure 43–A. Longitudinal section 0L.

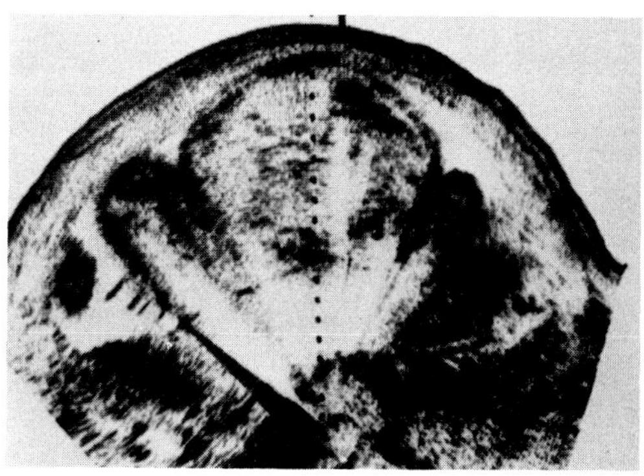

Figure 43–B. Transverse section +2.

DIAGNOSIS

Posterior placenta (not a previa).

ULTRASOUND FINDINGS

This placenta is very difficult to find; in fact we never see much of it at all. Although we cannot see where the placenta *is,* we can see where it *is not* — *not* on the anterior or side walls of the uterus nor in the fundus. Similarly, it is *not* covering the cervix since the baby's head is far down in the pelvis adjacent to the internal os. By elimination, therefore, we can deduce where the placenta *is* — on the posterior uterine wall. In advanced pregnancy it is usually not possible to see a posterior placenta because of acoustic shadowing from the baby's bones; in these situations we must check those portions of the uterus which are visible and then figure out where the placenta lies. One thing more: the baby's trunk is against the anterior wall of the uterus, pushed anteriorly by the placenta on the posterior wall. This, too, is a helpful sign in locating the placenta in later pregnancies.

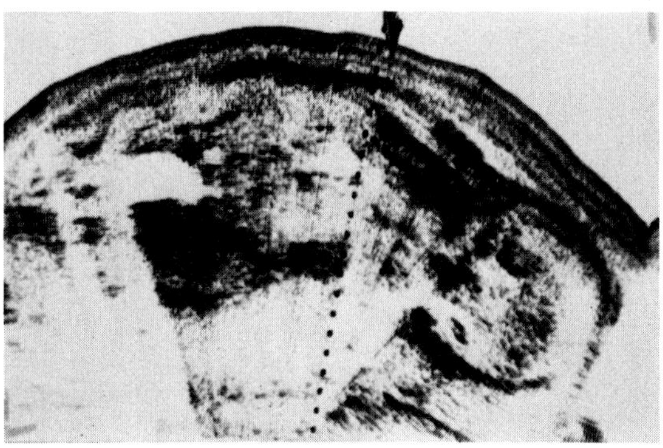

Figure 43–A. **Longitudinal section 0L.** Posterior placenta.

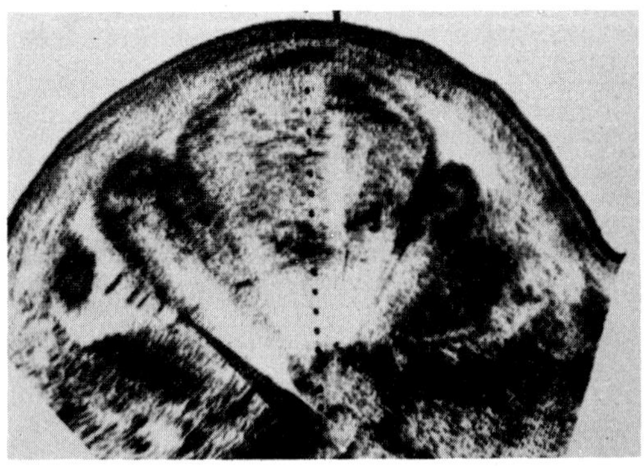

Figure 43–B. **Transverse section +2.** Posterior placenta.

Case 44

CLINICAL INFORMATION

This 47-year-old man has no complaints but on routine physical examination he was found to have a blood pressure of 156/102. An IVP at another hospital showed ". . . a ten cm. mass in the lower pole of the right kidney which could represent a renal tumor or a renal cyst. Ultrasound examination with possible cyst puncture is recommended." Since the outside hospital does not have an ultrasound machine, the patient was referred to our laboratory. Unfortunately, he did not bring his IVP films with him, so we will have to rely on the ultrasound scan alone. The technician thinks that the easily identified mass is probably a cyst, but she is worried because it cannot be shown to be completely echo-free.

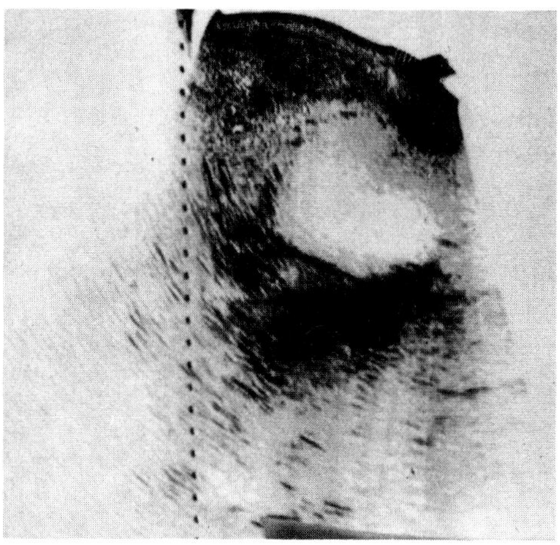

Figure 44–A. Transverse section +4. Prone scan of right kidney.

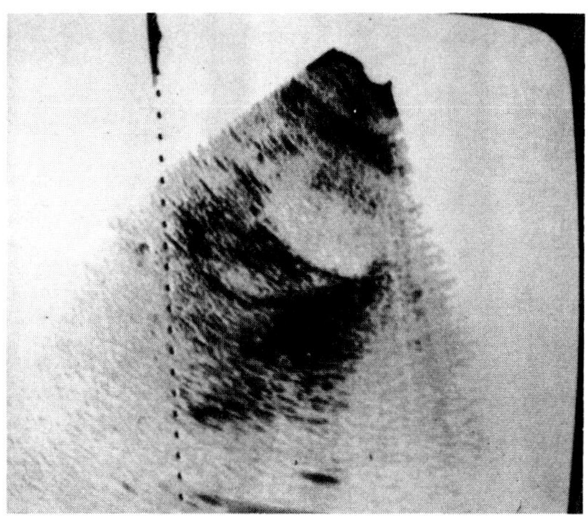

Figure 44–B. Longitudinal section 3R. Prone scan of right kidney.

DIAGNOSIS

Simple renal cyst with electrical interference artifacts.

ULTRASOUND FINDINGS

The renal mass is quite easy to identify and seems to be a straightforward renal cyst; two of the three criteria, a smooth back wall and good through transmission, are present. The only confusing aspect of this scan is the presence of the low-level, light gray "echoes" within the cyst cavity. These do not have the characteristics of reverberation artifacts (which we can see anteriorly in Figure 44–B), but do not look like real echoes either. We notice that they are rather widespread in our picture and seem to occur even in areas where there should be no echoes. We must be dealing with some type of artifact, but what is its source? Because of the confusion we performed a cyst puncture on the patient. This confirmed our impression that this is a benign renal cyst (we routinely perform cyst punctures on an outpatient basis).

Chasing down the origin of this low-level artifact was difficult. It required a couple of weeks and the services of several electronic consultants. The problem occurred episodically and was usually worse in the morning than in the afternoon. There would be long periods when the artifacts were not seen at all, and on occasion they would be present on nearly every examination. The puzzle was finally solved; we were picking up high-frequency electronic interference in the main power line to the scanner. Those long hours as a medical student on surgery finally paid off when the on-and-off pattern of the interference was recognized by one of us as that of the "Bovie" electrocautery unit used in the operating room. There are, however, other potential sources of this type of interference, such as automobile engines, airplanes, and electric arc welders. The cars and airplanes do not prove a problem for us because the steel girder construction of the hospital serves to shield the machine, and there are no electric arc welders in the hospital. The "Bovie" was used every day, however, and did present a problem. The solution: installation of an "isolation" transformer in the power source to the ultrasound scanner to eliminate the electrical interference problem, a relatively inexpensive maneuver.

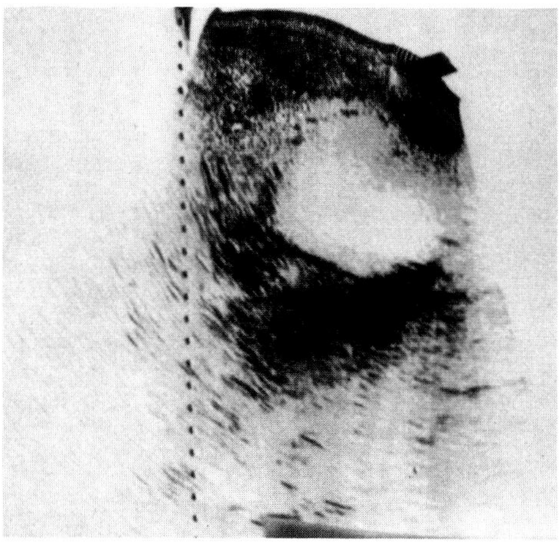

Figure 44–A. Transverse section +4. Renal cyst. Electrical interference artifacts. Prone scan of right kidney.

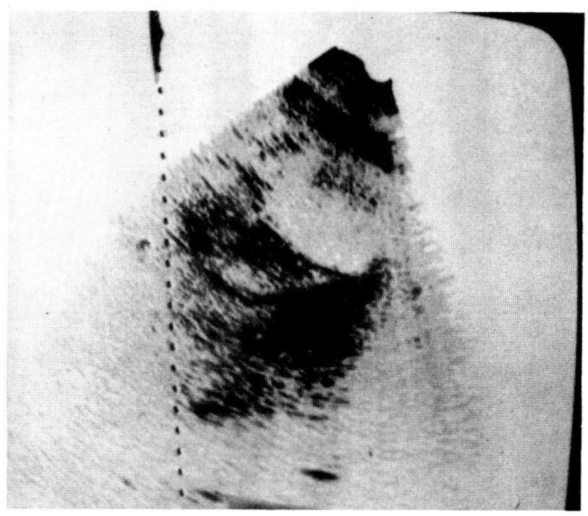

Figure 44–B. Longitudinal section 3R. Renal cyst. Electrical interference artifacts. Prone scan of right kidney.

Case 45

CLINICAL INFORMATION

This is a 23-year-old woman who has had several episodes of pelvic pain. The pain is sharp and steady and waxes and wanes over periods of one or two days at a time. It is not related to her menstrual cycles. She has never been pregnant and has been taking birth control pills for four and a half years. Her health is otherwise good. Physical examination discloses a firm, moderately tender, 5-centimeter mass in the left adnexal region; the left ovary cannot be palpated separately from the mass. The patient is afebrile. Routine laboratory studies, including pregnancy test and cervical culture, are normal. The patient was re-examined one week after the onset of her menstrual period and the mass was still present.

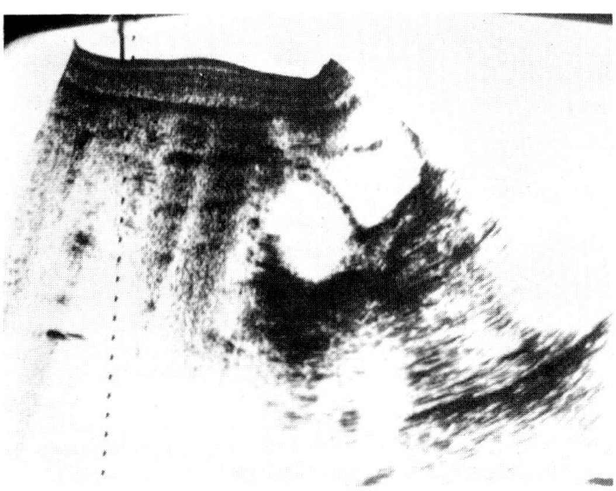

Figure 45–A. Longitudinal section 3L.

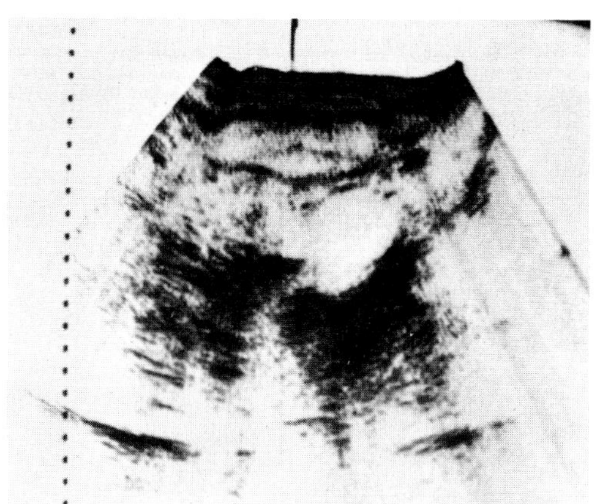

Figure 45–B. Transverse section –12.

DIAGNOSIS

Benign ovarian cyst.

ULTRASOUND FINDINGS

The longitudinal and cross-sectional scans show the pelvic mass in the left adnexal region; it is well-circumscribed and echo-free. The absence of internal echoes means the mass is homogeneous, or fluid-filled (or, in common parlance, a "cyst"). Cystic pelvic masses usually arise from the ovary, and there are many different pathologic types. In order to simplify things we classify ovarian cysts into three broad categories: physiologic, benign, and malignant. Physiologic cysts are those associated with the normal menstrual or pregnancy cycles and include simple follicle cysts and corpus luteum cysts (see Case 55). Benign cysts encompass a variety of pathologic conditions, such as giant follicular cysts, paraovarian cysts, cystic teratomas, endometriomas, and cystadenomas. These cysts have two common characteristics: they do not change significantly during the menstrual cycle and they are not malignant. Malignant cysts are those associated with a malignant tumor, such as ovarian carcinoma or malignant teratoma.

Physiologic cysts can usually be diagnosed because of their relatively small size and characteristic clinical behavior. It is very difficult to separate benign and malignant cysts on the basis of the ultrasound appearance alone because they both have protean manifestations and considerable overlap. Even when all the clinical and laboratory data are factored in, it is frequently impossible to exclude a malignant lesion, and the majority of cystic lesions ultimately come to surgery. Figures 45–C through 45–I are examples of the many faces of benign ovarian cysts.

Figure 45–C shows a small cyst that is not located in the adnexal region but is positioned inferior to the uterus (arrow) in the cul-de-sac. Figure 45–D is a very small cyst confined to the right ovary. The cyst in Figures 45–E and 45–F is huge and has spilled out of the pelvis into the abdomen. In such cases it may be difficult to separate the cyst (*Cy*) from the bladder (*Bl*); if confusion arises we ask the patient to void and we repeat the scan. The cyst in Figure 45–G is multilocular while that in Figure 45–H has an irregular back wall and possibly some internal echoes (most of these "echoes" are artifacts). Figure 45–I is a large cystic teratoma (dermoid). Some ultrasonographers have suggested that it is possible to diagnose a dermoid accurately by analyzing the echo pattern; we do not share this opinion.

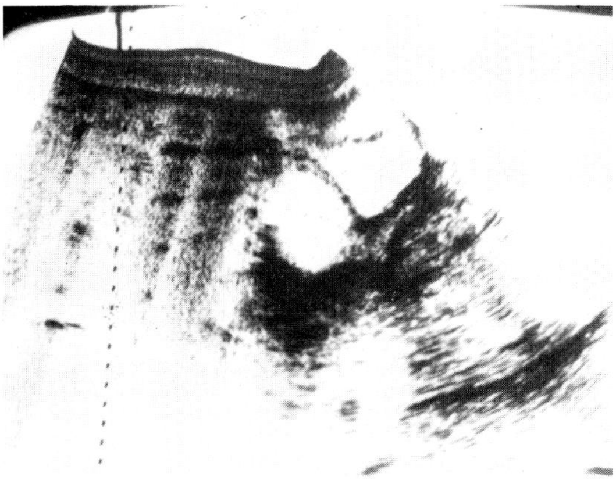

Figure 45–A. Longitudinal section 3L.
Ovarian cyst.

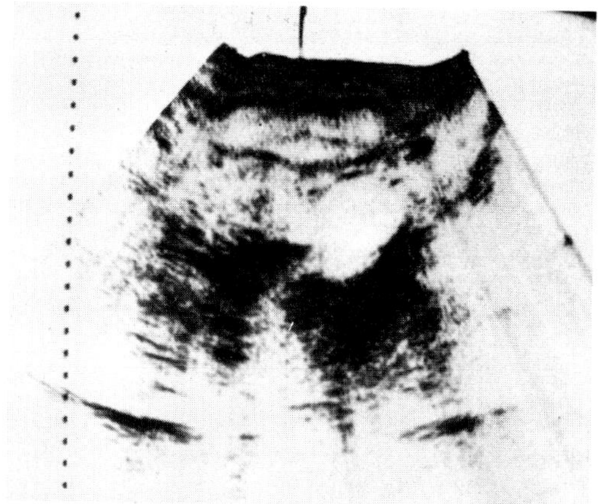

Figure 45–B. Transverse section –12.
Ovarian cyst.

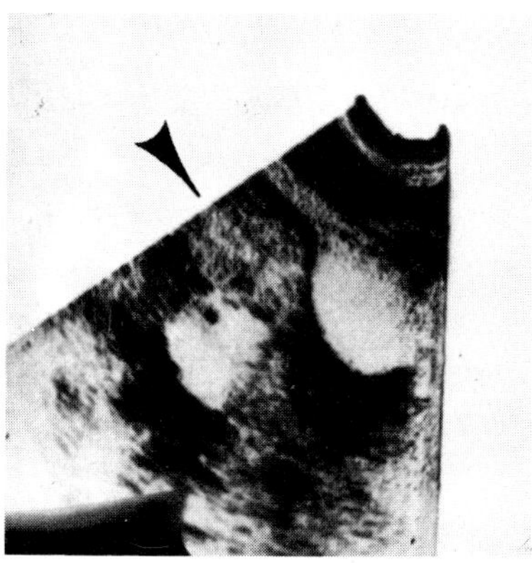

Figure 45–C. Longitudinal section 2L.
Ovarian cyst (arrowhead points to uterus).

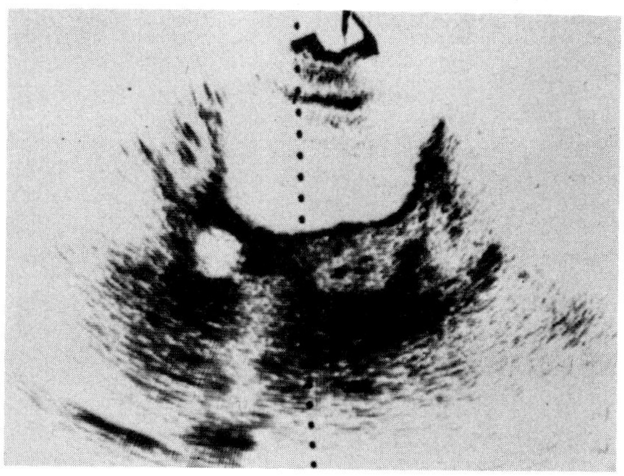

Figure 45–D. Transverse section –11.
Right ovarian cyst.

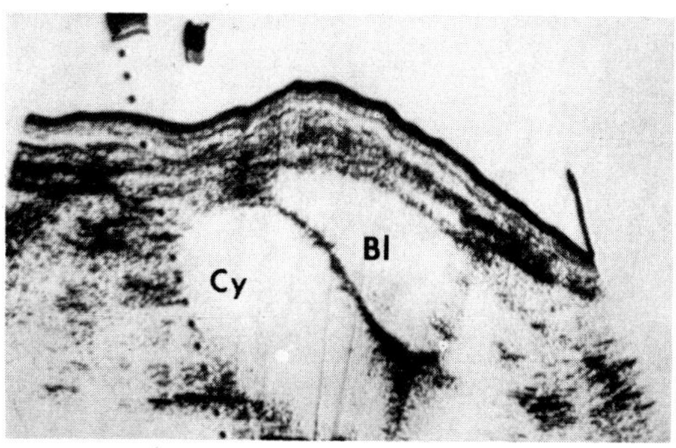

Figure 45–E. Longitudinal section 4L.
Massive ovarian cyst.

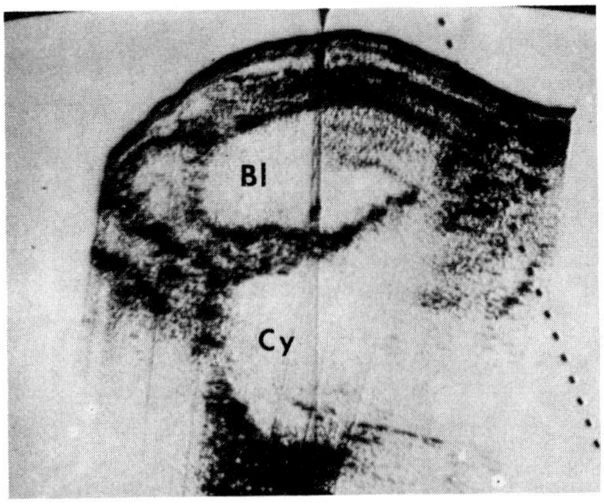

Figure 45–F. Transverse section −5. Massive ovarian cyst.

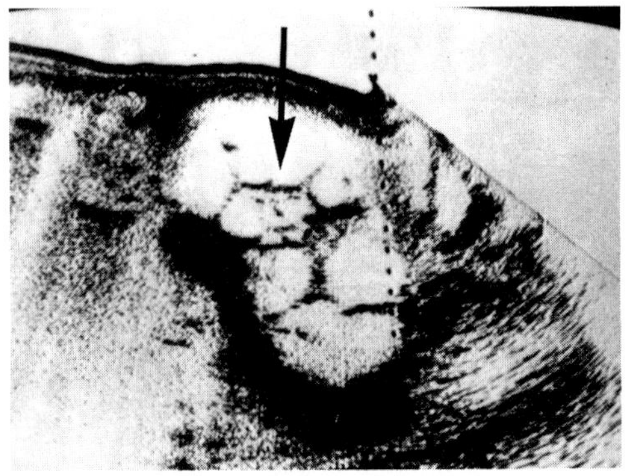

Figure 45–G. Longitudinal section 4R. Multilocular ovarian cyst.

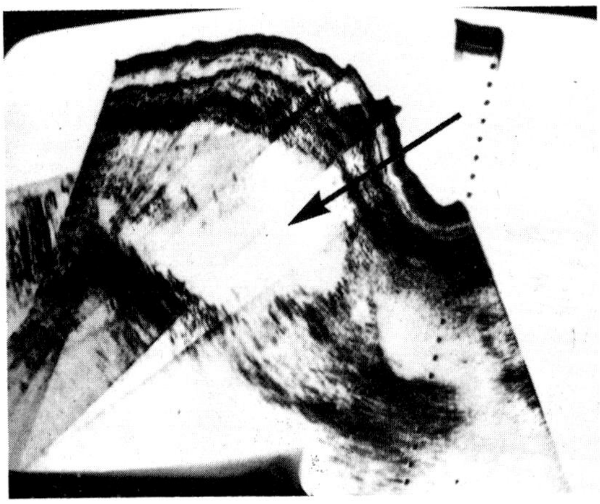

Figure 45–H. Longitudinal section 2R. Massive ovarian cyst.

Figure 45–I. Transverse section −2. Benign cystic teratoma.

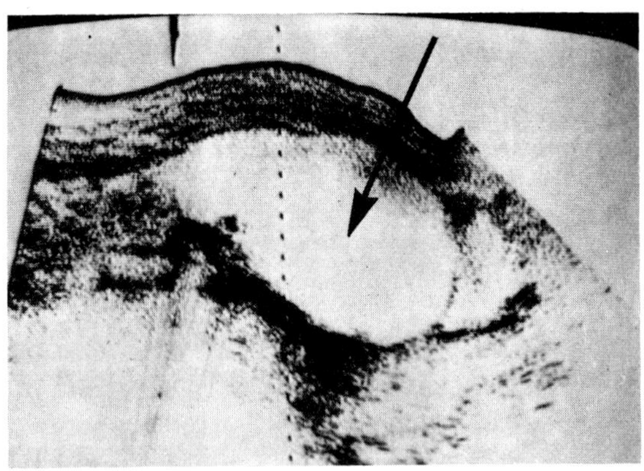

REFERENCES

Hobbins JC, Winsberg F: *Ultrasonography in Obstetrics and Gynecology*. Chapter 8. Williams & Wilkins Co., Baltimore (1977)

Morley P, Barnett E: The ovarian mass, Chapter 27 in *Ultrasonography in Obstetrics and Gynecology* Sanders and James, editors. Appleton-Century-Crofts, New York (1977)

Case 46

CLINICAL INFORMATION

This 68-year-old man has noticed swelling in his feet and legs which has been getting worse over the past few months. He is otherwise well and healthy, except for occasional back pains which have troubled him for several years. His physical examination reveals pitting edema of the legs and a firm, nontender, nonpulsatile mass in the central abdomen. The lungs are dry, and auscultation of the heart and an EKG are normal. Routine laboratory studies, including liver functions and serum amylase, are normal.

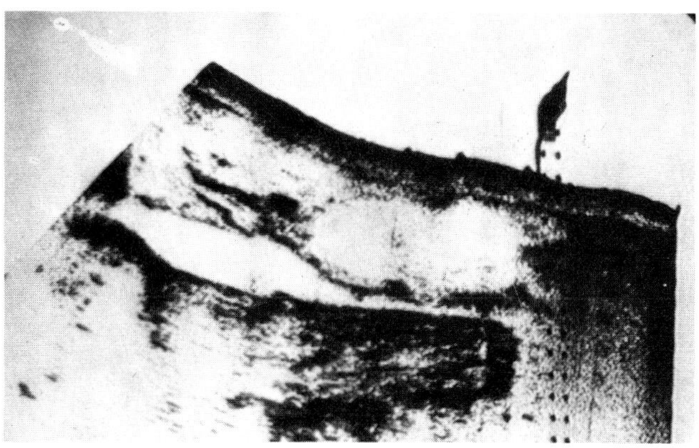

Figure 46–A. Longitudinal section 3R.

DIAGNOSIS

Compression of the inferior vena cava by a large aortic aneurysm.

ULTRASOUND FINDINGS

Figure 46–A is a longitudinal scan along the course of the inferior vena cava, made while the patient performed a Valsalva maneuver. A Valsalva maneuver should cause the inferior vena cava to dilate; and, indeed, the upper part of the vessel is quite distended. The lower part, however, is compressed by a large mass, which produces relatively weak echoes. There are several possibilities for the origin of this mass: enlarged retroperitoneal lymph nodes, a tumor arising either in the retroperitoneum or the overlying intraperitoneal structures, acute or chronic hemorrhage, or an aortic aneurysm. The cross-sectional scan (Figure 46–B) answers the question in this case and shows that we are dealing with an aortic aneurysm. If we had not recognized this as an aneurysm, we would have performed a percutaneous aspiration biopsy to obtain the diagnosis.

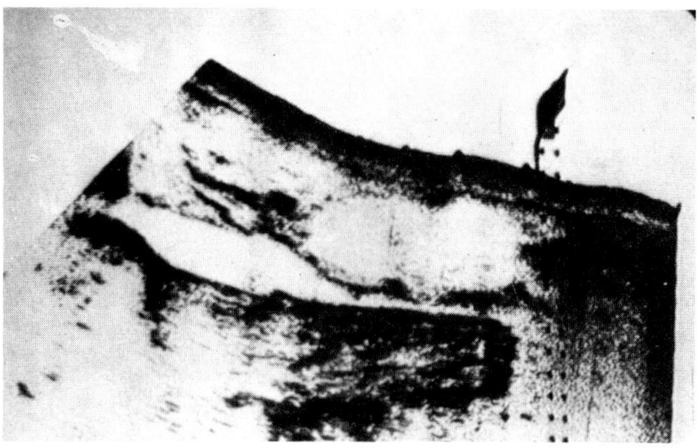

Figure 46–A. **Longitudinal section 3R.** Aortic aneurysm compressing vena cava.

Figure 46–B. **Transverse section +4.** Aortic aneurysm.

REFERENCES

Gosink BB: The inferior vena cava: mass effects. Am J. Radiol 130:533 (1978)

Walls WJ, Templeton AW: Ultrasonic demonstration of inferior vena caval compression: guide to pancreatic head enlargement. Radiology 123:165 (1977)

Case 47

CLINICAL INFORMATION

This 63-year-old man has symptoms of prostatism — frequency, nocturia, and difficulty with his urinary stream. He has recently had some episodes of pink urine and has, therefore, consulted his doctor. His physical examination is unremarkable except for an enlarged prostate, and routine laboratory studies show only hematuria. A transurethral prostatectomy is scheduled and an IVP ordered for preoperative evaluation. A film taken 45 minutes after the injection of contrast agent (Figure 47–A) shows a duplication of the right renal collecting system (which is otherwise normal), an enlarged prostate, and no visualization of the left kidney or collecting system. An ultrasound scan of the left kidney is requested; Figure 47–B is a typical longitudinal scan.

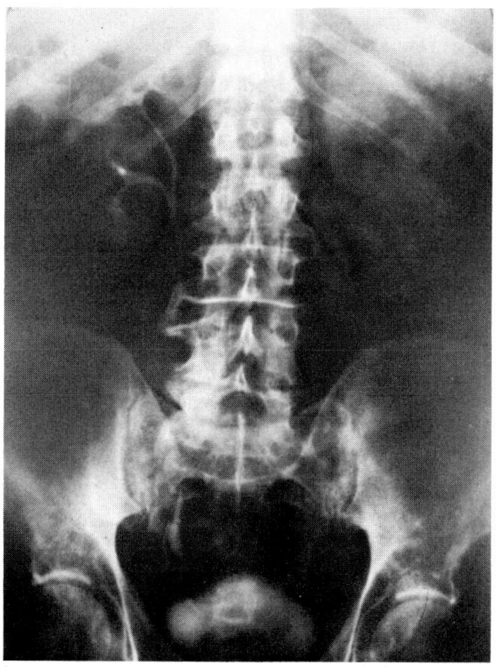

Figure 47–A.
IVP, 45-minute film.

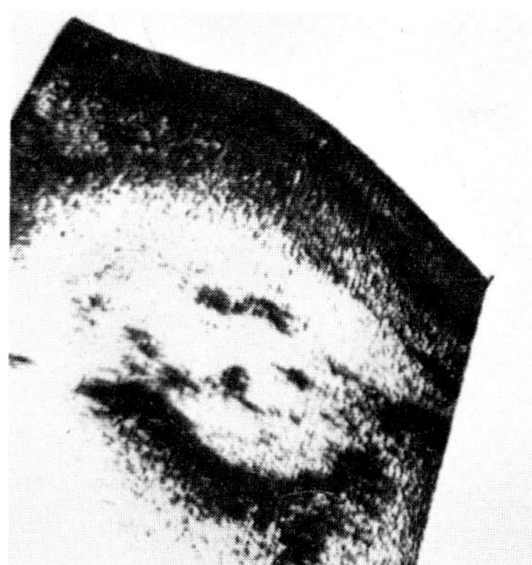

Figure 47–B.
Longitudinal section 1R.
Prone scan of left kidney.

DIAGNOSIS

Hydronephrosis.

ULTRASOUND FINDINGS

The clustered, high-intensity echoes arising from the normal renal sinus are not seen in this patient; instead there is an elongated irregular echo-free zone, bounded by a thin line of loud echoes, which represent the normal structures of the renal sinus (renal collecting system, fat, and blood vessels) displaced by the echo-free "mass." The absence of echoes tells us we are dealing with fluid, and there are only two causes for fluid in this location — hydronephrosis and parapelvic cysts. We can exclude a parapelvic cyst by the shape of the area (irregular, not round), and the IVP findings confirm this diagnosis. (In our experience parapelvic cysts are less common than hen's teeth; a fluid-filled mass in the renal sinus is almost always hydronephrosis.) The left ureter was obstructed at the uretero-vesical junction by a carcinoma of the prostate.

The diagnosis of hydronephrosis can be made easily and reliably by ultrasound. In the nonhydronephrotic kidney the loud echoes from the renal sinus are compact and closely grouped (Figure 47–C); if any of these echoes are displaced or separated by an echo-free zone, you should always think first of hydronephrosis. Figures 47–D through 47–F show other examples of hydronephrotic kidneys.

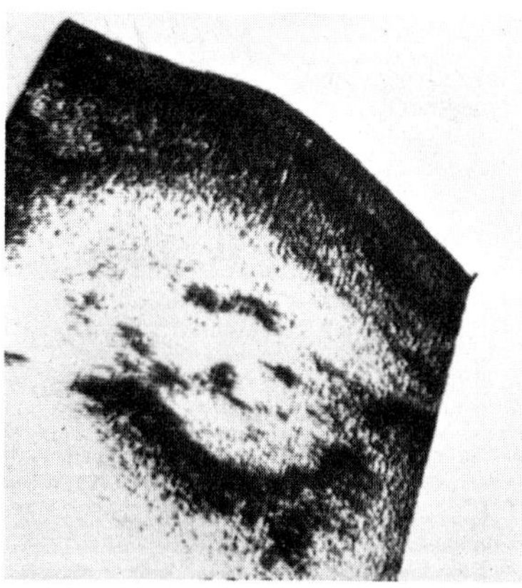

Figure 47–B. **Longitudinal section 1R.** Hydronephrosis. Prone scan of left kidney.

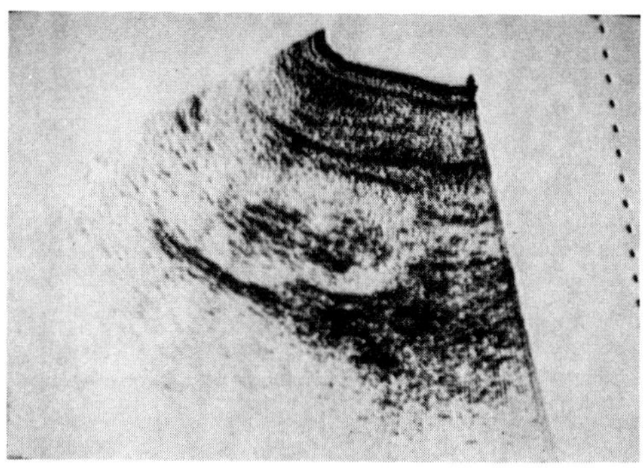

Figure 47–C. **Longitudinal section 0L.** Normal kidney. Prone scan.

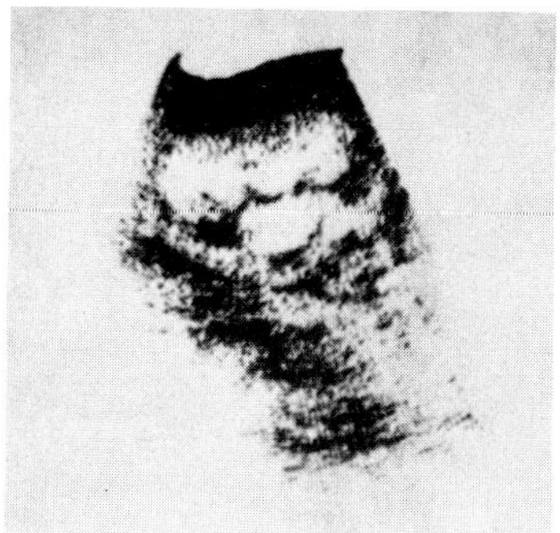

Figure 47–D. **Transverse section +5.** Hydronephrosis. Prone scan.

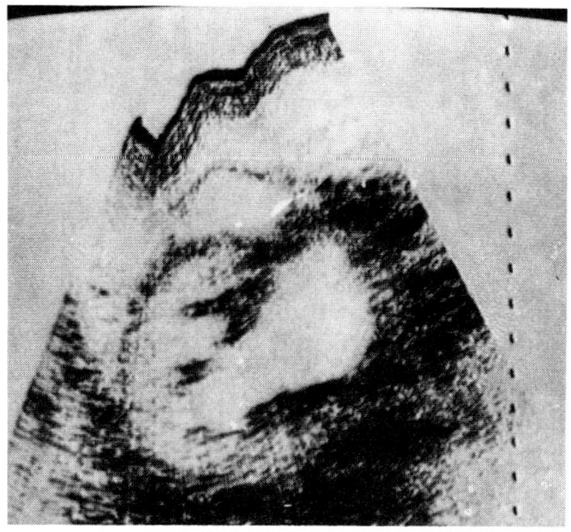

Figure 47–E. **Transverse section +7.** Hydronephrosis. Supine scan.

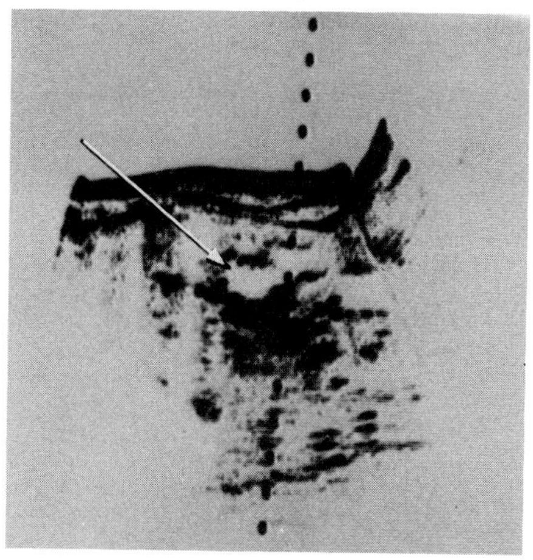

Figure 47–F. **Longitudinal section 0L.** Hydronephrosis in a 4-year-old. Prone scan.

REFERENCES

Ellenbogen PH, Scheible FW, Talner LB, et al: Sensitivity of gray scale ultrasound in detecting urinary tract obstruction. Am J Radiol 130:731 (1978)

Sanders RC, Conrad MR: The ultrasonic characteristics of the renal pelvicalyceal echo complex. J Clin Ultra 5:372 (1977)

Case 48

CLINICAL INFORMATION

This 6-month-old boy seems to be healthy, but his mother is worried that he is not eating enough because he does not always finish his bottle and "spits up" every couple of days. His height and head circumference are in the fiftieth percentile for his age, but his weight is in only the fifth percentile. The pediatric resident is suspicious that there may be an abdominal mass in the right upper abdomen and requests an ultrasound examination.

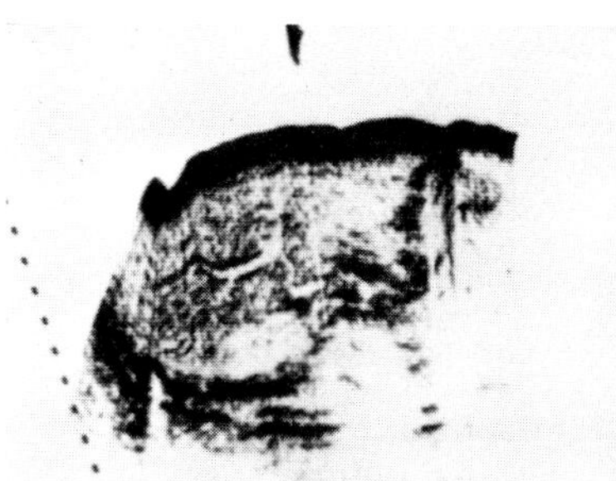

Figure 48–A.
Transverse section +3.5.

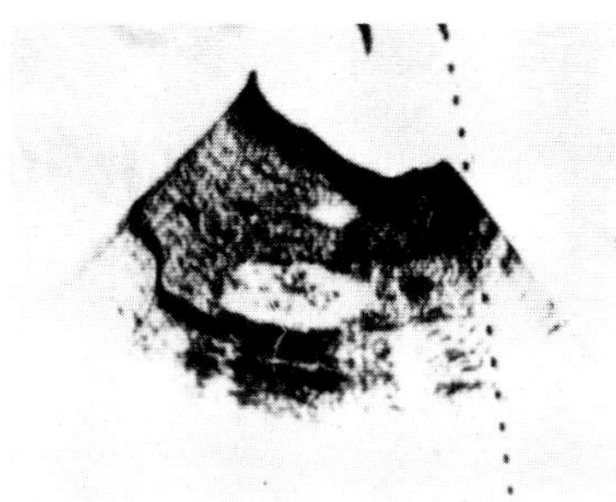

Figure 48–B.
Longitudinal section 2R.

DIAGNOSIS

Normal infant abdomen.

ULTRASOUND FINDINGS

The right upper quadrant is nicely demonstrated and is normal. We can clearly see the right kidney and suprarenal region as well as the liver and tiny gallbladder. There are no abnormal masses present, and the texture of the liver is uniform, with nothing to suggest a focal lesion. The palpable "mass" is the edge of the liver.

Scanning small infants can be a technical challenge. Because the anatomic structures of interest all lie within 5 centimeters of the transducer face, it is important to use a transducer that is well focused in the near field. Most of the transducers designed for adult use do not have good near-field resolution and produce poor pictures in infants. It is usually possible to go to higher frequencies, such as 5 or 7.5 MHz, because penetrating power is unnecessary in these small patients. The biggest problem the technician will encounter is keeping the patient still; a squirmy subject defies scanning. We try to schedule all these examinations to follow immediately after a feeding in the hope that the baby will fall asleep. If necessary, we sedate them with chloral hydrate (30 mg./kg.). We have found that trying to restrain a wriggling infant while attempting to make a scan is impossible.

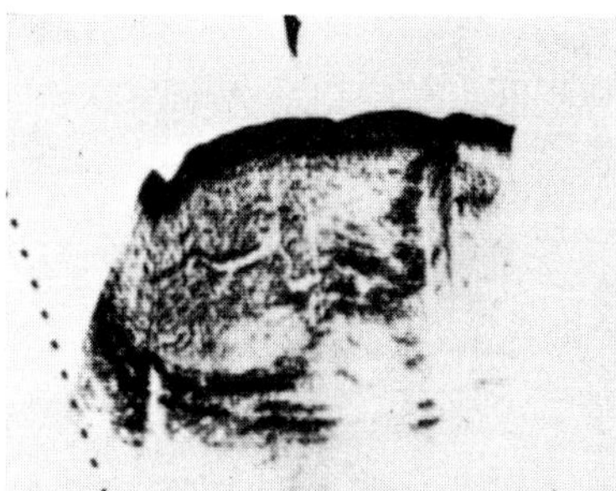

Figure 48–A. **Transverse section +3.5.** Normal abdomen.

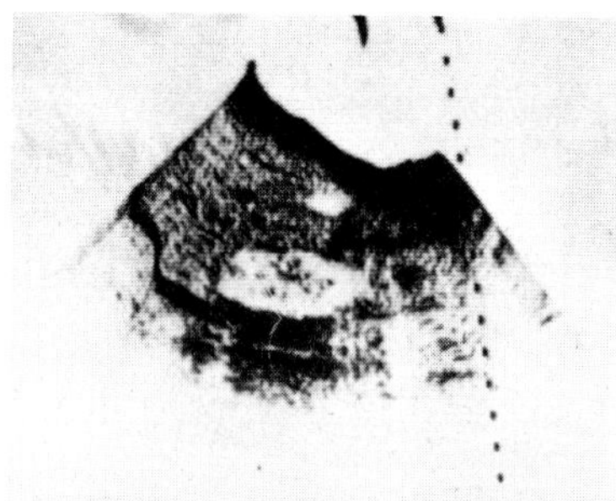

Figure 48–B. **Longitudinal section 2R.** Normal abdomen.

REFERENCE

Walls WJ, Roberts FF, Templeton AW: B-scan diagnostic ultrasound in the pediatric patient. Am J Radiol 120:431 (1974)

Case 49

CLINICAL INFORMATION

This 52-year-old woman has a flock of complaints — abdominal pain, diarrhea, fever, lack of pep. Most of these symptoms are chronic but the abdominal pain is rather new and seems to be worsening. It is centered in the left lower quadrant and is best described as a dull aching. She has never had a major illness or surgical operation. Her general physical examination is unremarkable except for the pelvic examination, which reveals a firm, slightly tender mass in the left adnexal region. Laboratory studies, including liver function tests and a Pap smear, are normal. Skull, chest, and plain abdominal radiograms are normal. A barium enema shows a few small diverticula but no evidence of diverticulitis or tumor; there is no evidence of an extrinsic pelvic mass. An IVP is normal. Figure 49–A is a longitudinal scan through the left adnexal region and Figure 49–B is a cross-section through the pelvis.

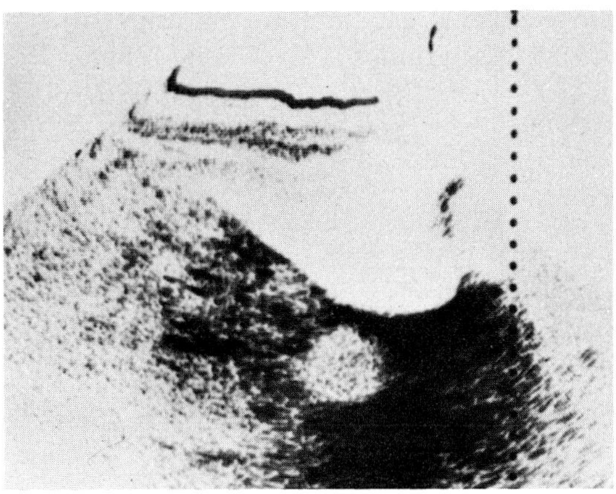

Figure 49–A. Longitudinal section 2L.

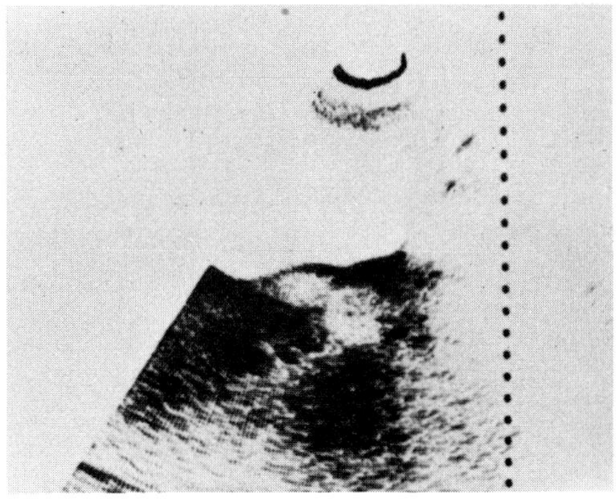

Figure 49–B. Transverse section −12.

DIAGNOSIS

Carcinoma of the ovary.

ULTRASOUND FINDINGS

The ultrasound scans clearly demonstrate the palpable pelvic mass: it is about 3.5 centimeters in diameter, filled with echoes (and hence not fluid-filled), and separate from the uterus (Figure 49–B). A solid ovarian mass in a patient of this age is cancer until proven otherwise. Tumors of the ovary vary greatly in composition, ranging from completely solid tumors such as this to nearly totally cystic masses, such as the carcinoma shown in Figure 49–C. Sometimes they are well circumscribed, but they can also invade surrounding tissues and obscure organ boundaries. Figure 49–D shows an ovarian tumor growing into the cul-de-sac and uterus (black arrows); the fundus of the uterus (open arrow) cannot be separated from the tumor. If there is ascites present we can be almost certain that we are dealing with a malignant tumor. Figure 49–E shows an ovarian carcinoma (arrow) and a large amount of ascites (*A*).

We do not think it is possible to tell the histologic type of tumor from the ultrasound scan appearance. Figures 49–F through 49–I are scans from different patients with pelvic tumors; the scans are nearly identical yet the tumors are not. Figure 49–F is an ovarian carcinoma, Figure 49–G is a malignant teratoma (malignant dermoid), Figure 49–H is squamous cell carcinoma of the vagina, and Figure 49–I is Hodgkin's disease. Figure 49–J is from another postmenopausal patient with pelvic pain and a mass: the findings are the same as the preceding pictures but this is not a tumor at all — it is a pelvic abscess.

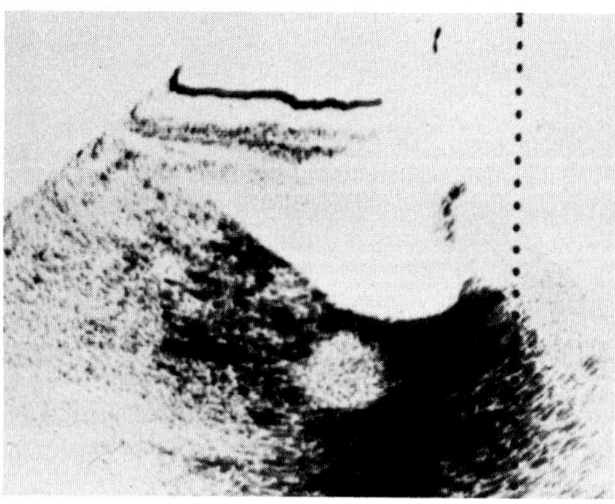

Figure 49–A. Longitudinal section 2L. Carcinoma of the ovary.

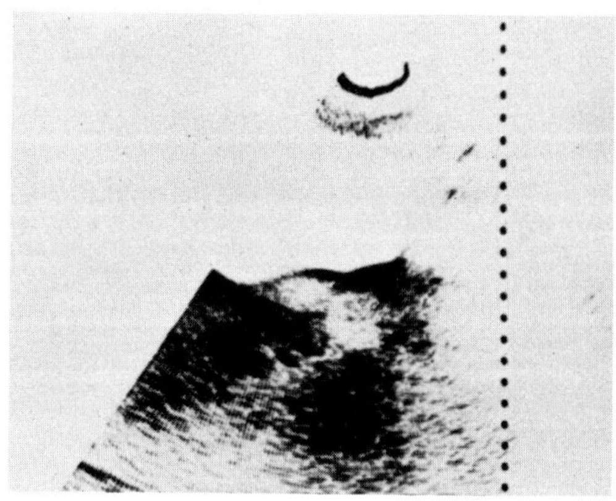

Figure 49–B. Transverse section −12. Carcinoma of the ovary.

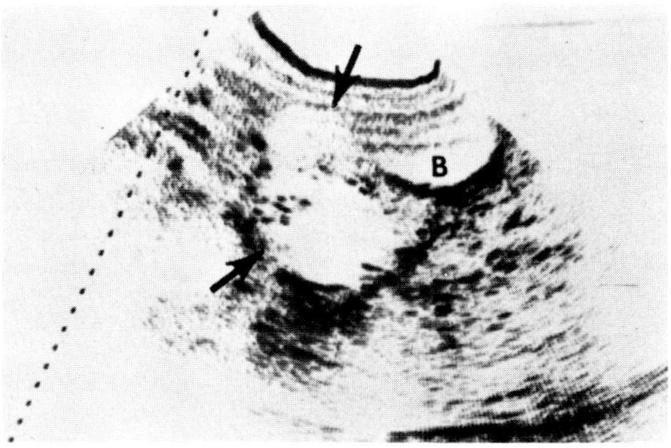

Figure 49–C. Longitudinal section .5L. Carcinoma of the ovary. B=Bladder.

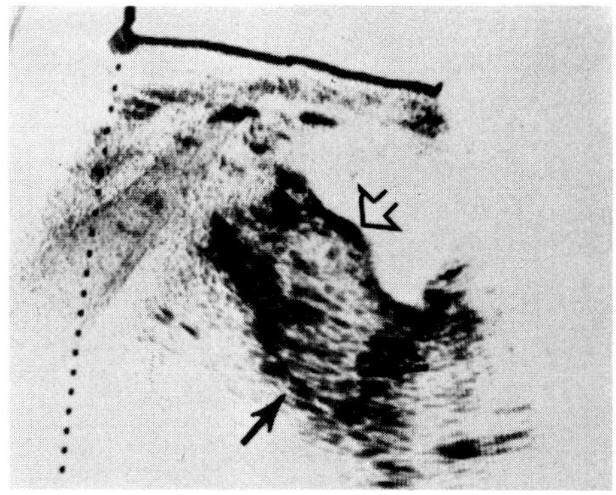

Figure 49–D. Longitudinal section 0L. Carcinoma of the ovary.

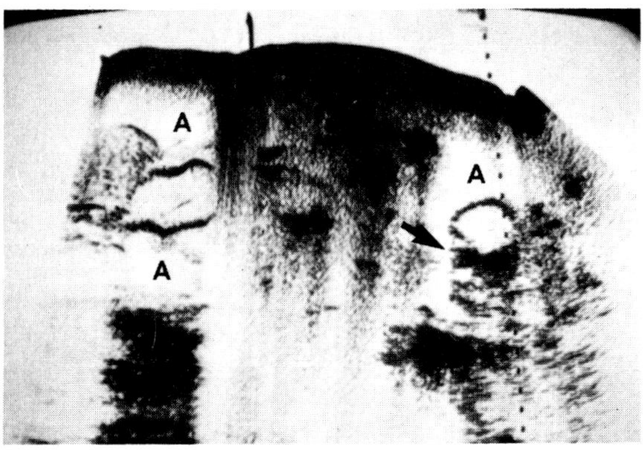

Figure 49–E. Longitudinal section 5R. Ovarian carcinoma and malignant ascites.

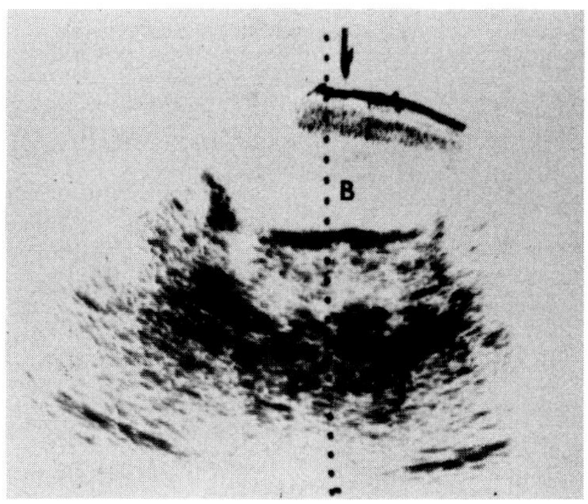

Figure 49–F. Transverse section −13. Carcinoma of the ovary. B=Bladder.

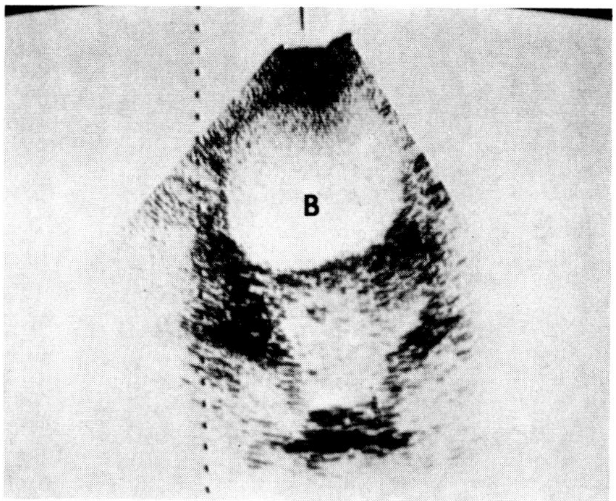

Figure 49–G. Transverse section −14. Malignant teratoma. B=Bladder.

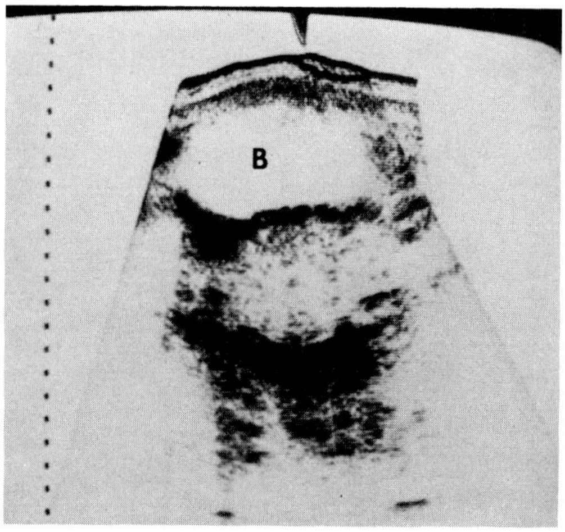

Figure 49–H. Transverse section −15. Carcinoma of the vagina. B=Bladder.

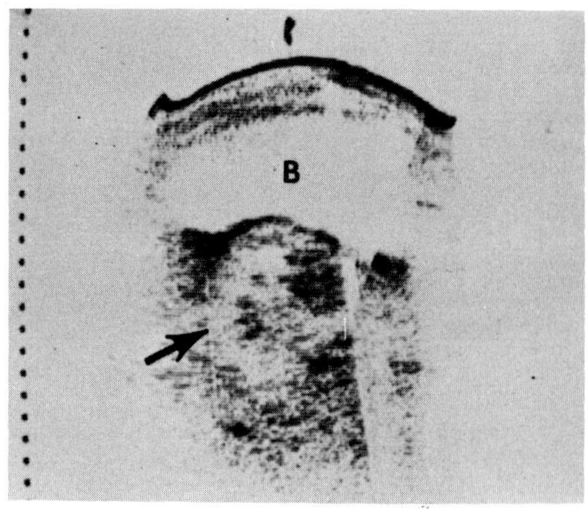

Figure 49–I. Transverse section −14. Hodgkin's disease. B=Bladder.

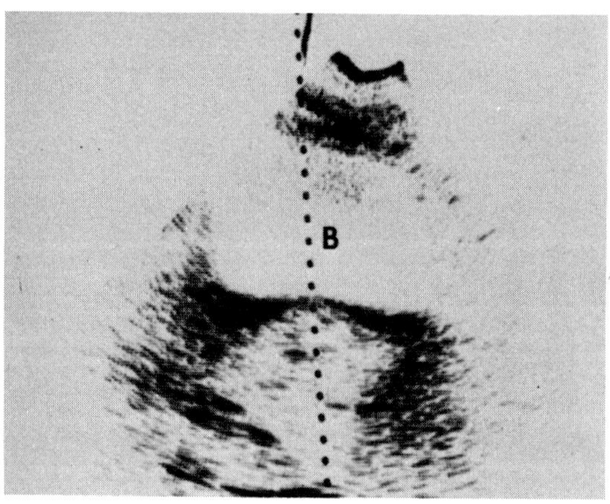

Figure 49–J. Transverse section −14. Pelvic abscess. B=Bladder.

REFERENCES

Lawson TL, Albarelli JN: Diagnosis of gynecologic pelvic masses by gray scale ultrasonography. Am J Radiol 128:1003 (1977)

Morley P, Barnett E: The ovarian mass, Chapter 27 in *Ultrasonography in Obstetrics and Gynecology* Sanders and James, editors. Appleton-Century-Crofts, New York (1977)

Bowie JD: Ultrasound of gynecologic pelvic masses: The indefinite uterus and other patterns associated with diagnostic error. J Clin Ultra 5:323 (1977)

Case 50

CLINICAL INFORMATION

This 22-year-old woman comes to the Emergency Room complaining of severe abdominal pain of sudden onset. The pain is centered in the midabdomen, does not radiate, and is not colicky. She has never had a similar episode and does not have any other medical problems. Her abdomen shows diffuse mild rebound tenderness. A pelvic examination and cervical smear are normal. There are no white cells or blood in the urine, but she has a fever of 101 degrees, and her white blood cell count is 14,000. Stat blood chemistries are sent off, and she is readied for surgery for probable appendicitis, pending the result. When the serum amylase turns out to be 800, the operation is cancelled and the patient sent to the ultrasound laboratory.

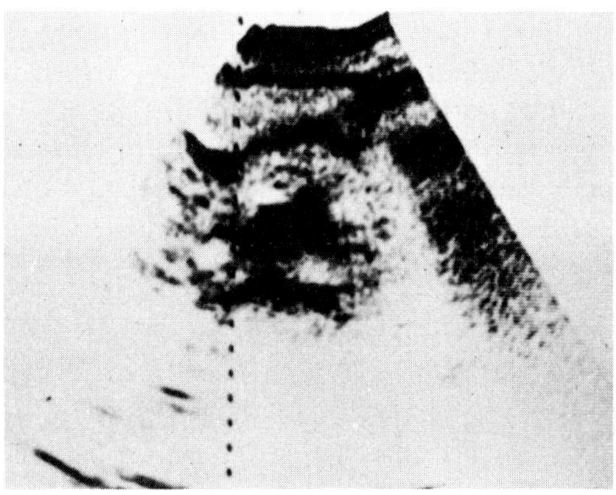

Figure 50–A. Transverse section +8.

DIAGNOSIS

Acute pancreatitis.

ULTRASOUND FINDINGS

The striking aspect of this examination is the ease with which the pancreas can be seen. It is plump, relatively echo-free, and seems to dominate the central abdomen. This is the classic appearance of acute pancreatitis — a swollen, edematous organ, enlarged throughout. Figures 50–B through 50–D are scans of other patients with the same disease. The normal pancreas (Figure 50–E) is usually quite echogenic and can be difficult to separate from the surrounding retroperitoneal fat. Usually it cannot be identified as a discrete organ, but can be located only by defining the neighboring vessels. When the pancreas is inflamed, however, it becomes very edematous; and this increased fluid content decreases the echogenicity and allows the gland to stand out clearly from its surroundings. On transverse or cross-sectional scans it appears as if a bratwurst had been draped over the spine and aorta.

When these ultrasonic findings are associated with the proper clinical setting, as in this case, the diagnosis can be made with certainty. If the clinical picture is less clear, the diagnosis can usually be suggested as long as the pancreas is diffusely involved *throughout its entire length.* In most, but not all, cases of pancreatitis, the gland is totally involved, whereas tumors of the pancreas are almost always *focal* lesions involving only a portion of the gland and sparing the remainder. Differential diagnosis is rarely a problem. The only potential pitfall is a retroperitoneal lymphoma, particularly Hodgkin's disease, which not infrequently mimics pancreatitis (Figure 22–F). The clinical setting and extent of the abnormal area usually makes differentiation easy. Two very rare conditions which can also look like this are horseshoe kidney and retroperitoneal fibrosis.

(Some of the latest generation of scanners have signal-processing options, which display the normal pancreas in light shades of gray rather than the dark echoes shown in Figure 50–E. The ultrasonographer must become familiar with the appearance of the normal pancreas on the specific machine which he is using.)

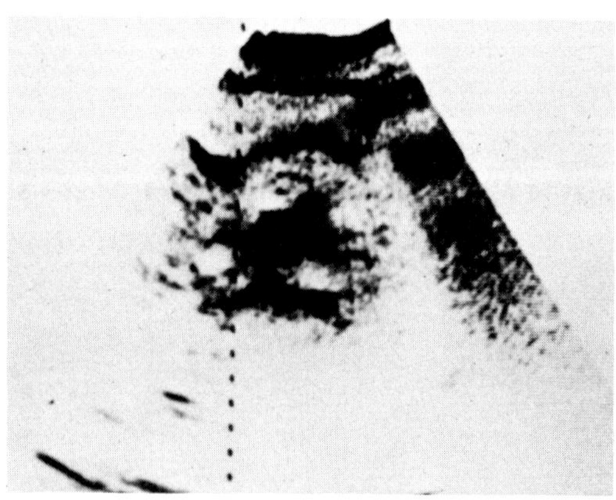

Figure 50–A. Transverse section +8.
Acute pancreatitis.

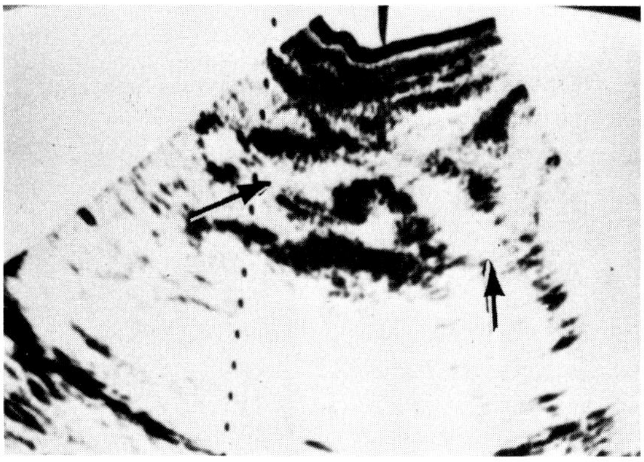

Figure 50–B. Transverse section +11. Acute pancreatitis.

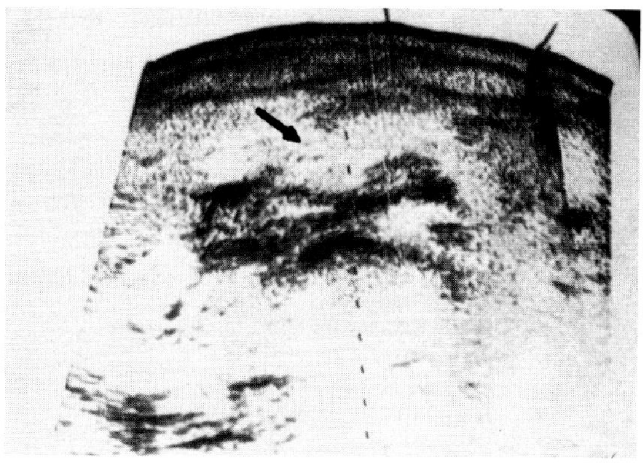

Figure 50–C. Transverse section +7. Acute pancreatitis.

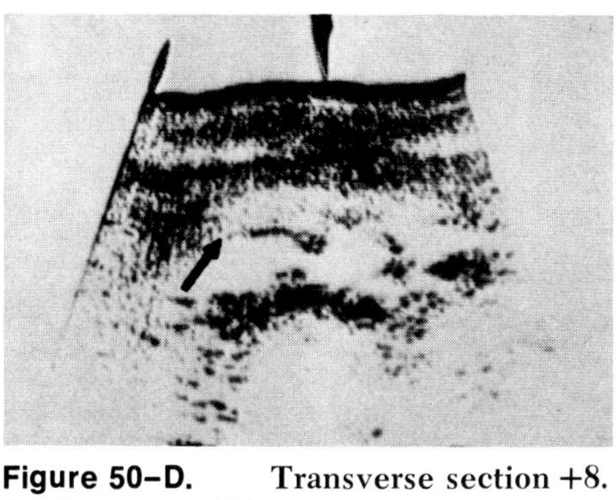

Figure 50–D. Transverse section +8. Acute pancreatitis.

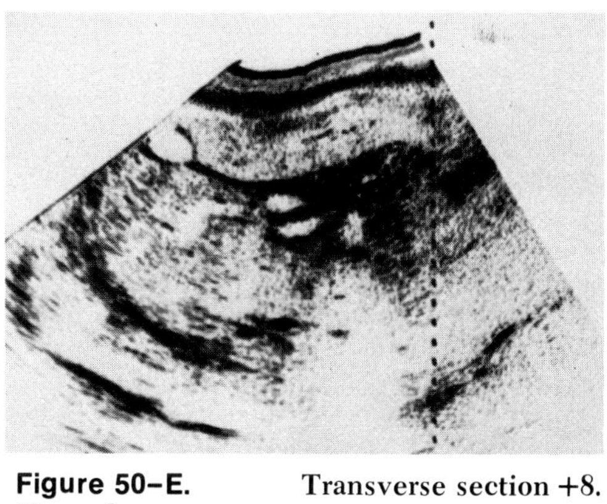

Figure 50–E. Transverse section +8. Normal pancreas.

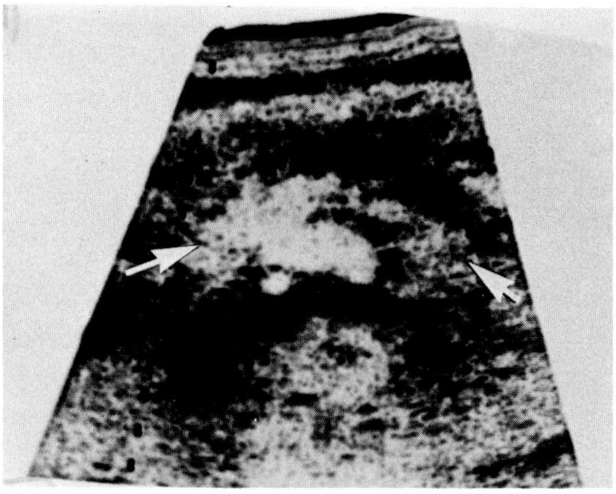

Figure 50–F. Transverse section +7. Hodgkin's disease.

REFERENCES

Perlmutter GS: Chapter 7, Pancreas. in *Abdominal Gray Scale Ultrasonography.* BB Goldberg, editor. John Wiley & Sons. New York 1977

Doust BD: Ultrasonic examination of the pancreas. Rad Clin N Am 13:467 (1975)

Mindell HJ, Kupic EA: Horseshoe kidney: ultrasonic demonstration. Am J Radiol 129:526 (1977)

Sanders RC, Duffy T, McLaughlin MG, et al: Sonography in the diagnosis of retroperitoneal fibrosis. J Urol 118:944 (1977)

Case 51

CLINICAL INFORMATION

This 73-year-old woman has always been in excellent health and, despite her age, is quite active. She even skis regularly. She consulted her doctor because of a recent aching in her back that occurred particularly after skiing. Her physical examination and basic blood and urine studies were normal. Her doctor wondered whether she might have a renal cause for these symptoms and ordered an IVP (Figure 51–A). The radiologist is uncertain of the findings. He suspects that the kidneys are normal, but is somewhat concerned about the possibility of a parapelvic mass in the right kidney and recommends ultrasound to help evaluate this. The ultrasound technician has scanned the right kidney carefully, using an expanded (2:1) scale (*not* the "zoom") and has obtained transverse (Figure 51–B) and longitudinal (Figure 51–C) pictures through the renal pelvis. What are we dealing with?

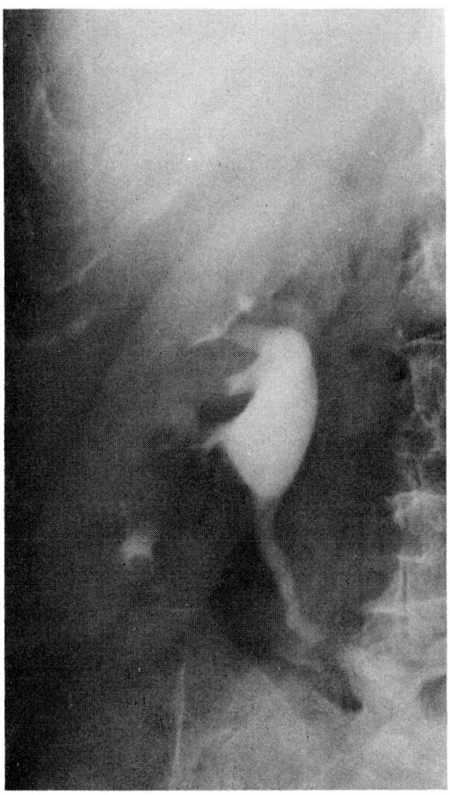

Figure 51–A.
IVP, 8-minute compression film of right kidney.

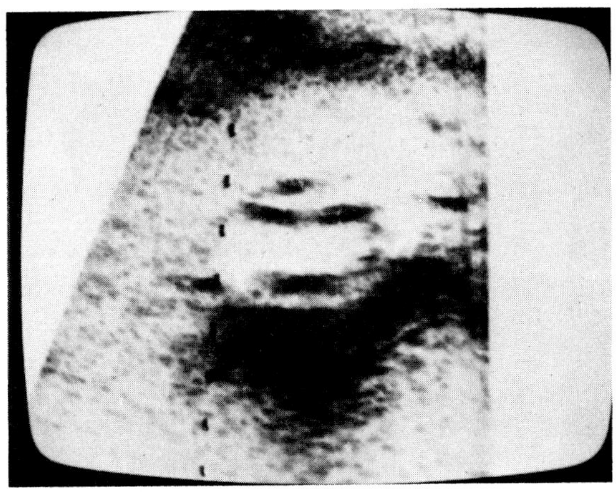

Figure 51–B. **Transverse section +3.**
Prone scan of right kidney.

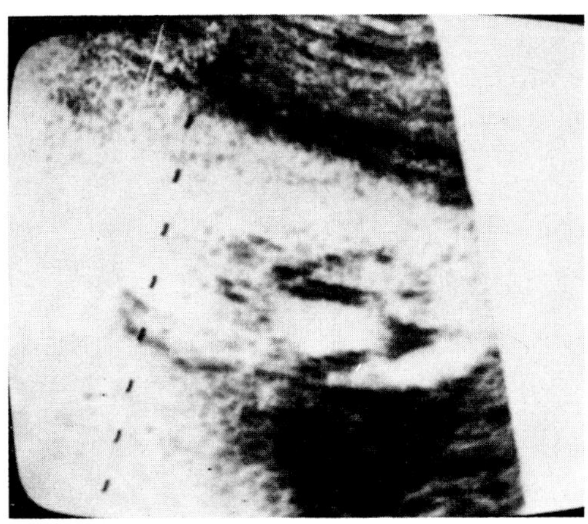

Figure 51–C. **Longitudinal section 2L.**
Prone scan of right kidney.

DIAGNOSIS

Extrarenal pelvis.

ULTRASOUND FINDINGS

There is an echoless area in the middle of the kidney where we normally expect to see only the clustered, high-amplitude echoes characteristic of the normal renal sinus. (These echoes arise from the multiple interfaces in the sinus — blood vessels, calyces, fat, and so on.) Because the mass is echoless, we know it is homogeneous and, therefore, probably fluid-filled; what fluid-filled masses could be present in this region?

One possibility is a parapelvic renal cyst. The shape of the area argues against this, however; cysts are round, and this structure is rather oval and elongated. (A multiloculated cyst may not be round, but we find no evidence of a septum, which should be present in any multiloculated cyst.) A parapelvic abscess is unlikely on the basis of the clinical findings and relatively normal IVP. A calyceal diverticulum is excluded by the IVP. Could this, then, represent a part of the renal collecting system, by far the most common cause of fluid-filled "masses" in the renal sinus? Well, actually, this is a typical appearance for a slightly enlarged extrarenal pelvis. On the basis of the ultrasound picture alone, we cannot say whether the renal pelvis is large because of hydronephrosis or whether this is physiologic dilatation secondary to the fact that the pelvis lies outside the renal sinus — the so-called "extrarenal pelvis." Because we do not see any evidence of dilated calyces, we suspect that this is in fact an extrarenal pelvis, and the IVP certainly confirms our suspicion. (The patient's doctor was a man of little faith; he requested an arteriogram to verify the ultrasound diagnosis.)

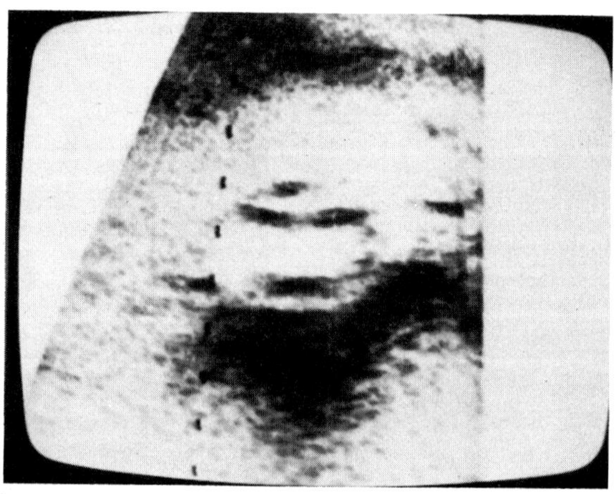

Figure 51–B. Transverse section +3. Extrarenal pelvis. Prone scan of right kidney.

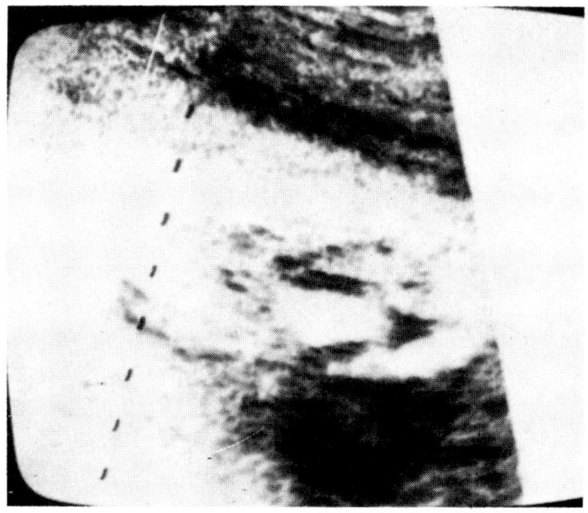

Figure 51–C. Longitudinal section 2L. Extrarenal pelvis. Prone scan of right kidney.

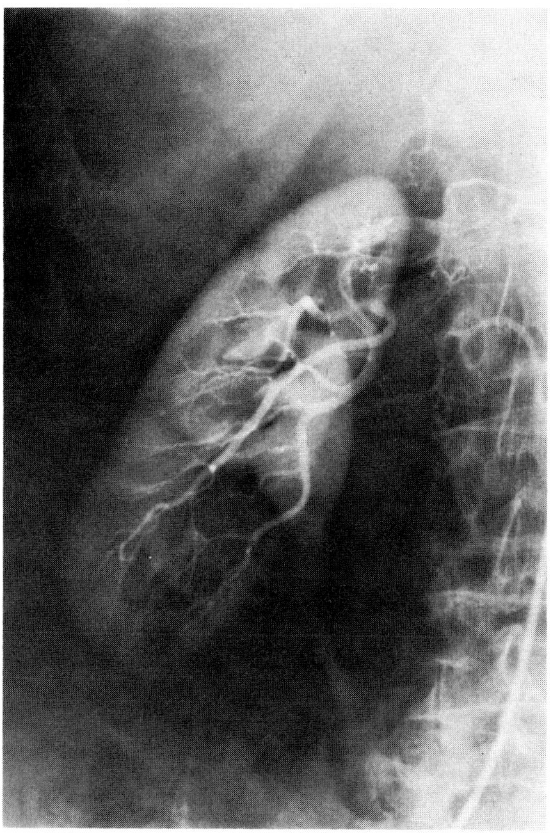

Figure 51–D.
Arteriogram of kidney in Figures 51–A through 51–C.

REFERENCES

Sanders R C, Conrad M R: The ultrasonic characteristics of the renal pelvicalyceal echo complex. J Clin Ultra 5:372 (1977)

Masatello V J, Smith E H, Carrera G F, et al: Ultrasonic evaluation of the obstructed duplex kidney. Am J Radiol 129:113 (1977)

Case 52

CLINICAL INFORMATION

This 53-year-old man has had a "cold" for the past six months which he "just hasn't been able to shake off." He frequently feels nauseated, tires easily, has abdominal cramps, and has lost 20 pounds. The physical examination does not turn up any definite abnormalities. There are a few scattered wheezes and rhonchi in the chest (he is a heavy smoker) and the liver is slightly enlarged. The laboratory results show a mild anemia, slightly elevated blood glucose, normal liver function tests, and occult blood in the stool. Chest and plain abdominal radiograms are negative. Barium studies have not been performed.

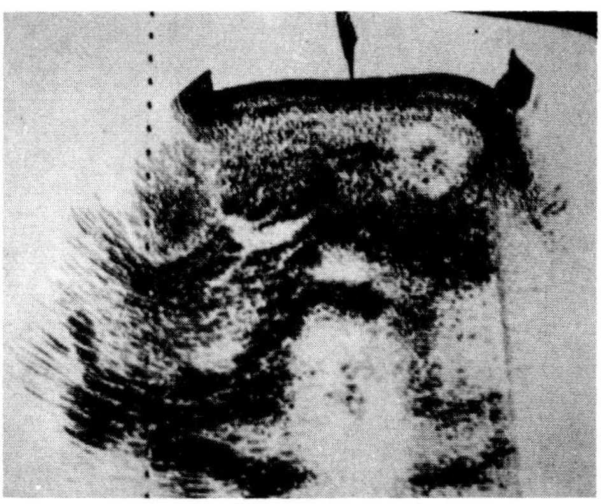

Figure 52–A. Transverse section +12.

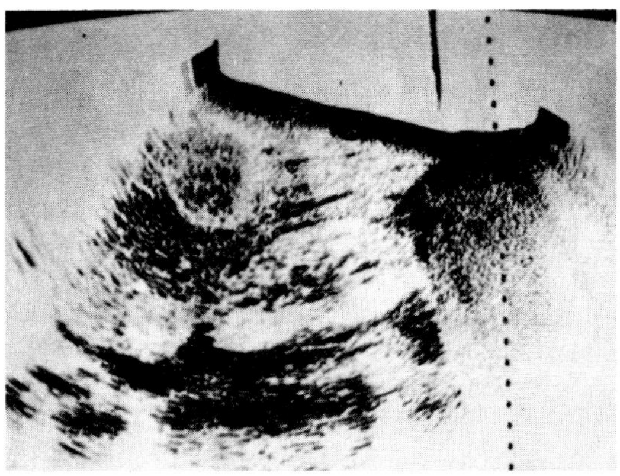

Figure 52–B. Longitudinal section 8R.

DIAGNOSIS

Large liver metastases (carcinoma of the colon).

ULTRASOUND FINDINGS

There are three large focal defects in the liver. The cross-section (Figure 52–A) shows one relatively echo-free area in the tip of the left lobe and a second area with an echo-free ring and dense center in the right lobe. The longitudinal section (Figure 52–B) again demonstrates the ring lesion as well as a third echo-free area deep in the right lobe just above the kidney. Large, multiple, solid lesions in the liver are almost pathognomonic for liver metastases. (Focal regenerative hyperplasia can also give this appearance, but this disorder is rare and the clinical setting is different. Multiple liver cysts do not contain internal echos.)

In order to recognize metastases we must be familiar with the appearance of the normal liver (Figures 52–C and 52–D), which is homogeneously inhomogeneous. Although this sounds like self-contradictory gobbledygook, it actually describes the liver texture quite well. The liver normally generates echoes of several different amplitudes, corresponding to the many different internal interfaces; hepatocytes, bile ducts, blood vessels, and supporting stroma all combine to generate echoes of different amplitudes, which generally fall into the middle shades of gray. The liver texture is, therefore, inhomogeneous. The composition of the liver is the same throughout all the lobes, and so the distribution of the many interfaces is also uniform and the overall appearance of the liver texture is homogeneous. This pattern is analogous to the texture of burlap cloth; the coarseness of burlap produces a homogeneous inhomogeneity.

Metastases cause focal distortion and disruption of the normal liver texture (just as a hole disrupts the pattern in the burlap). The monotonous homogeneous liver texture is interrupted by focal areas of either loud or low-level echoes, or a combination of the two. Figures 52–E through 52–J are examples of the many different ways metastases can distort the normal liver pattern. Some are dense, loud echoes (Figures 52–E, 52–G, and 52–H), while others are very weak echoes or echoes are absent altogether (Figures 52–H, 52–I, and 52–J). Sometimes the lesions contain both loud and weak echoes (Figure 52–F). On occasion, different parts of the liver will contain different types of lesions (Figures 52–A, 52–B, and 52–H).

When there are unequivocal focal areas of abnormal echoes in the liver, we can diagnose metastatic liver disease with an accuracy approaching 100 per cent. Our true positive rate will be very high and the false positive rate very low. Unfortunately, excluding liver metastases is not such an accurate business. Very small or diffuse liver metastases may not alter the internal texture enough to be appreciated on the ultrasound picture; therefore, we believe very strongly that the presence of a normal ultrasound scan of the liver *does not exclude liver metastases*. Some ultrasonographers do not agree with us on this point. The problem of the accuracy of ultrasound in diagnosing liver metastases is discussed in greater detail in Case 88.

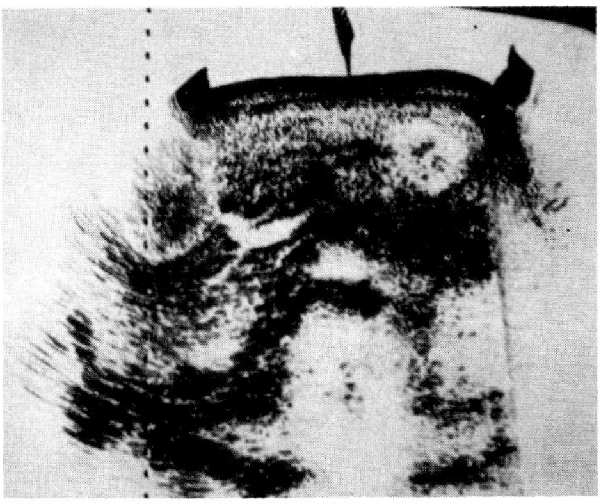

Figure 52–A. Transverse section +12.
Liver metastases.

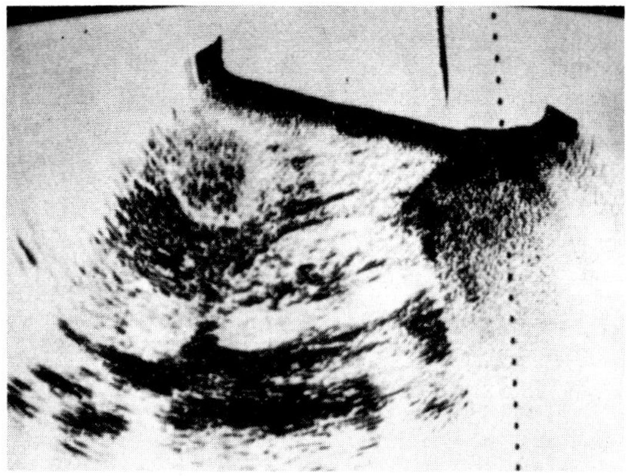

Figure 52–B. Longitudinal section 8R.
Liver metastases.

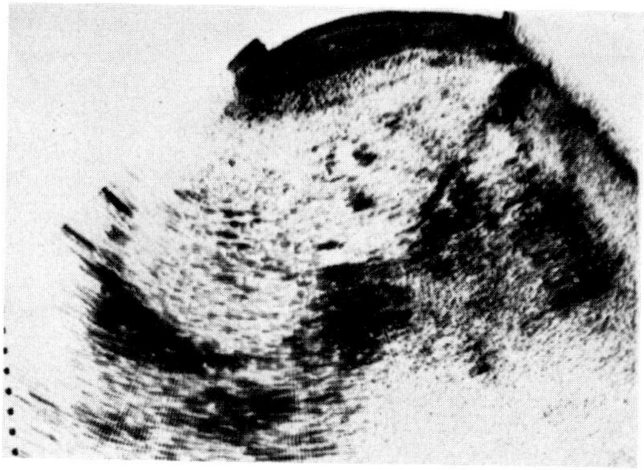

Figure 52–C. Transverse section +14.
Normal liver.

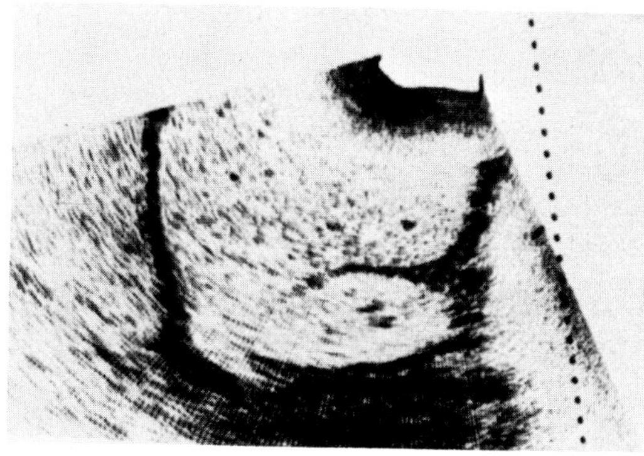

Figure 52–D. Longitudinal section 8R.
Normal liver.

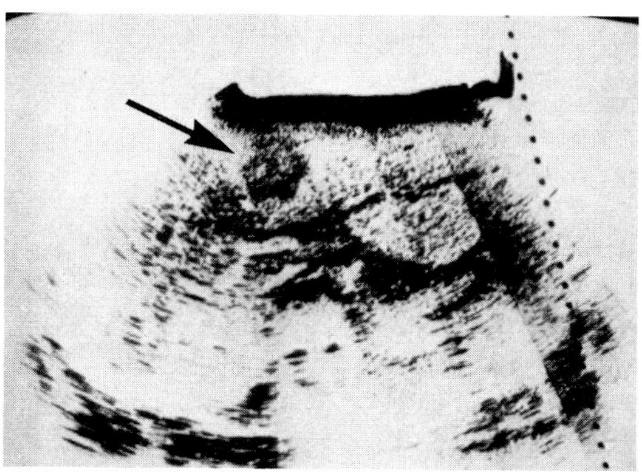

Figure 52–E. Transverse section +16.
Liver metastases.

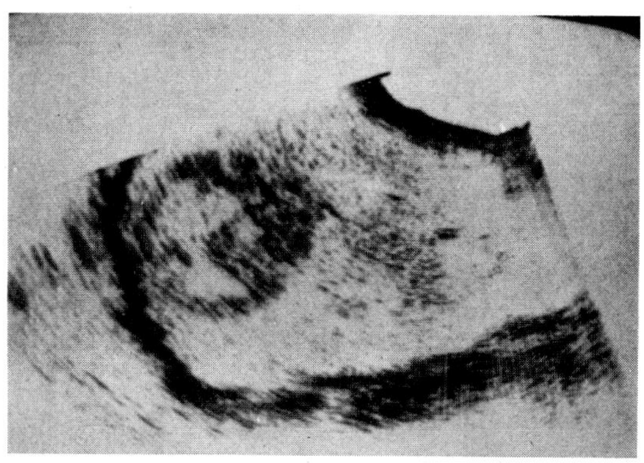

Figure 52–F. Longitudinal section 10R.
Liver metastases.

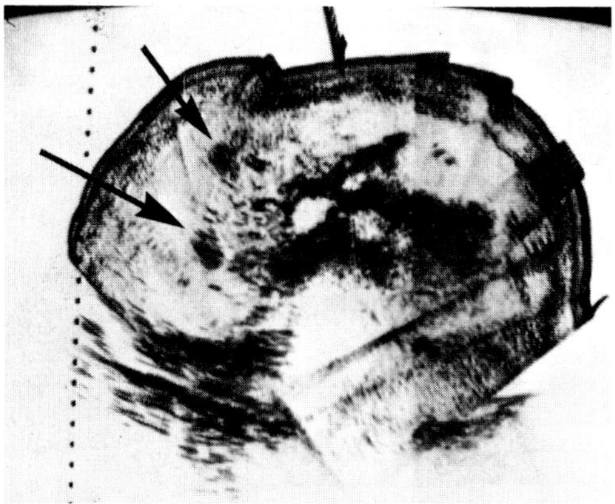

Figure 52–G. Transverse section +8.
Liver metastases.

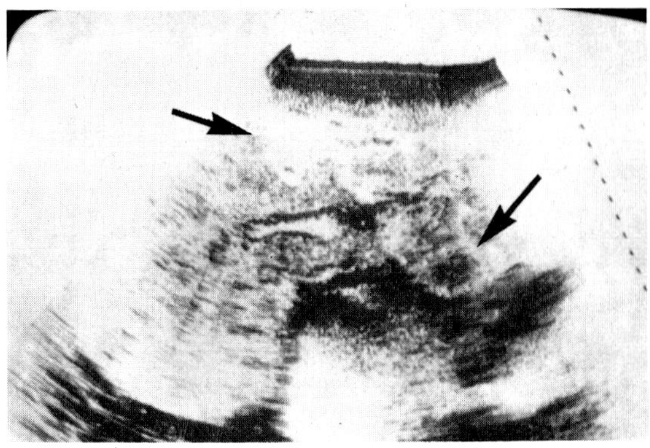

Figure 52–H. Transverse section +13.
Liver metastases.

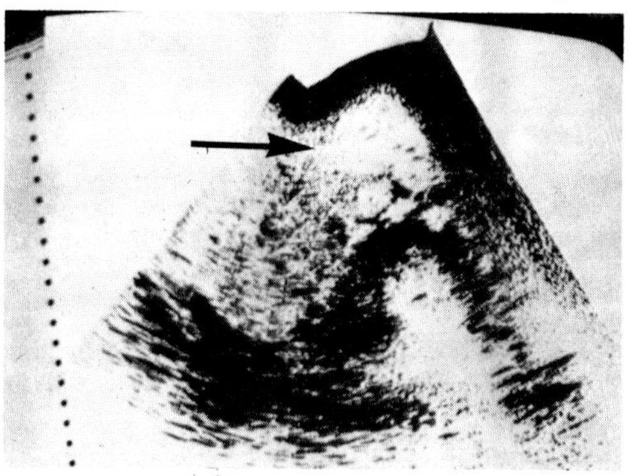

Figure 52–I. Transverse section +10.
Liver metastases.

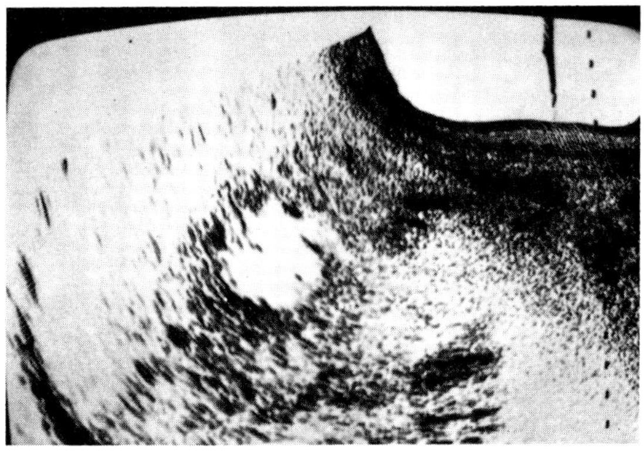

Figure 52–J. **Longitudinal section 10R.** Cystic liver metastasis.

REFERENCES

Scheible W, Gosink B B, Leopold G R: Gray scale echographic patterns of hepatic metastatic disease. Am J Radiol 129:983 (1977)

Zatz L M, Gouldin J A, Hanley G A: A comparison of gray scale ultrasound and radionuclide for the detection of focal hepatic lesions. J Clin Ultra 5:178 (1977)

Green B, Bree R L, Goldstein H M, et al: Gray scale ultrasound evaluation of hepatic neoplasms. Radiology 124:203 (1977)

Taylor K J W, Sullivan D, Rosenfield A T, et al: Gray scale ultrasound and isotope scanning: complementary techniques for imaging the liver. Am J Radiol 128:277 (1977)

Case 53

CLINICAL INFORMATION

This 74-year-old man has had a steady, dull aching in his left flank and back. The pain has been present for many months and seems to be worsening. He has also noticed some episodes of pinkish urine. Many years ago he had tuberculosis and was once told that it had infected his kidneys; he has never been given any medicine for TB, however, and had always assumed that the infection had "burned itself out." Upon questioning, he admits to a decreasing appetite and small (5 lb.) weight loss in the past year. His urinalysis shows many red blood cells, but it is otherwise normal. An IVP is not of good quality; the kidneys are only faintly visualized, and the collecting systems are not seen in their entirety. Despite the poor visualization the radiologist feels that the left kidney is displaced anteriorly and somewhat laterally, but he cannot tell whether there is a renal mass. These two longitudinal scans (Figures 53–A and 53–B) and one transverse scan (Figure 53–C) of the left kidney were performed with the patient prone.

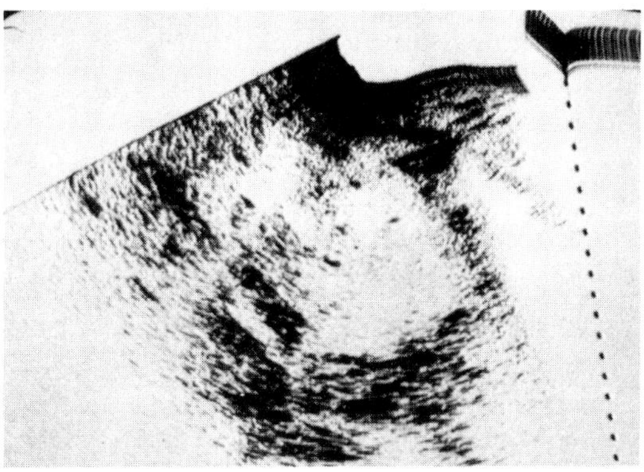

Figure 53–A. Longitudinal section 2L. Prone scan of left kidney.

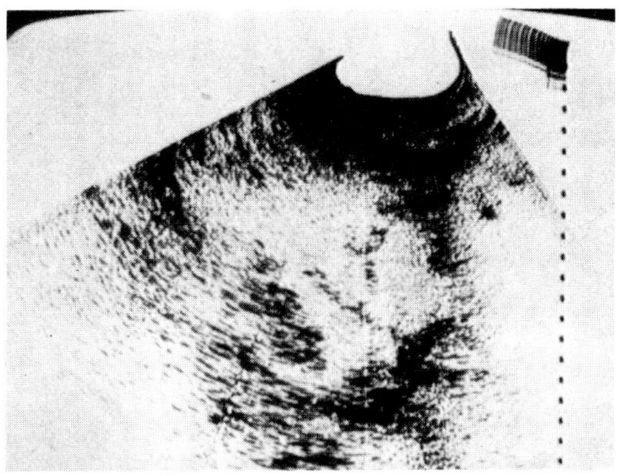

Figure 53–B. Longitudinal section 0L. Prone scan of left kidney.

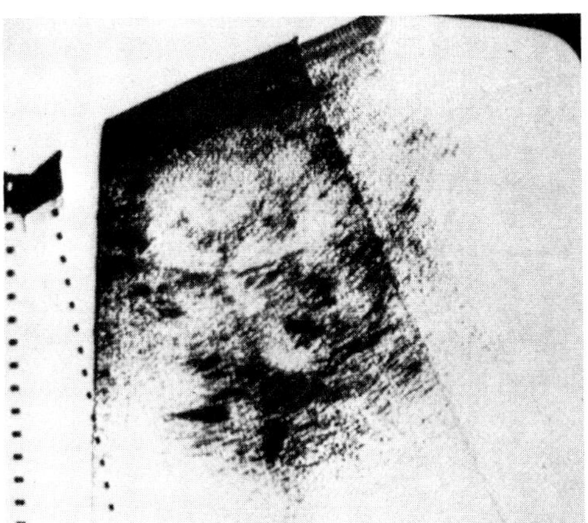

Figure 53–C. Transverse section +3. Prone scan of left kidney.

DIAGNOSIS

Massive hypernephroma (renal cell carcinoma).

ULTRASOUND FINDINGS

This is a rather straightforward case; the mass of tumor is apparently arising in the posterior portion of the kidney and is seen most clearly there. The lobular, irregular margins and multiple internal echoes tell us that we are dealing with a solid mass, not a cyst. Although it is theoretically possible that a perinephric abscess or granuloma could produce this many internal echoes, such an occurrence is rare. The only problem for the ultrasonographer is to decide if this tumor is of renal or extrarenal origin (i.e., retroperitoneal sarcoma). Although in Figures 53–B and 53–C we can identify a partial boundary between the tumor and the kidney, this boundary is never complete. Figure 53–A clearly shows that the tumor is arising from the renal parenchyma.

As we have said before, we do not think it is possible to make a histologic diagnosis of tumor type from the appearance of the ultrasound scan alone; this requires either an angiogram or a biopsy. Look at the scan in Figure 53–D. This is a longitudinal scan of the right kidney in a different patient — the orientation is similar to that of Figure 53–A. The kidney in this patient has been almost entirely replaced by tumor. (The normal, uninvolved liver is seen clearly anterior to the tumor.) Although its appearance is similar to that of the hypernephroma, this tumor is a massive *transitional cell carcinoma.*

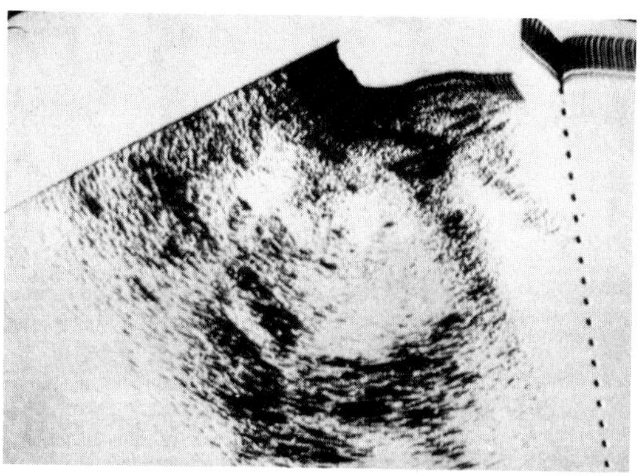

Figure 53–A. Longitudinal section 2L. Hypernephroma. Prone scan of left kidney.

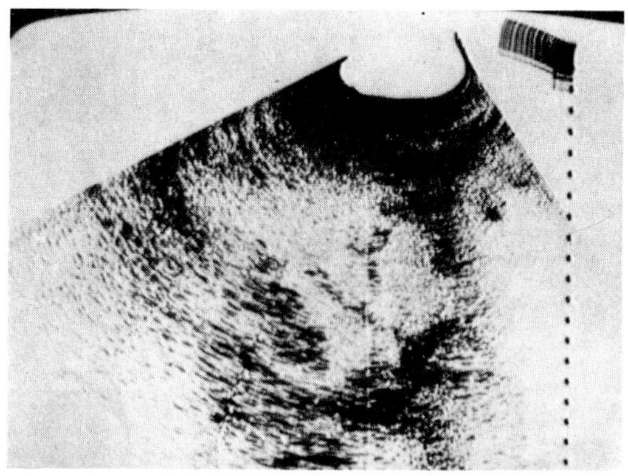

Figure 53–B. Longitudinal section 0L. Hypernephroma. Prone scan of left kidney.

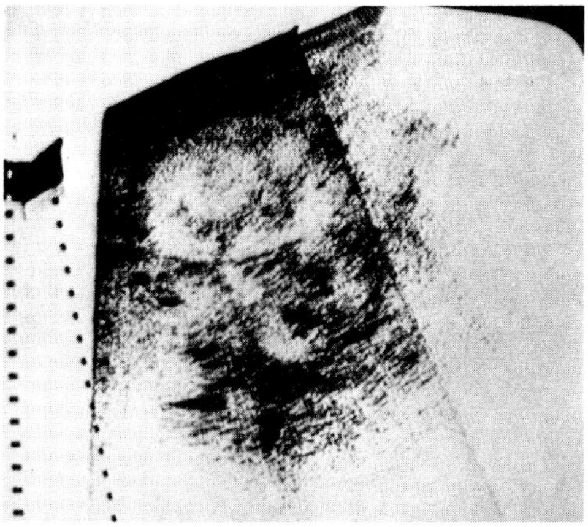

Figure 53–C. **Transverse section +3.** Hypernephroma. Prone scan of left kidney.

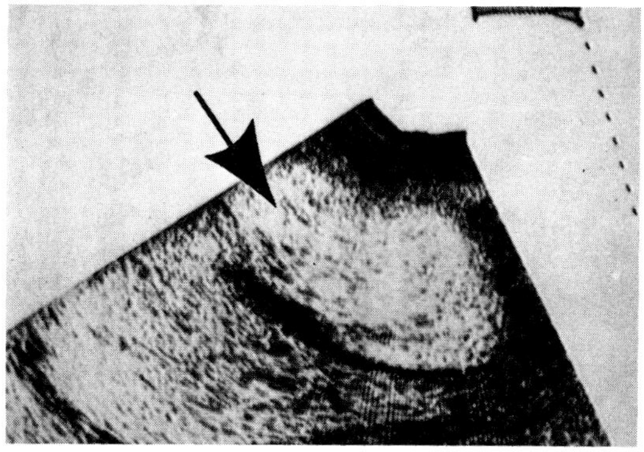

Figure 53–D. **Longitudinal section 0L.** Transitional cell carcinoma (arrow). Prone scan of right kidney.

REFERENCES

Maklad N F, Chuang V P, Doust B D, et al: Ultrasonic characterization of solid renal lesions. Radiology 123:733 (1977)

Sanders R C: Renal ultrasound. Rad Clin N. Am 13:417 (1975)

Case 54

CLINICAL INFORMATION

This 33-year-old woman has had episodic upper abdominal pains, which her doctor suspects are caused by gallstones. An oral cholecystogram shows only faint visualization of the gallbladder so she is sent for ultrasonic cholecystography. The technician keeps getting this picture. Are there gallstones? What can you do to verify your diagnosis?

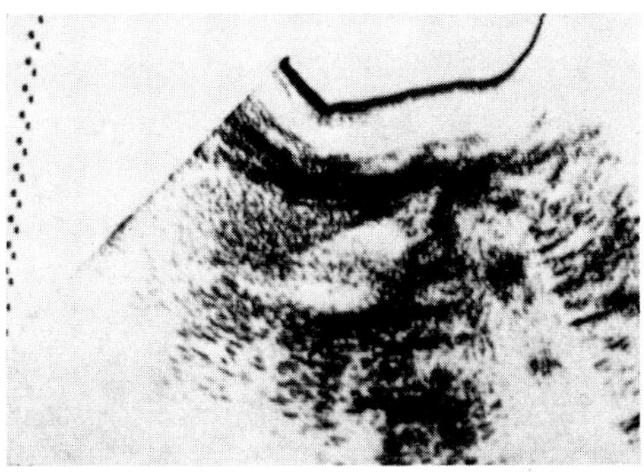

Figure 54–A.
Normal gallbladder with reverberation artifact. Longitudinal section through gallbladder.

DIAGNOSIS

Normal gallbladder (reverberation artifact simulating floating gallstones).

ULTRASOUND FINDINGS

There is a band of echos (straight arrow) running through the center of the gallbladder lumen on the supine scan (Figure 54–A). Although these echoes could represent a layer of floating gallstones, they are rather atypical. The band of echoes is wide, there is no acoustic shadow, and the echoes seem to parallel the loud echoes arising from the peritoneum and superior surface of the liver (curved arrow). Perhaps we are seeing the first generation reverberation artifact of these loud echoes? To test this hypothesis we will scan the patient in the upright position: floating gallstones will shift position so that they remain parallel to the floor while reverberation artifacts will stay parallel to the abdominal wall. The upright scan (Figure 54–B) confirms our suspicion — the echoes stay aligned with the abdominal wall and are artifacts, not stones.

Figures 54–C and 54–D are other examples of reverberation artifacts masquerading as floating gallstones: upright scanning quickly solves the dilemma in these situations. Figures 54–E, 54–F, and 54–G are scans of real floating stones. Their appearance is a little different from the artifacts; they are more sharply defined, they frequently, but not always (Figure 54–E), cast an acoustic shadow, and they are oriented horizontally. If the scan is performed upright, as in Figure 54–G, the stones shift to keep their horizontal orientation.

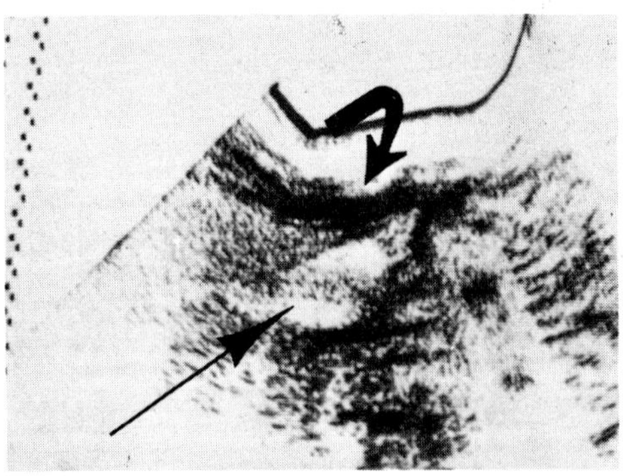

Figure 54–A.
Longitudinal section through gallbladder.

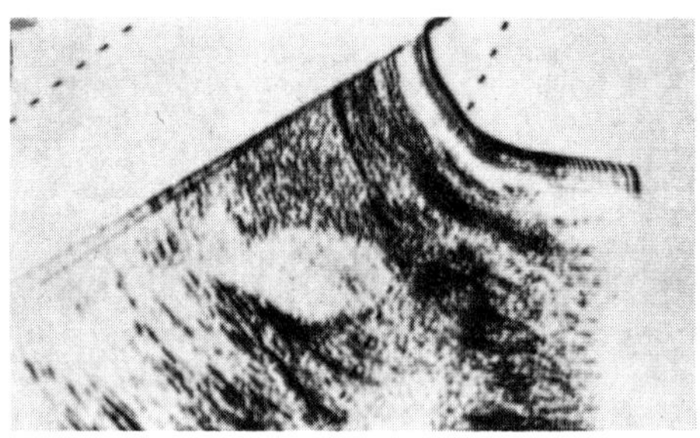

Figure 54–B.
Normal gallbladder. Upright scan of patient in Figure 54–A.

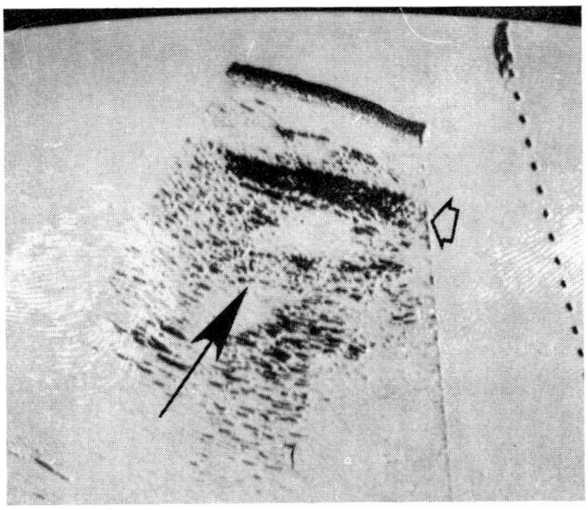

Figure 54–C. **Longitudinal section 4R.** Reverberation artifact (solid arrow) from anterior abdominal wall (open arrow).

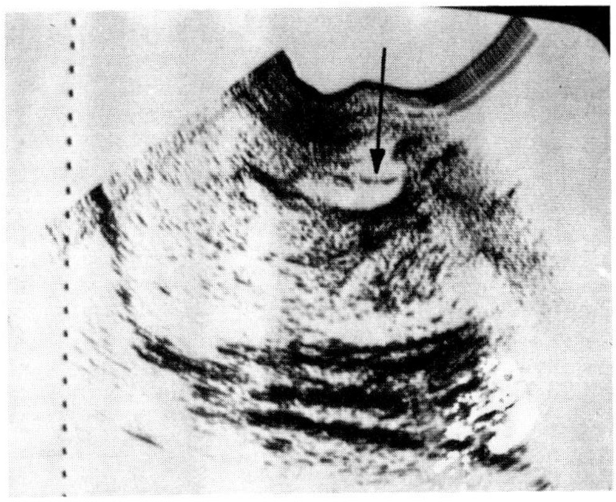

Figure 54–D. **Longitudinal section 7R.** Reverberation artifact in gallbladder.

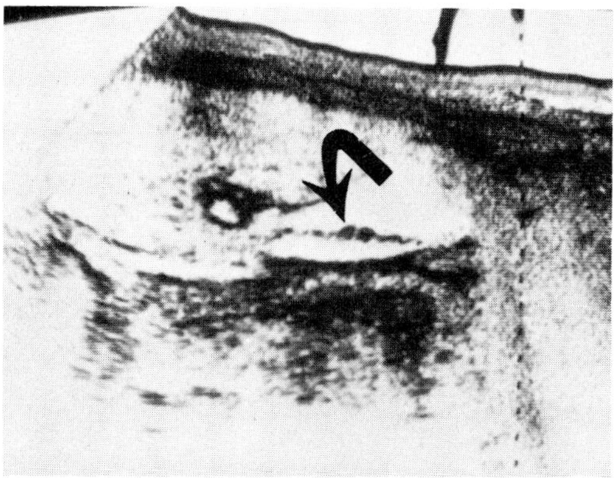

Figure 54–E. **Longitudinal section 4R.** Floating gallstones.

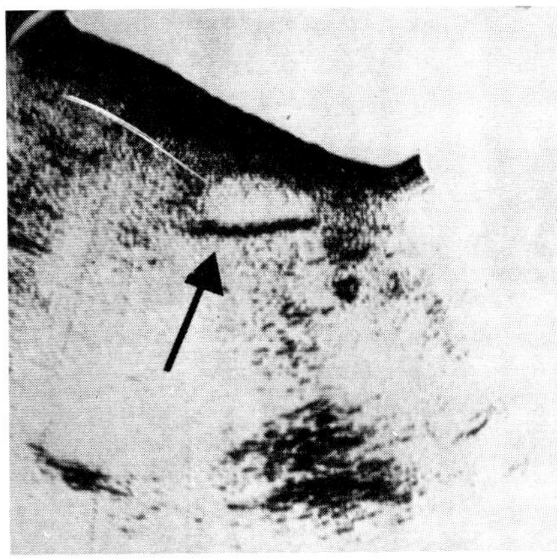

Figure 54–F. **Longitudinal section 5R.** Floating gallstones.

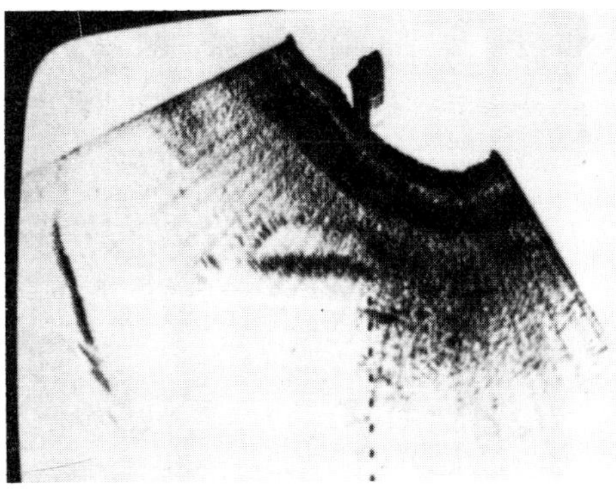

Figure 54–G.
Upright scan, longitudinal section 5R. Floating gallstones.

REFERENCES

Bartrum R J Jr, Crow H C: *Gray-Scale Ultrasound: A Manual for Physicians and Technical Personnel.* Chapters 5 and 6. W B Saunders Co, Philadelphia (1977)

Crow H C, Bartrum R J Jr, Foote S R: Expanded criteria for the ultrasonic diagnosis of gallstones. J Clin Ultra 4:289 (1976)

Case 55

CLINICAL INFORMATION

This 30-year-old asymptomatic woman is having a routine physical examination before having an IUD inserted. The doctor finds that her uterus is normal in size but deviated to the right, and there is a small, firm mass in the left ovary.

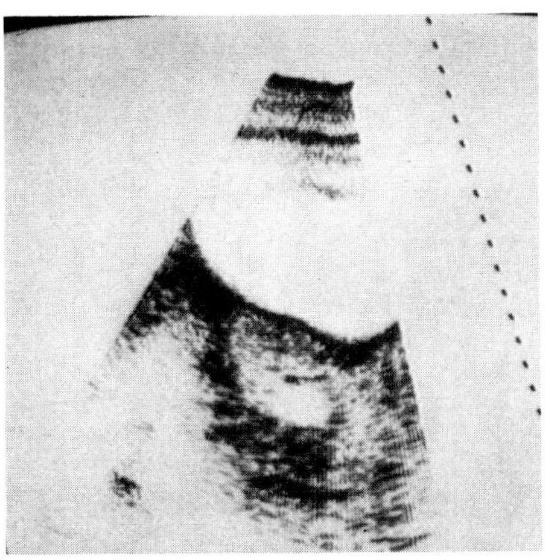

Figure 55–A. Longitudinal section 1L.

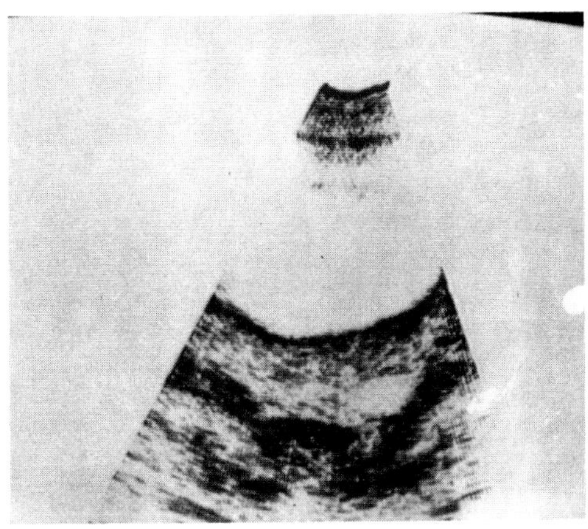

Figure 55–B. Transverse section −10.

DIAGNOSIS

Corpus luteum cyst.

ULTRASOUND FINDINGS

There is a small echo-free area in the region of the left ovary. It is difficult in this case to separate the ovary from the uterus, and the dextroflexion of the uterus causes it to appear large on the cross-section (Figure 55–B). The echoless mass has a smooth back wall and is quite typical of a corpus luteum cyst. These benign cysts are the most common ovarian mass in premenopausal women. They arise in the second half of the menstrual cycle from the corpus luteum and often grow to 2 centimeters before regressing during the menstrual period. If the patient is re-examined during the first half of her cycle no mass will be found. When the levels of estrogen and progesterone are maintained, a corpus luteum cyst may continue to grow and become quite formidable. Figure 55–C shows a 4-centimeter cyst in the cul-de-sac (arrow) in a woman who is 14 weeks pregnant. Figure 55–D shows a huge, multilocular corpus luteum cyst in a woman with a choriocarcinoma.

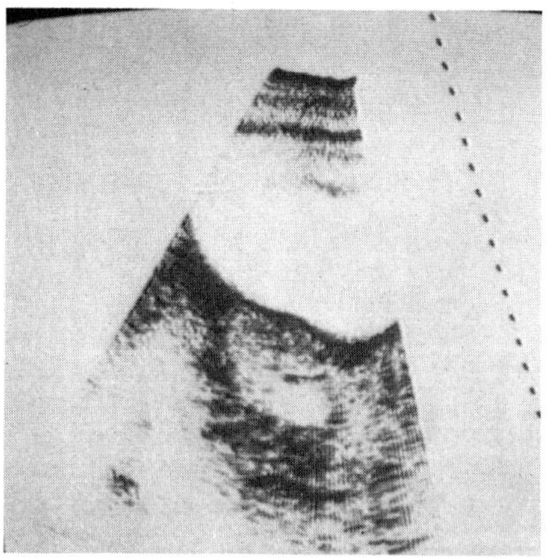

Figure 55–A. Longitudinal section 1L. Corpus luteum cyst.

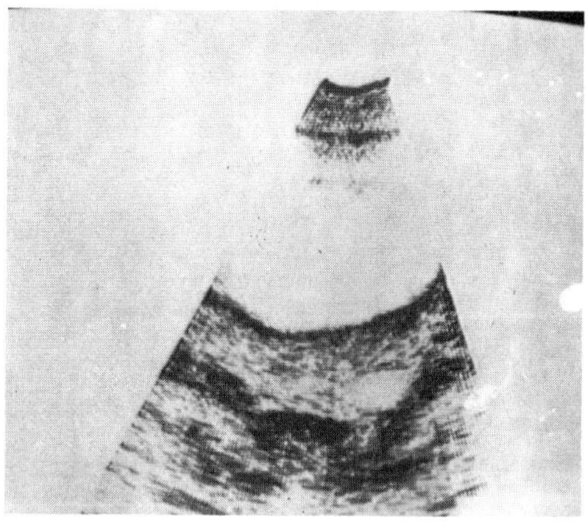

Figure 55–B. Transverse section −10. Corpus luteum cyst.

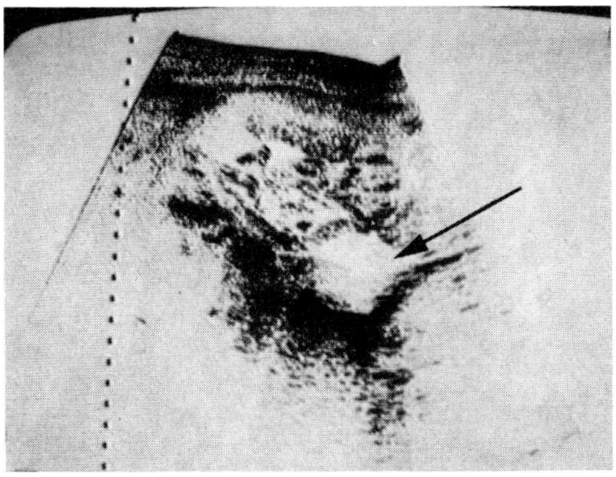

Figure 55–C. **Longitudinal section 1R.** Corpus luteum cyst in pregnancy.

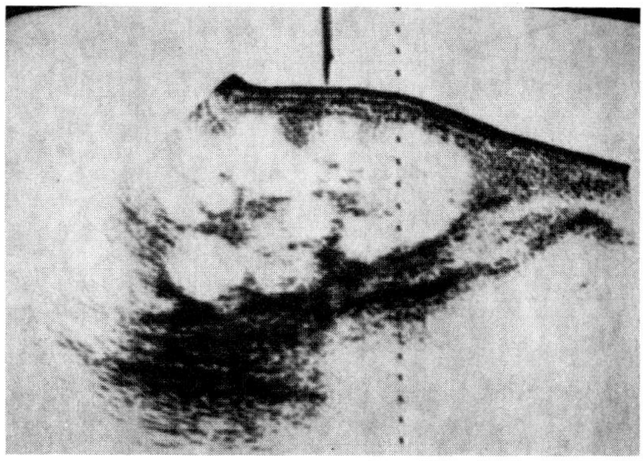

Figure 55–D. **Longitudinal section 6L.** Multilocular corpus luteum cyst (theca lutein cyst) in patient with choriocarcinoma.

REFERENCES

Hobbins J C, Winsberg F: *Ultrasonography in Obstetrics and Gynecology.* Chapter 8. Williams & Wilkins Co, Baltimore (1977)

Morley P, Barnett E: The ovarian mass, Chapter 27 in *Ultrasonography in Obstetrics and Gynecology* Sanders and James, editors. Appleton-Century-Crofts, New York (1977)

Scheer K, Goldstein D P: Use of ultrasonography to follow regression of theca lutein cysts. Radiology 108:673 (1973)

Case 56

CLINICAL INFORMATION

Figures 56–A, 56–B, and 56–C are all transverse scans at level +3 in three different patients. The clinical question is the same in each case — is there metastatic tumor in the bowel or mesentery? Which, if any, of these patients have intra-abdominal metastases? What can you do to confirm your diagnosis?

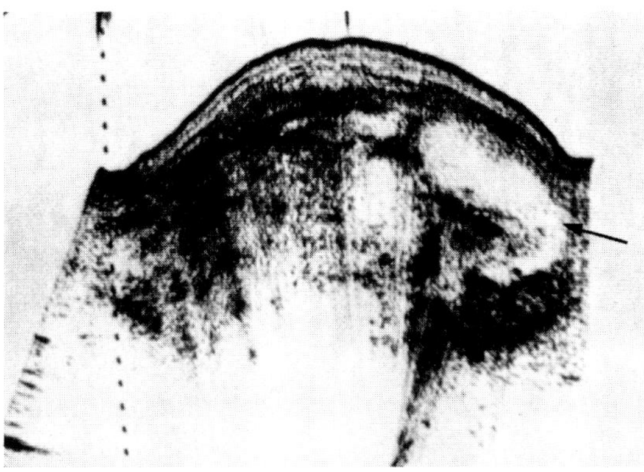

Figure 56–A. Transverse section +3.

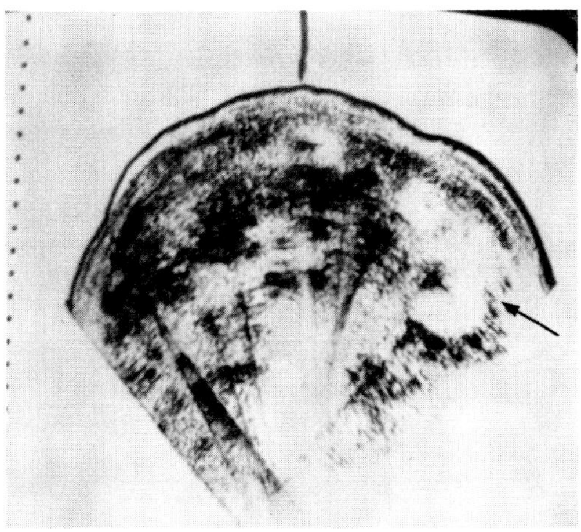

Figure 56–B. Transverse section +3.

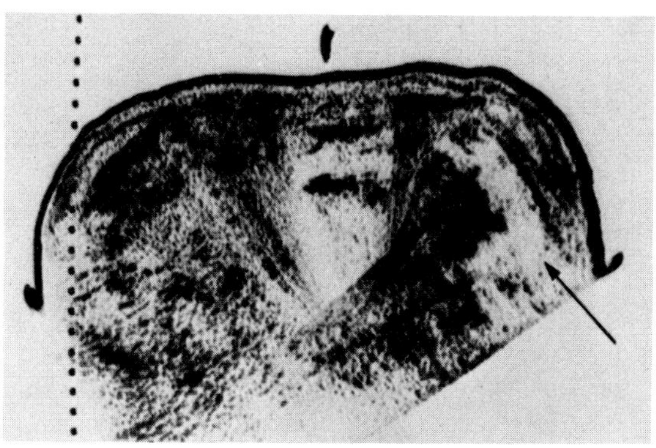

Figure 56–C. Transverse section +3.

DIAGNOSIS

Patients A and B have mesenteric metastases; patient C is normal.

ULTRASOUND FINDINGS

All these scans are essentially the same: there is a semicircular relatively echo-free mass in the left abdomen (arrows). It is unusual to see this appearance in the lower abdomen; the more common picture is a diffuse, amorphous collection of echoes and artifacts which arise from partially air-filled small and large bowel. Whenever a well-defined structure is encountered in the lower abdomen we have to be on guard. Could this represent a pathologic mass or is it simply a loop of normal bowel filled with fluid or feces?

This is a difficult problem to sort out. First, we make certain that the "mass" we are seeing is a constant finding. Fluid in the bowel is changed by peristalsis and if our mass is not reproducible on multiple scans over several minutes we dismiss it as fluid-filled bowel. Feces in the colon does not move as rapidly and it is necessary to scan the patient on a different day following an enema to eliminate this type of "mass." (This is what was done with patient C; she was given an enema and re-examined — the "mass" was gone.) Secondly, we examine the patient: a real mass should be palpable and have a feeling quite different from bowel. Finally, if we are still confused, we can perform a percutaneous biopsy of the suspicious "mass." While this may seem to be a bit extreme, it is an innocuous procedure and is certainly preferable to an exploratory operation, laparoscopy, or further expensive and complicated diagnostic imaging procedures.

Patient A had known ovarian carcinoma and patient B a malignant carcinoid tumor. Aspiration of the "masses" in these patients confirmed the clinical suspicion that the tumor had spread to the mesentery.

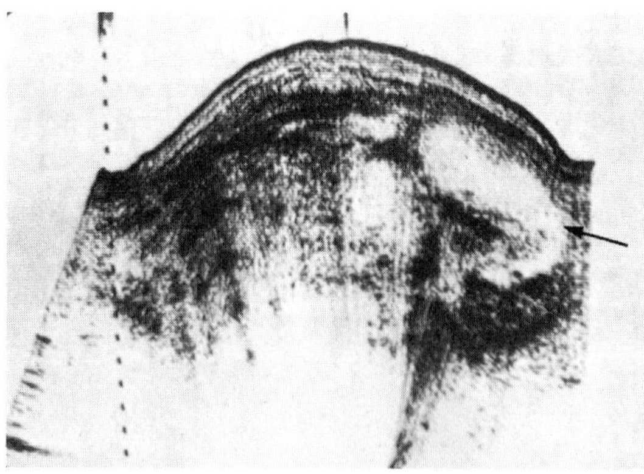

Figure 56–A. Transverse section +3. Metastatic ovarian carcinoma.

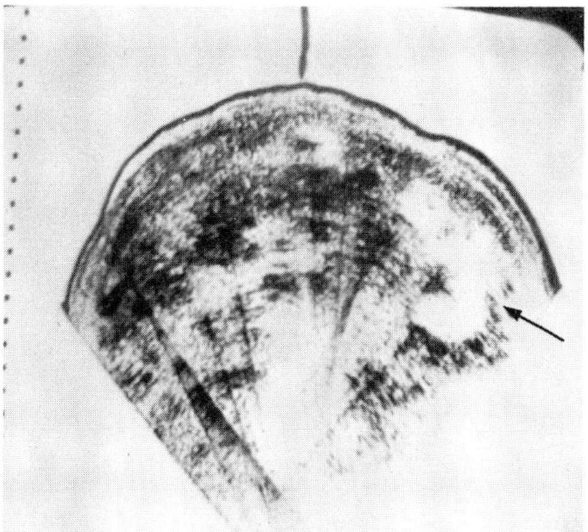

Figure 56–B. Transverse section +3. Malignant carcinoid.

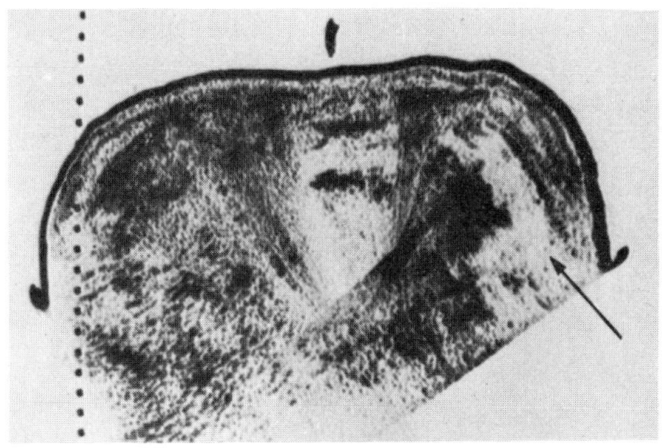

Figure 56–C. Transverse section +3. Normal, feces-filled colon.

REFERENCES

Bartrum R J Jr, Crow H C: *Gray-Scale Ultrasound: A Manual for Physicians and Technical Personnel.* Chapter 13. W B Saunders Co, Philadelphia (1977)

Goldstein H M, Zornoza J, Wallace S, et al: Percutaneous fine needle aspiration biopsy of pancreatic and other abdominal masses. Radiology 123:319 (1977)

Case 57

CLINICAL INFORMATION

This 42-year-old man has a chronic disease (we won't tell you what it is yet) and is sent for ultrasound examination to evaluate an abdominal mass. The patient, while not robust, is not particularly ill and has no general complaints.

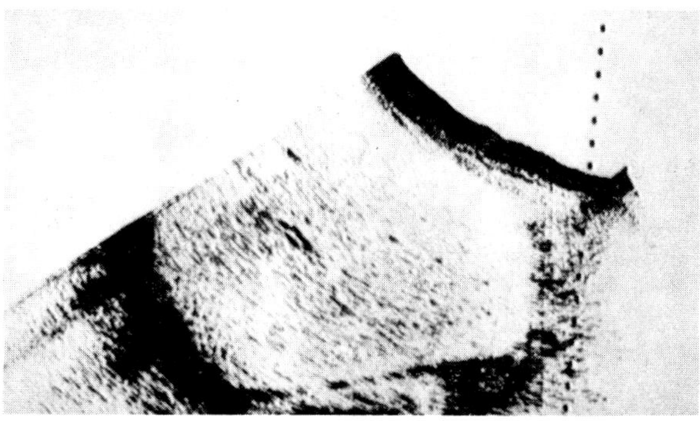

Figure 57–A. Longitudinal section 12L.

DIAGNOSIS

Massive splenomegaly (myelofibrosis).

ULTRASOUND FINDINGS

Did you call this an enlarged liver? It certainly looks like it but since the scan was performed in the *left* upper quadrant the liver is not a likely candidate. (When confronted with this problem we must always check with the technicians to be certain that the scan is labeled correctly and check the patient to make sure there is not situs inversus.) This is the spleen, massively enlarged and containing higher-level echoes than normally seen. There is not much to be said about splenomegaly; ultrasound is helpful in documenting that a palpable mass in the left upper quadrant is indeed the spleen and in getting a good estimate of its size. Sometimes, as in this case, ultrasound is used to follow the advance or regression of the spleen in chronic disease.

We do not feel comfortable trying to make a specific diagnosis as to the cause of the splenomegaly: it seems that all enlarged spleens look pretty much the same and there are no good ultrasonic signs on which to base a specific diagnosis. Fortunately, this is not often a *clinical* problem. Figures 57–B and 57–C are scans from a patient with idiopathic splenomegaly and Figure 57–D is an enlarged spleen secondary to leukemia; this patient also had a simple cyst (arrow) in the left kidney.

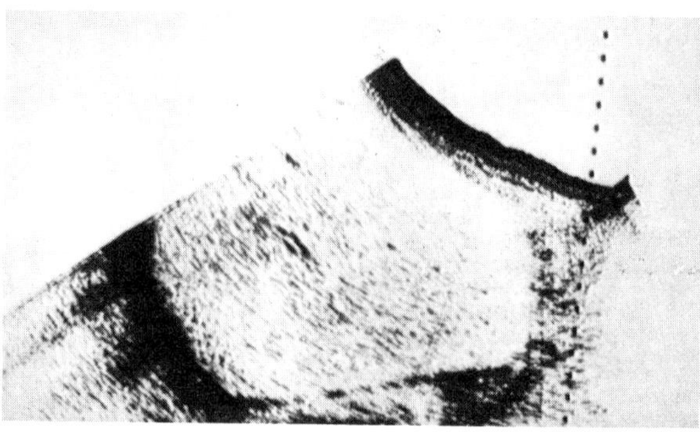

Figure 57–A. **Longitudinal section 12L.** Splenomegaly (myelofibrosis).

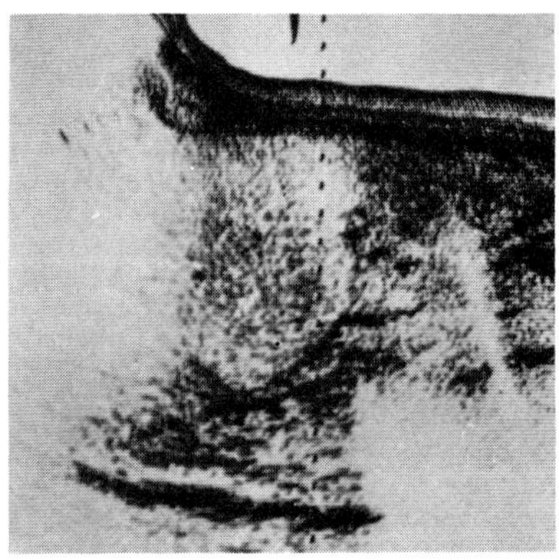

Figure 57–B. **Longitudinal section 6L.** Splenomegaly (idiopathic).

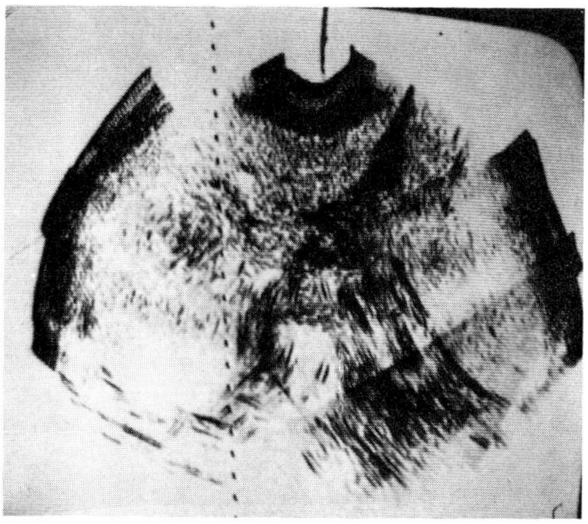

Figure 57–C. Transverse section +9.5. Splenomegaly (idiopathic).

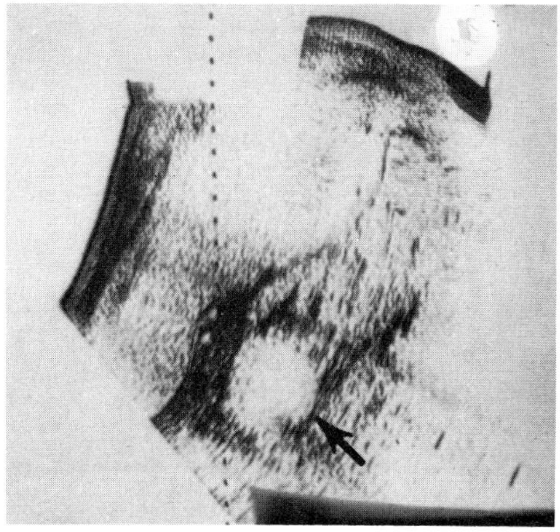

Figure 57–D. Splenomegaly (leukemia). Right lateral decubitus scan; section through a rib interspace. Notice the small cyst in the left kidney.

REFERENCES

Hunter T B, Haber K: Unusual sonographic appearance of the spleen in a case of myelofibrosis. Am J Radiol 128:138 (1977)

Taylor K J W, Milan J: Differential diagnosis of chronic splenomegaly by gray scale ultrasonography. Br. J Radiol 49:519 (1976)

Case 58

CLINICAL INFORMATION

This 73-year-old woman is 16 years out from the diagnosis and treatment of Stage I carcinoma of the cervix. At the time of the initial diagnosis she received both internal and external radiation followed by a simple abdominal hysterectomy. There was no evidence of metastatic disease at that time, nor has there been any suggestion of recurrence. During the past six months she has had some episodes of back pain and a lumbosacral spine radiograph is interpreted as showing a mass in the right lower quadrant of the abdomen (the bones are normal). An IVP (Figure 58–A) suggests that this mass is intimately associated with the lower portion of the right kidney, but it is difficult to see and the radiologist cannot characterize it further. The mass is not easily palpable (although some observers insist that it is definitely there), and careful review of previous films fails to turn up any suggestion of its prior existence. The urinalysis is normal. The patient is not particularly easy to scan; she is moderately obese and does not hold her breath well. This scan, performed with the patient supine, is typical of those obtained.

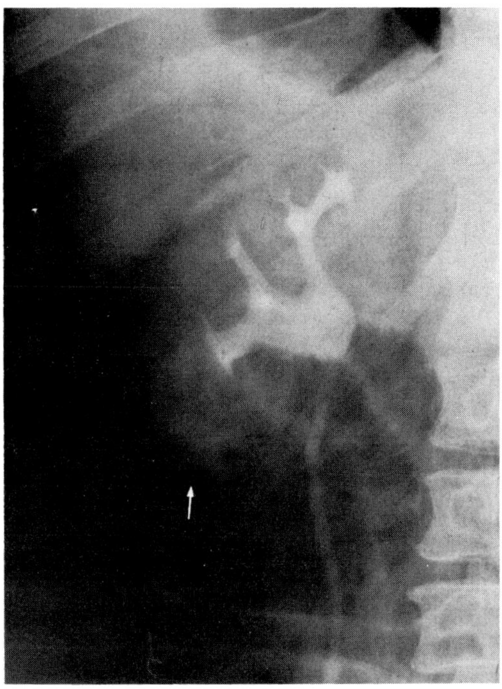

Figure 58–A.
IVP, 8-minute compression film of right kidney.

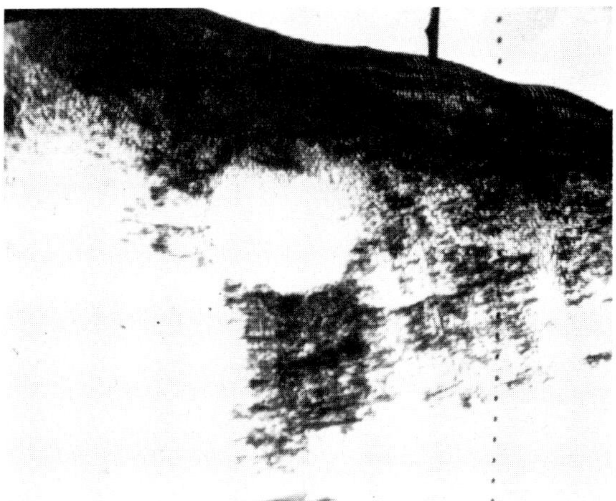

Figure 58–B. Longitudinal section 6R.

DIAGNOSIS

Benign renal cyst.

ULTRASOUND FINDINGS

The mass has two of the three characteristics of a benign renal cyst: no internal echoes and increased through transmission. The third and most important sign is absent, however: the back wall is not smooth but contains two small lumps. The lack of a smooth back wall is sometimes the only finding in a cystic renal tumor or renal abscess. While we felt that the most likely diagnosis in this case was renal cyst, we recommended that because of the irregular wall the diagnosis be confirmed by cyst puncture. (Another diagnostic possibility, massive hydronephrosis, was excluded by IVP.) Puncture was performed, the fluid was drained and contained no red blood cells or tumor cells, and the patient's backaches lessened.

Figure 58–C is a nearly identical scan obtained on another patient. There is a large echo-free mass in the kidney, and there is increased through transmission. The back wall, however, is not a smooth, continuous line. This mass was a cystic hypernephroma.

Figure 58–B. Longitudinal section 6R. Renal cyst.

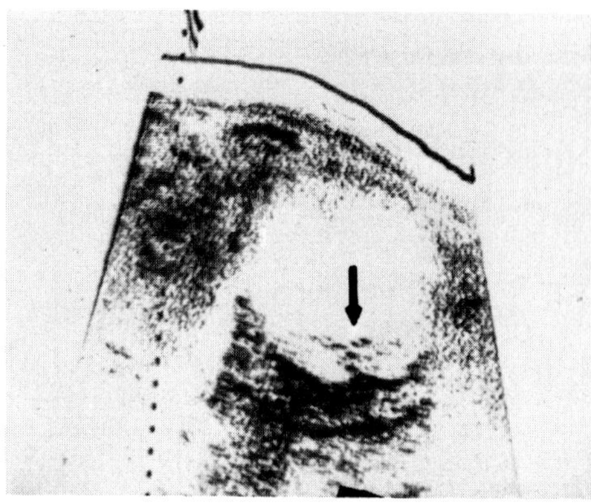

Figure 58–C. Transverse section +2. Prone scan; cystic hypernephroma.

REFERENCE

Green W M, King D L, Cassarella W J: A reappraisal of sonolucent renal masses. Radiology 121:163 (1976)

Case 59

CLINICAL INFORMATION

This is a 63-year-old man who has cancer of the pancreas. The diagnosis was made three months earlier (by ultrasound!) and, as the tumor was small and there was no evidence of metastatic disease, a "curative" surgical procedure — total pancreatectomy — was performed. Although he weathered the immediate postoperative period without serious difficulty, he continued to have pain and tenderness in his abdomen, associated with fevers and abnormal laboratory tests. His appetite is poor and he has continued to lose weight. His doctor is concerned that there is an intra-abdominal abscess or recurrence of the tumor and requests ultrasound to help sort out the differential diagnostic possibilities. Can you make a specific diagnosis? What would you recommend as the next diagnostic procedure?

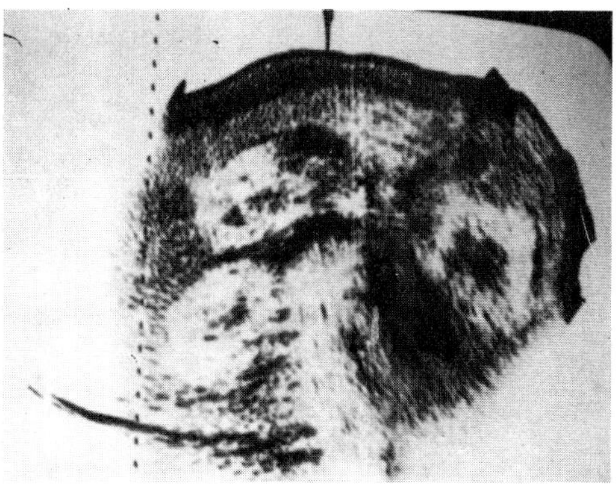

Figure 59–A. Transverse section +6.

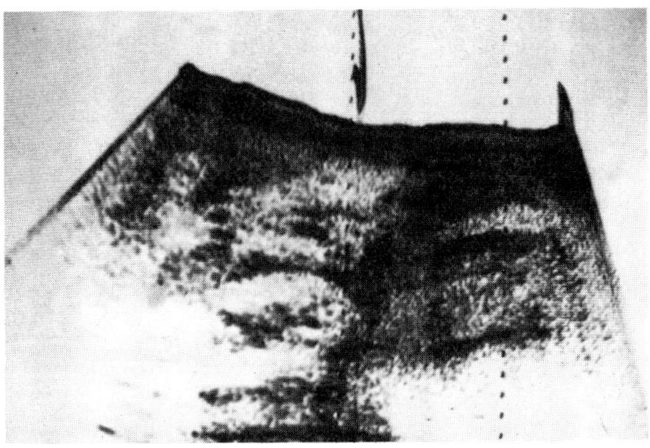

Figure 59–B. Longitudinal section 4R.

DIAGNOSIS

Extensive intra-abdominal abscess.

ULTRASOUND FINDINGS

The cross-section (Figure 59–A) shows two unusual echo collections (arrows) in the midabdomen. One is located in the porta hepatis and the other in the left flank. Figure 59–B is a longitudinal scan through the collection on the right, showing that it extends down over the kidney. Although there are some echo-free portions, the bulk of these masses is composed of fairly high-level echos. This echo pattern is often labeled "complex" to indicate that it is characteristic of neither a fluid-filled lesion nor a solid mass. Unfortunately, recognizing the complex echo pattern is not much use in differential diagnosis because so many different lesions can give this appearance. All that we can say is that the mass is very heterogeneous; we cannot distinguish tumor from necrotic tumor from abscess from hematoma from dilated bowel. Percutaneous biopsy is necessary to make these distinctions and should be the next diagnostic procedure. (There are many other imaging techniques which might help, such as CT scan, gallium scan, arteriography, barium studies, and so on. While any of these may give useful information, none of them is as simple, quick, or accurate as the biopsy.) Biopsy in this patient yielded necrotic, infected material; surgical exploration was performed, and an extensive abscess evacuated. No new tumor was found. The patient, however, subsequently died of metastatic disease.

Figures 59–C through 59–G are other examples of intra-abdominal abscess. There is no characteristic appearance: some are echo-free (Figures 59–C and 59–D) while others have various degrees of "complexity." In most situations the clinical setting is so suggestive that an ultrasonic confirmation of the diagnosis of abscess is not difficult. In confusing cases we do not hesitate to aspirate.

Ultrasound is a good method of following abscesses. The patient shown in Figure 59–G has an incompletely drained subhepatic abscess. The cause is also apparent: the drainage tube (wavy arrow) is not in the most dependent part of the collection.

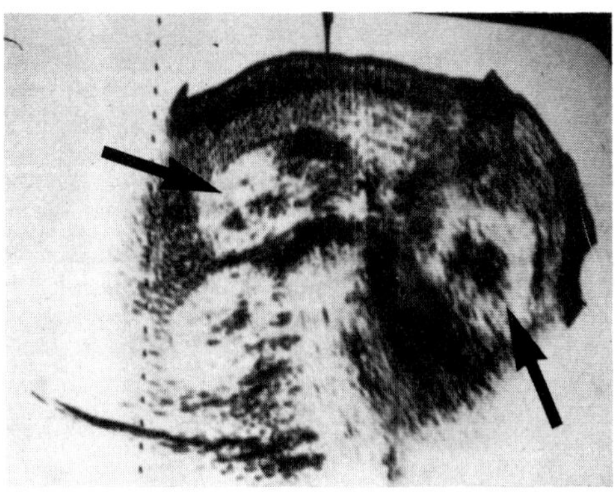

Figure 59–A. Transverse section +6. Abscess.

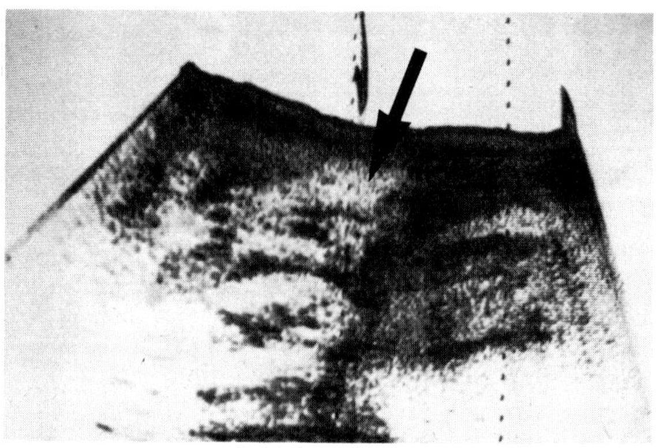

Figure 59–B. Longitudinal section 4R. Abscess.

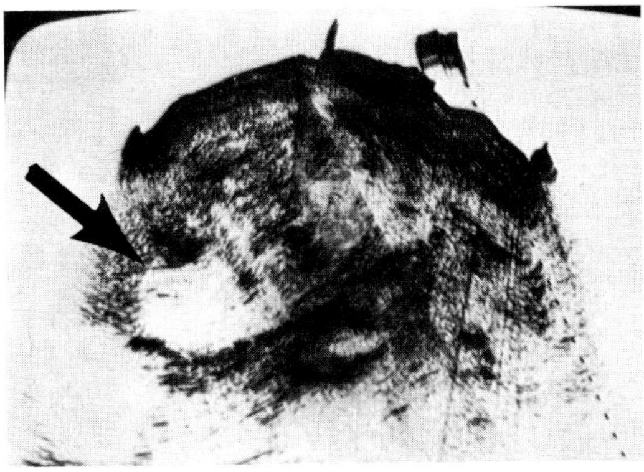

Figure 59–C. Transverse section +12.
Subhepatic abscess.

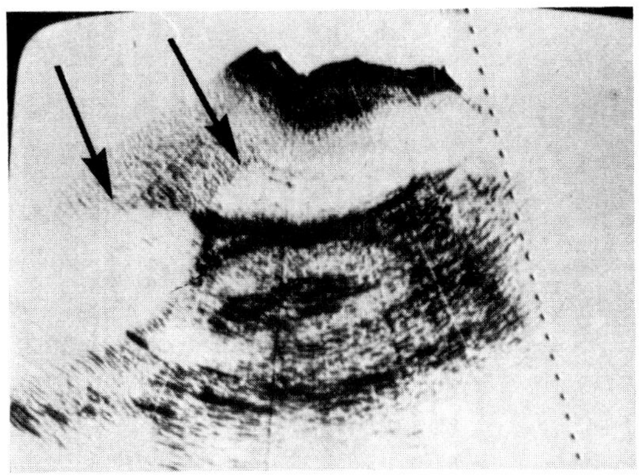

Figure 59–D. Longitudinal section 10R.
Subhepatic abscess.

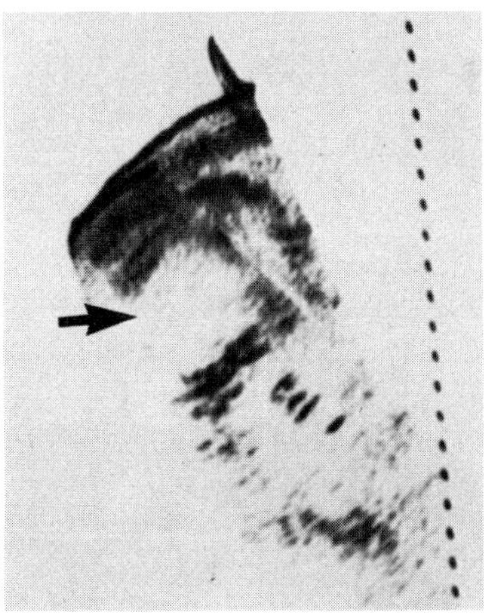

Figure 59–E. Transverse section 0.
Right lower quadrant abscess.

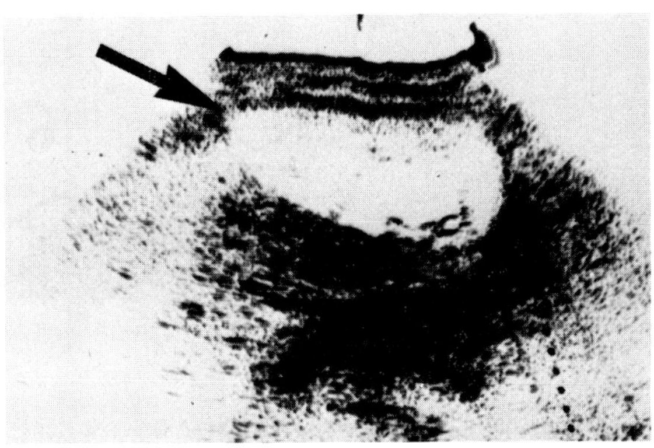

Figure 59–F. Transverse section +14.
Suprahepatic abscess.

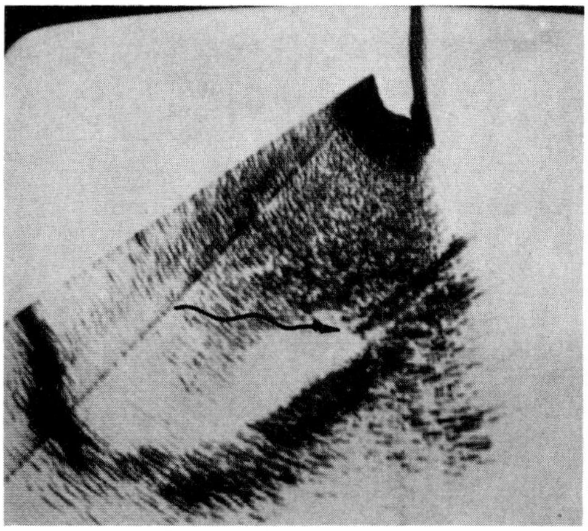

Figure 59–G. **Longitudinal section 12R.** Subhepatic abscess with drain.

REFERENCES

Doust B D, Quiroz F, Stewart J M: Ultrasonic distinction of abscesses from other intra-abdominal fluid collections. Radiology 125:213 (1977)

Laing F C, Jacobs R P: Value of ultrasonography in the detection of retroperitoneal inflammatory masses. Radiology 123:169 (1977)

Smith E H, Bartrum R J Jr: Ultrasonically-guided percutaneous aspiration of abscesses. Am J Radiol 122:308 (1974)

Conrad M R, Sanders R C, Mascadero A D: Perinephric abscess aspiration under ultrasonic guidance. Am J Radiol 128:459 (1977)

Case 60

CLINICAL INFORMATION

This 26-year-old woman is 16 weeks pregnant by dates and has had minor vaginal bleeding for the past two days.

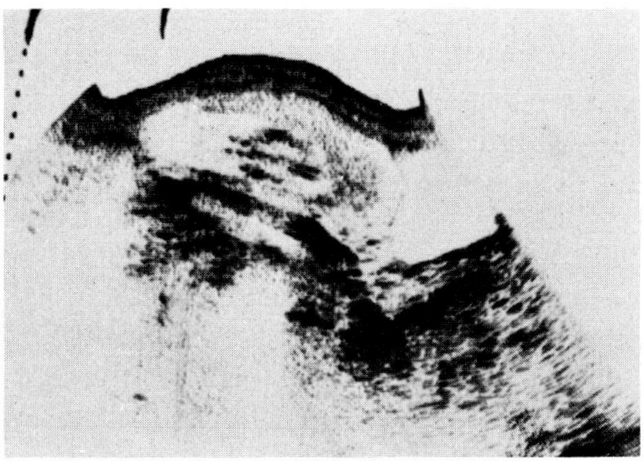

Figure 60–A. Longitudinal section 1L.

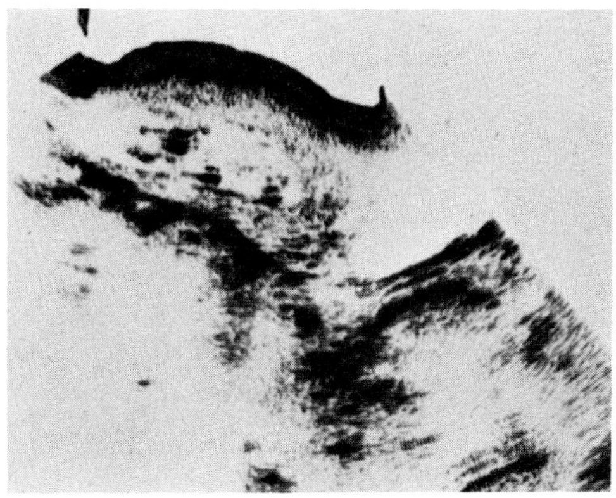

Figure 60–B. Longitudinal section 0L.

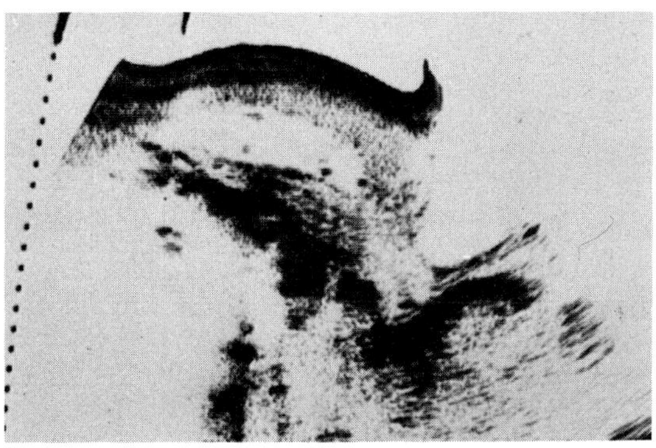

Figure 60–C. Longitudinal section 1R.

DIAGNOSIS

Placenta previa.

ULTRASOUND FINDINGS

Figures 60–A, 60–B, and 60–C are longitudinal sections through the left, central, and right portions of the cervix respectively. On all the scans the placenta covers the internal os completely and extends from high on the anterior wall of the uterus well onto the posterior wall. This is an unequivocal placenta previa. There is no way this placenta will "migrate" to uncover the os.

The diagnosis of placenta previa is somewhat more difficult than one would expect. The problem comes in locating the *internal* os of the cervix, something which cannot be done precisely with ultrasound. Using waterfilled doughnut pessaries, it has been shown that the *external* os of the cervix is in very close approximation to the *lowest part* of the urinary bladder in the midline longitudinal section (straight arrow in Figure 60–D). The internal os is located a variable distance cranial to this and we can only guess at its location. Many placentas extend down low on the uterine neck and could well cover a portion of the os: we label these cases "marginal placenta previa" since we are unable to tell whether the cervix is actually covered by the placenta. We reserve the term "complete (or unequivocal) placenta previa" for those cases in which the placenta is definitely seen on both sides of the os, not only on the midline section but on scans to the left and right as well. The curved arrow in Figure 60–D, a scan from another patient, shows the edge of the placenta. It clearly extends onto the anterior uterine wall: even though we are uncertain of the exact location of the cervix, in this case we know it has to be covered.

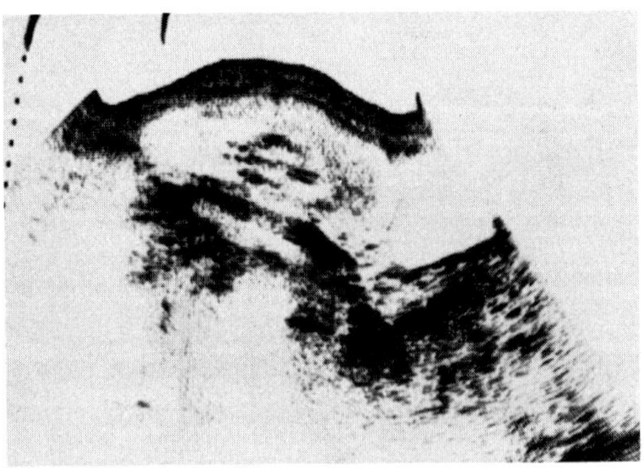

Figure 60–A. Longitudinal section 1L. Placenta previa.

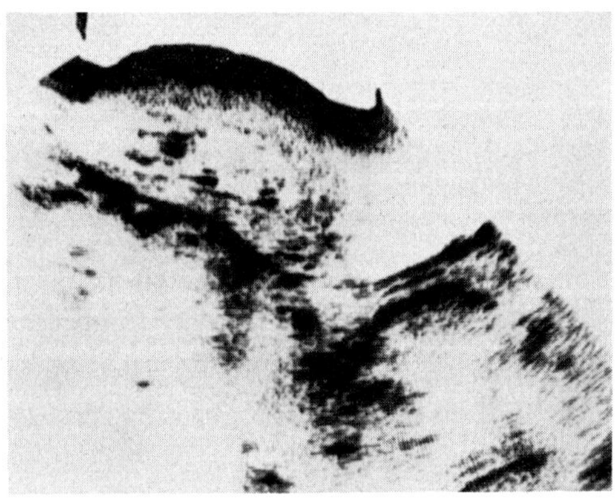

Figure 60–B. Longitudinal section 0L. Placenta previa.

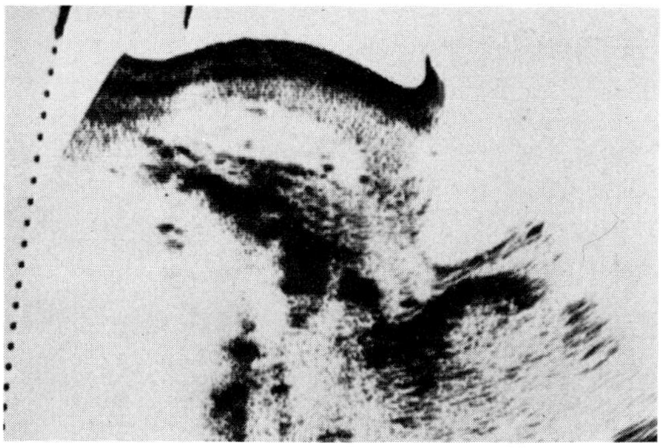

Figure 60–C. Longitudinal section 1R. Placenta previa.

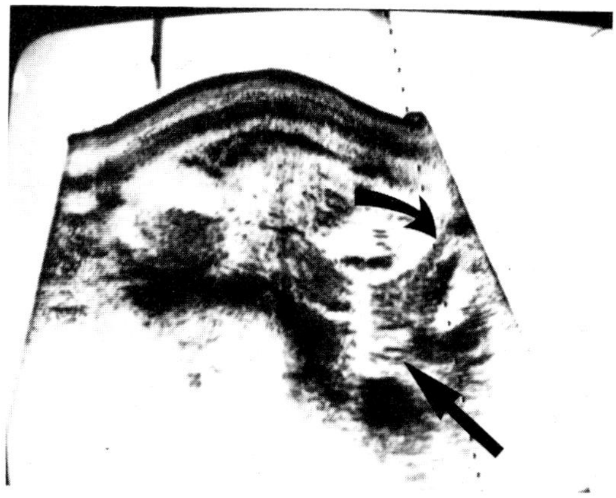

Figure 60–D. Longitudinal section 0L. Placenta previa.

REFERENCES

Williamson D, Bjorgen J, Baier B: Ultrasonic diagnosis of placenta previa: value of a post-void scan. J Clin Ultra 6:58 (1978)

Reed M F: Ultrasonic placentography. Br J Radiol 46:255 (1973)

Case 61

CLINICAL INFORMATION

This 58-year-old woman has been having pains in her abdomen and back, which have increased slowly in severity over the past three months. They have been associated with a decreasing appetite and periods of nausea with occasional vomiting. She is uncertain whether she has lost any weight. Of more concern, however, is the presence of a "lump" under her left rib cage which has been increasing in size. This mass can be easily palpated on physical examination and is firm but not particularly tender.

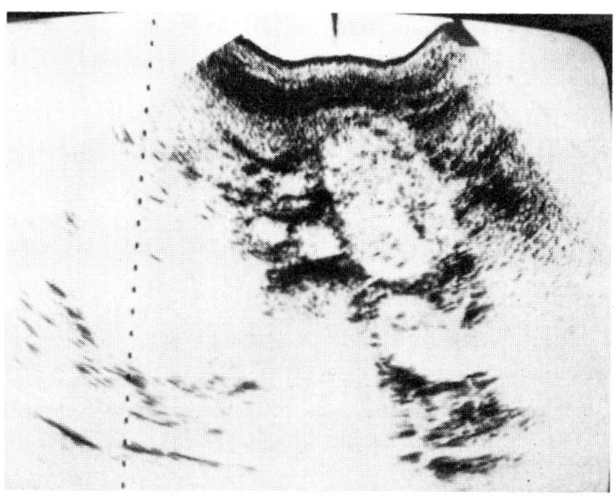

Figure 61–A. Transverse section +11.

DIAGNOSIS

Cystadenocarcinoma of the pancreas.

ULTRASOUND FINDINGS

When confronted with this clinical setting — a palpable abdominal mass of unknown etiology — we must consider two questions. First, from what organ is the mass arising? Providing the answer to this involves not only scanning the mass to locate its boundaries but also careful scrutiny of the adjacent abdominal organs to determine whether they are actively involved within the mass or merely passively displaced by it. In this case there is no problem locating the mass because it occupies nearly all of the left upper quadrant, but we must decide what organs are involved. The left kidney and the liver can be seen and are clearly separate from the mass. Obviously both the pancreas and retroperitoneal lymph nodes may be involved; and we can never clearly see the spleen as a distinct organ. Similarly, we cannot say whether or not the stomach is involved; this will require an upper GI series.

The second question is whether or not the mass is fluid-filled or, in more scientific terms, "homogeneous." A completely echo-free mass is almost always cystic — occasionally a lymphoma or sarcoma can mimic a fluid-filled lesion — and this can be very helpful in differential diagnosis. If the mass does contain internal echoes, it is not a simple cyst; and more complex lesions such as tumor, necrotic tumor, abscess, hematoma, or pseudocyst must be considered. The mass in question here falls into the latter category: Although there are some echo-free areas, which may represent fluid, for the most part this lesion is inhomogeneous, or solid. Although some ultrasonographers feel that the nature of the echo pattern in a mass can be used to make a more specific histiologic diagnosis, this has not been the experience of the vast majority of ultrasound laboratories. We feel that trying to make a specific diagnosis based on echo pattern alone is little better than guesswork and is a disservice not only to the patient and referring physician but also to the credibility of the ultrasound examination itself.

The histiologic diagnosis in this instance was cystadenocarcinoma of the pancreas — a rare, low-grade form of malignancy. Perhaps in retrospect the relatively echo-free component of the mass should suggest the diagnosis, but we believe this is stretching things a bit. We have seen cases with nearly identical appearance (some of which appear elsewhere in this workbook) in which the final diagnoses have been such things as (1) necrotic adenocarcinoma of the pancreas, (2) pancreatitis, (3) metastatic tumor in retroperitoneal lymph nodes, and (4) Hodgkin's disease. Figure 61–B is a scan of a retroperitoneal fibrosarcoma. For practical purposes it is preferable to give a relatively nonspecific diagnosis such as: "Large, generally solid tumor, possibly containing fluid-filled regions; the origin of this tumor could be any of the organs in the left upper quadrant of the abdomen with the exception of the liver and kidney. Percutaneous aspiration biopsy is recommended for more specific diagnosis."

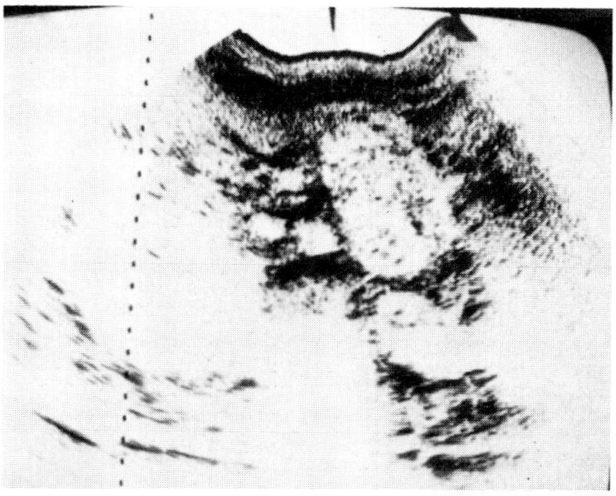

Figure 61–A. Transverse section +11. Cystadenocarcinoma of the pancreas.

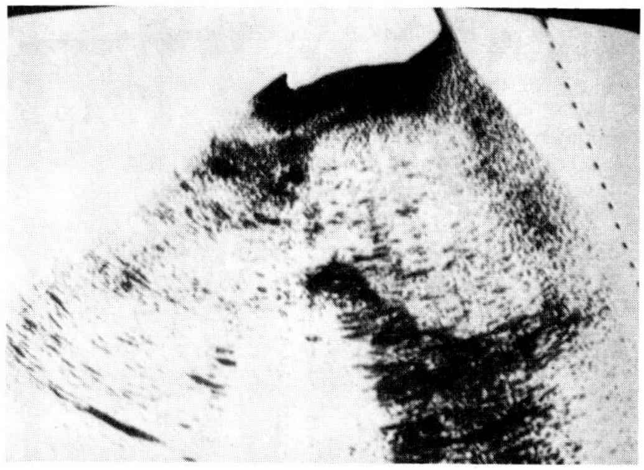

Figure 61–B. Transverse section +6. Retroperitoneal fibrosarcoma.

REFERENCE

Wolson A H, Walls W J: Ultrasonic characteristics of cystadenoma of the pancreas. Radiology 119:203 (1976)

Case 62

CLINICAL INFORMATION

This 45-year-old woman has had periodic episodes of right flank pain. The pain is not severe, nor does it last for more than one or two hours. The episodes are infrequent, and she does not consider this to be much of a problem. Because she is seeing her doctor for her annual checkup, however, she thinks she should mention it. She has no other medical problems except for idiopathic epilepsy, well controlled with phenytoin (Dilantin) for many years. Her physical examination is normal, but a urinalysis reveals a few red cells, many white cells, and a suggestion of white cell casts. While awaiting a quantitative urine culture (which subsequently showed a colony count of greater than 100,000), her doctor ordered an IVP. The radiologist who interpreted this study was concerned that there might be a parapelvic mass in the right kidney (Figure 62–A). This ultrasound scan was made through the long axis of this kidney with the patient supine.

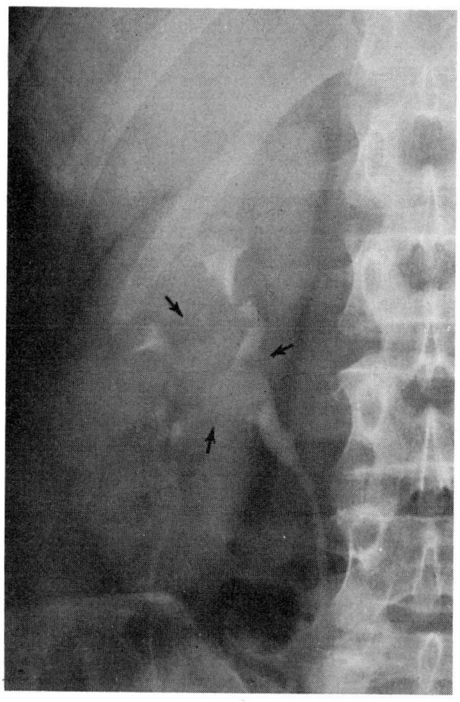

Figure 62–A. IVP, 8-minute compression film of right kidney.

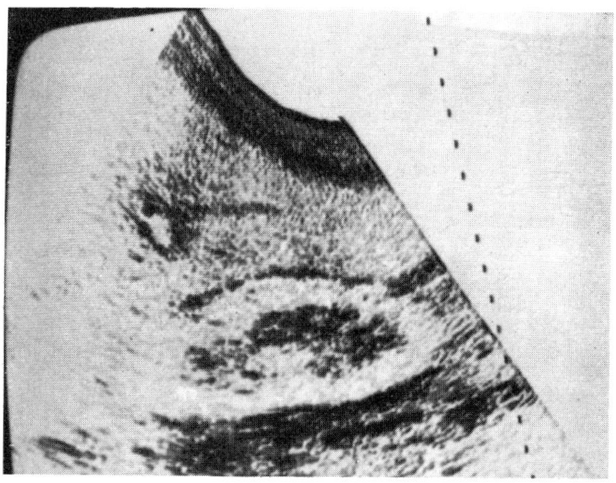

Figure 62–B. Supine scan. **Longitudinal section 7R.**

DIAGNOSIS

Column of Bertin (infolding of renal parenchyma).

ULTRASOUND FINDINGS

The kidney is very well demonstrated in this scan. Whenever possible, we like to scan the right kidney with the patient supine, using the liver as an acoustic window. In this way we can avoid the posterior ribs. The renal cortex in this kidney is normal, as is most of the renal sinus. There is, however, a small mass in the central portion of the renal sinus, which has displaced the high intensity echoes normally found there. The echoes from this mass are of the same intensity as those of the renal parenchyma, and the mass blends smoothly with the adjacent cortical tissue. Since the mass does not have a smooth wall and does contain low-level echoes we can exclude a fluid-filled lesion, such as a parapelvic cyst or calyceal diverticulum (Figure 62–E). The continuity with the renal cortex and the fact that the echo strength is the same as that of the normal renal parenchyma suggest that we may be dealing with a benign infolding of the renal cortex, a so-called "column of Bertin." We cannot exclude a tumor completely, however, particularly an adenoma, adenocarcinoma, or even a transitional cell tumor — all of which might be found in this location. We recommend, therefore, a renal arteriogram; and this confirms our diagnosis of a column of Bertin (Figure 62–C). Figure 62–D is a scan of a transitional cell carcinoma; we defy you to distinguish it from that of the column of Bertin.

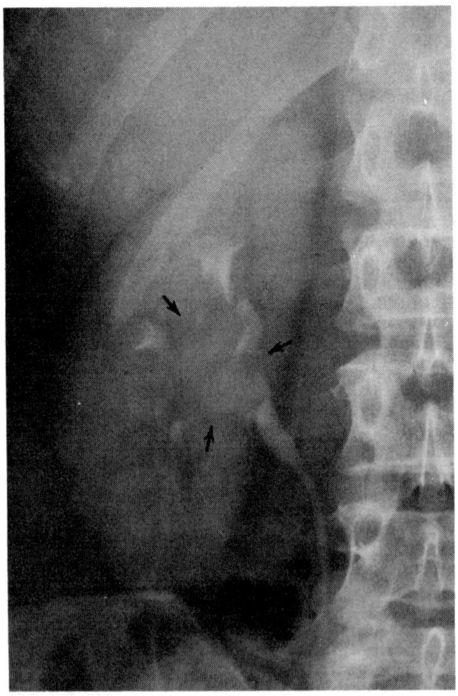

Figure 62–A. Column of Bertin. IVP, 8-minute compression film of right kidney.

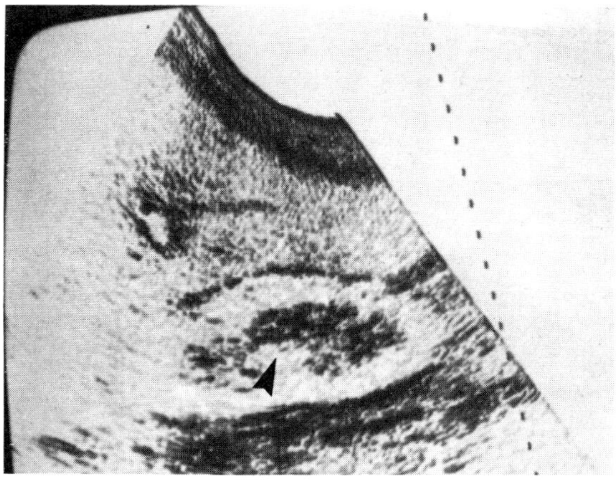

Figure 62–B. Longitudinal section 7R. Supine scan. Column of Bertin.

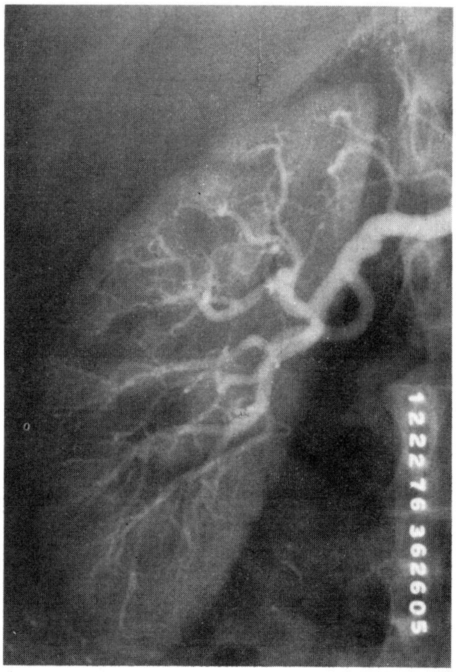

Figure 62–C. Column of Bertin. Arteriogram of kidney in Figures 62–A and 62–B.

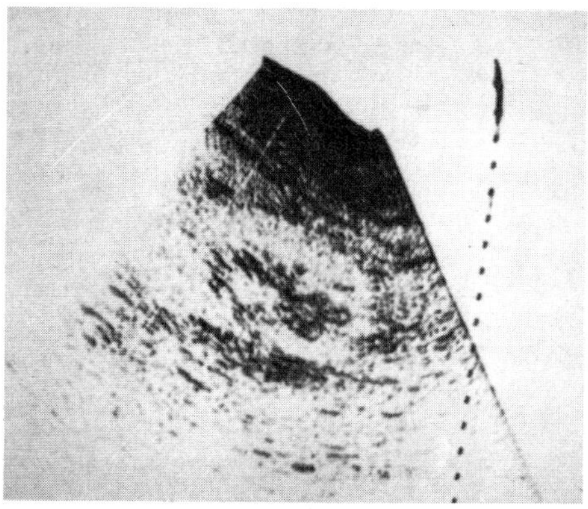

Figure 62–D. Longitudinal section 1R. Prone scan of right kidney. Transitional cell carcinoma.

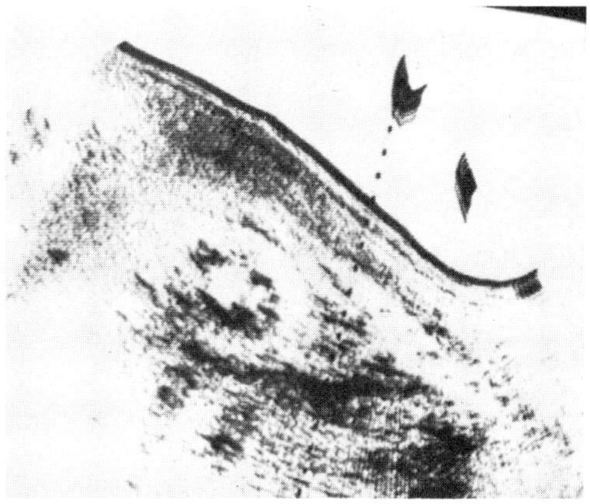

Figure 62–E. Longitudinal section 1R. Prone scan of right kidney. Parapelvic cyst.

REFERENCES

Pollack M M, Goldberg B B: Chapter 10, Kidney, in *Abdominal Gray Scale Ultrasonography*, Goldberg B B, ed. John Wiley & Sons, New York (1977)

Davidson A J: *Radiologic Diagnosis of Renal Parenchymal Disease*. Chapter 2. W B Saunders Co, Philadelphia (1977)

Case 63

CLINICAL INFORMATION

This 63-year-old woman arrived at the emergency ward complaining of severe right upper quadrant pain, nausea, and vomiting. Her abdomen was extremely tender and there was guarding of the right upper quadrant with mild rebound tenderness. Several examiners felt a mass in this area as well. She had a fever of 102.8 and a white blood cell count of 18,000 with a marked left shift, but liver function tests and urinalysis were normal. Her only previous hospital admission was for an appendectomy 37 years before.

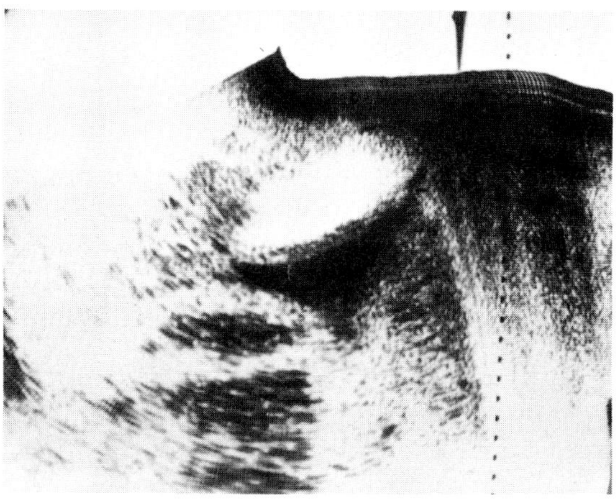

Figure 63–A. Longitudinal section 10R.

DIAGNOSIS

Acute cholecystitis.

ULTRASOUND FINDINGS

The palpable mass in the right upper quadrant is the gallbladder, which is enlarged and has a thickened wall. (When evaluating the wall of any fluid-filled viscus we must always confine our observations to the *posterior* wall, that is, the wall behind the fluid. For unexplained reasons the anterior wall of a fluid-filled organ or mass is never clearly or accurately seen.) This is the classic appearance of acute cholecystitis and should cause few diagnostic problems. However the clinical presentation is *more important* than the ultrasound findings in these cases. The presence of an enlarged gallbladder or a thick gallbladder wall *is not sufficient grounds* for diagnosing either acute or chronic cholecystitis; the clinical setting also must strongly suggest the diagnosis.

Judging the thickness of the wall of the gallbladder can be tricky business. Figure 63–B is a scan of the gallbladder in a patient with midepigastric and substernal pain: the wall of the gallbladder is diffusely thickened (this was verified on multiple scans). On the basis of this finding, we made a diagnosis of chronic cholecystitis. Unfortunately the patient died of a myocardial infarction nine hours after this scan was performed: at autopsy the gallbladder was considered to be completely normal. We usually hesitate to diagnose gallbladder disease by ultrasound when no stones are present; the assessment of the wall of the gallbladder has proved for us to be much less reliable than the recognition of gallstones.

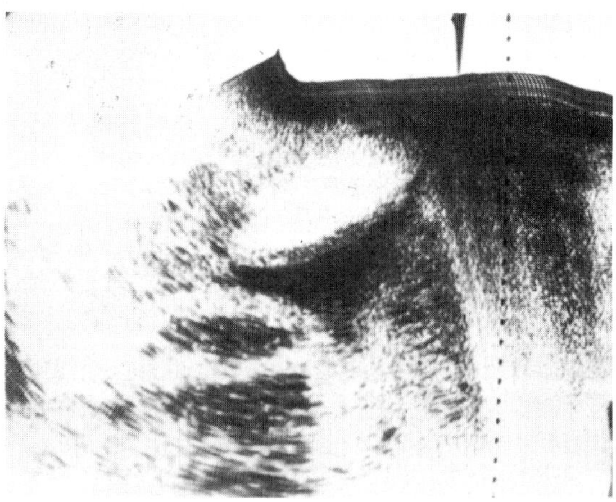

Figure 63–A. **Longitudinal section 10R.** Acute cholecystitis.

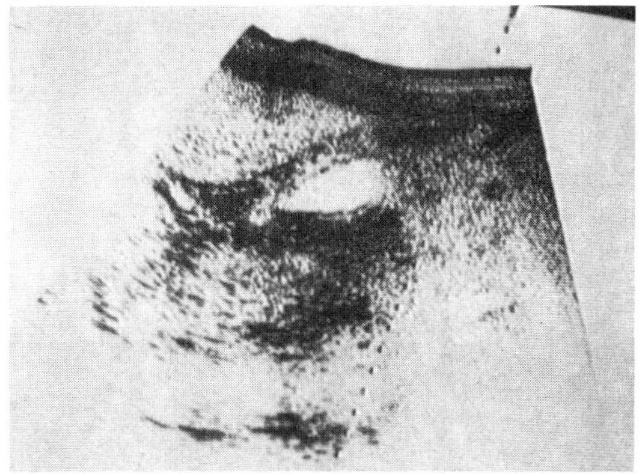

Figure 63–B. Longitudinal section 4R. Normal gallbladder.

REFERENCE

Berger M, Smith E, Holm H H, et al: Utility of ultrasound in the differential diagnosis of acute cholecystitis. Arch Surg 112:273 (1977)

Case 64

CLINICAL INFORMATION

This 23-year-old woman is 21 weeks pregnant and is having a routine ultrasound scan for gestational dating, placental localization, and so on.

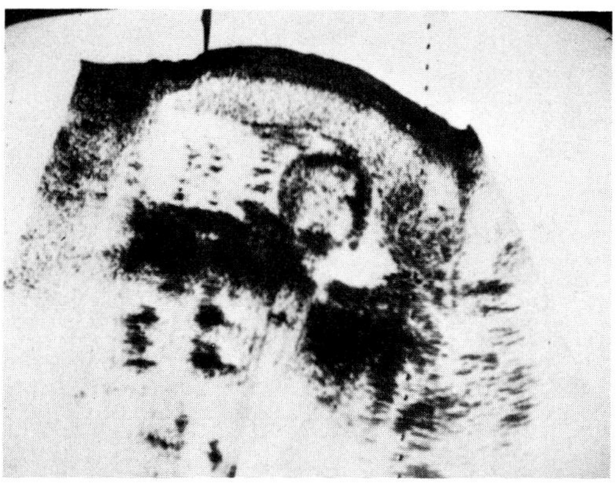

Figure 64–A. Longitudinal section 0L.

DIAGNOSIS

Marginal placenta previa.

ULTRASOUND FINDINGS

The placenta is anterior but extends down quite low onto the neck of the uterus. We cannot tell the exact location of either the placental margin or the internal os but it is certainly possible that a portion of the placenta covers the cervix. We call this a "marginal placenta previa" and recommend a follow-up examination around 30 to 32 weeks of gestation. As the uterus elongates with advancing pregnancy, the placenta is usually carried cephalad, away from the cervical os. Figure 64–B is a scan of this same patient at 31 weeks: the placenta is not going to cause problems.

This phenomenon of the change from a marginal placenta previa to normal placental position has been referred to as placental "migration." There is no agreement as to its cause; differential lengthening of the uterine fundus relative to the neck and atrophy of the lower placental margin are the most widely accepted theories. Some cases are undoubtedly due to technical factors. The internal cervical os is not always located in the midsagittal scan plane, particularly if the bladder is over-distended, and the midline scan may appear to show a marginal placenta previa when the actual cervix is off to the right or left. This problem can be minimized by rescanning the suspected "marginal placenta previa" after the patient has voided and obtaining sections not only in the midline but also 2 centimeters to the right and 2 centimeters to the left. If you are still suspicious of a placenta previa ask to re-examine the patient in the third trimester.

Figure 64–C shows a posterior marginal placenta previa at 27 weeks. Figures 64–D and 64–E are scans of a different patient at 18 and 35 weeks. The placenta is very low and seems to cover the os on the early scan (Figure 64–D), yet in the third trimester (Figure 64–E), it is well out of the way.

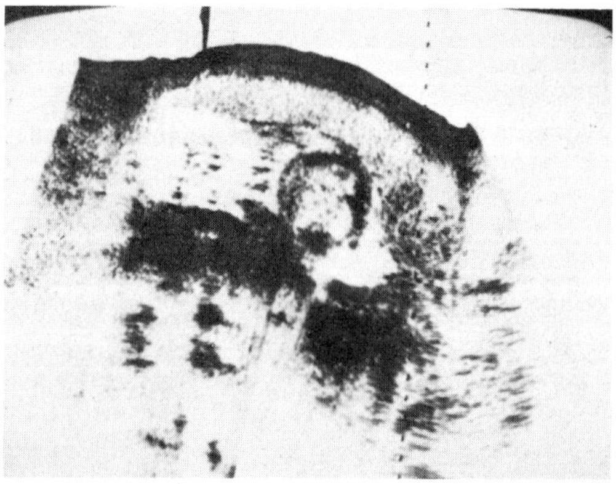

Figure 64–A. Longitudinal section 0L. Marginal placenta previa at 21 weeks.

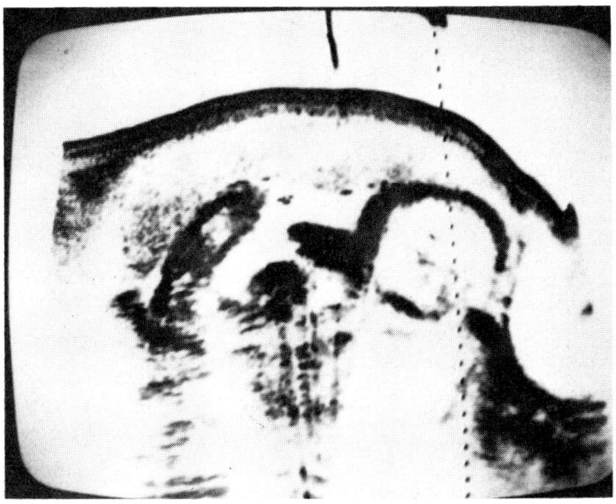

Figure 64–B. Longitudinal section 0L. Normal placentation; same patient as in Figure 64–A at 31 weeks.

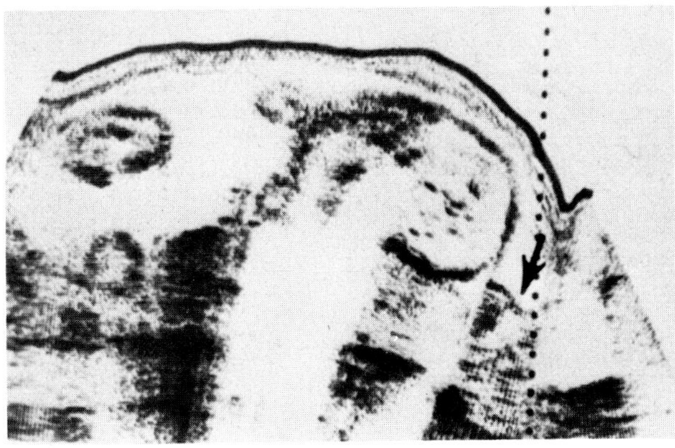

Figure 64–C. Longitudinal section 0L. Posterior marginal placenta previa.

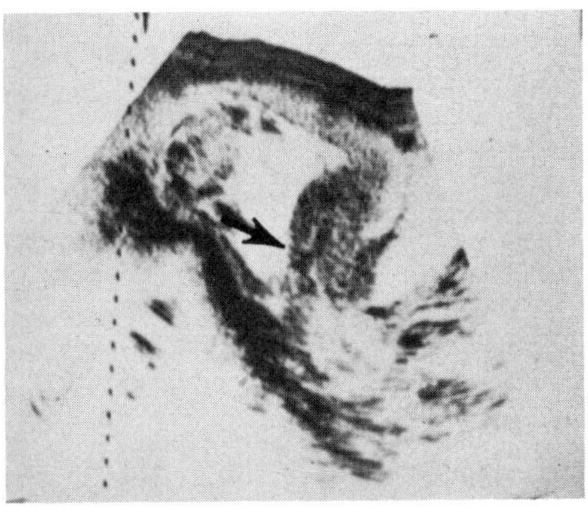

Figure 64–D. Longitudinal section 0L. Anterior marginal placenta previa at 18 weeks.

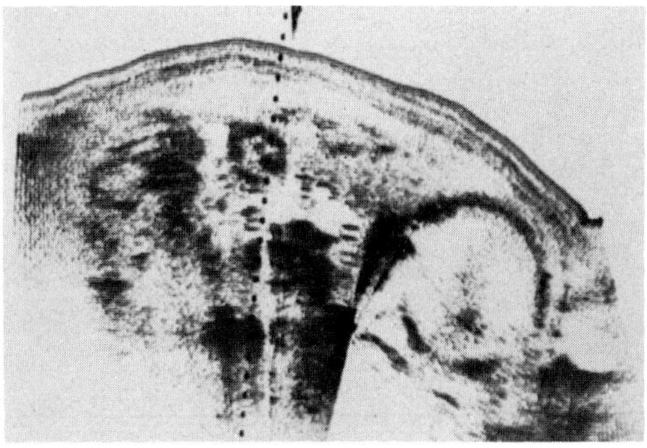

Figure 64–E. **Longitudinal section 0L.** Normal placentation; same patient as in Figure 64–D at 35 weeks.

REFERENCE

King D L: Placental migration demonstrated by ultrasonography. Radiology 109:167 (1973)

Case 65

CLINICAL INFORMATION

This 69-year-old widow has had a cold; and because she is planning to depart for her yearly jaunt to Florida next week, she wants to make certain it is not developing into pneumonia or "anything serious which would delay my trip." Her doctor is not worried about the cold but is concerned about a prominent, pulsatile mass in her central abdomen; he thinks it may represent an aortic aneurysm.

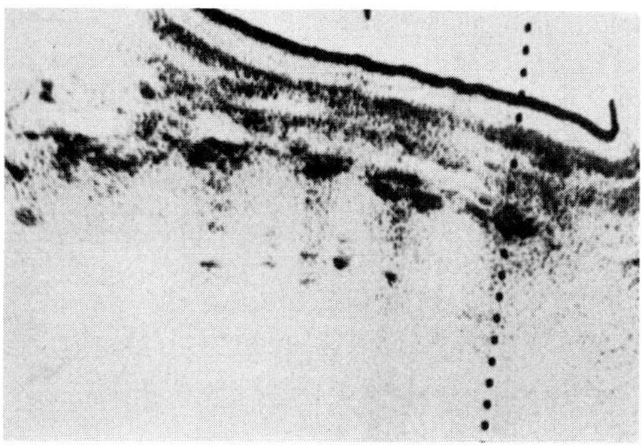

Figure 65–A. Longitudinal section 2R.

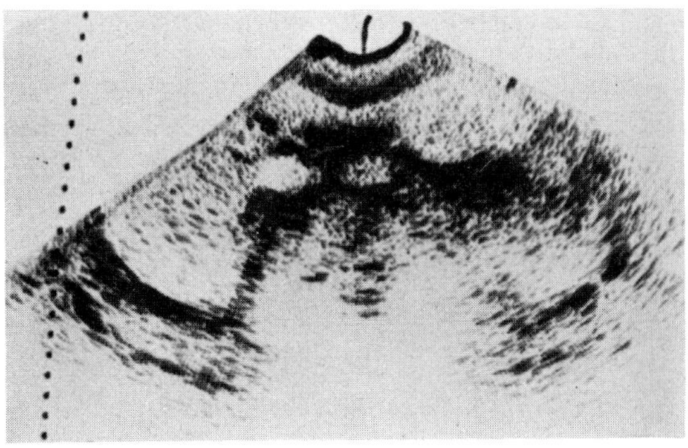

Figure 65–B. Transverse section +9.

DIAGNOSIS

Aorta of normal caliber with probable arteriosclerosis.

ULTRASOUND FINDINGS

Figure 65–A is a longitudinal scan through the aorta showing that it is of normal caliber throughout its length. (In some tortuous aortas it is not always possible to include the entire vessel on a single scan; it is important to make sure that its entire abdominal course has been visualized even if this requires piecing it together from several different longitudinal scans.) Selected cross-sectional scans (Figure 65–B) assure us that we are not missing a small sacular aneurysm projecting laterally.

What is "normal" caliber for the abdominal aorta? Most surgeons consider a diameter of 3 cm. or less, measured from outer wall to outer wall, as definitely normal. Above 3 cm. things get more confusing and terms such as "ectasia," "early aneurysm," "asymptomatic aneurysm," and "low-risk aneurysm" begin to appear. Once the diameter exceeds 5 cm., most people feel that there is a medically significant aneurysm with considerable risk to the patient. (These figures were derived from measurement of the aortic wall calcifications as seen on radiographs, and, because of magnification factors, do not correspond exactly to the measurements obtained by ultrasound. We have found that the ultrasonic diameter of the aorta is roughly 0.5 cm. *less* than the radiographic diameter.)

We might point out another feature of the appearance of this aorta which indicates that the vessel is not entirely normal: Note that the walls are slightly irregular, and there are short areas where the echoes from the walls are unusually strong. This pattern of echo production is caused by arteriosclerotic change within the wall with plaque formation and calcification.

Why was the referring physician, an experienced clinician, misled in this case? We suspect the answer is that the aorta is very close to the surface in this thin patient and the lack of intervening tissue made its palpations unusually prominent.

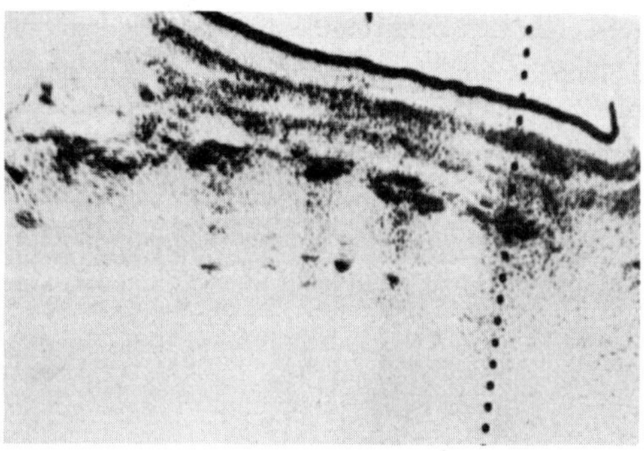

Figure 65–A. **Longitudinal section 2R.** Normal caliber aorta.

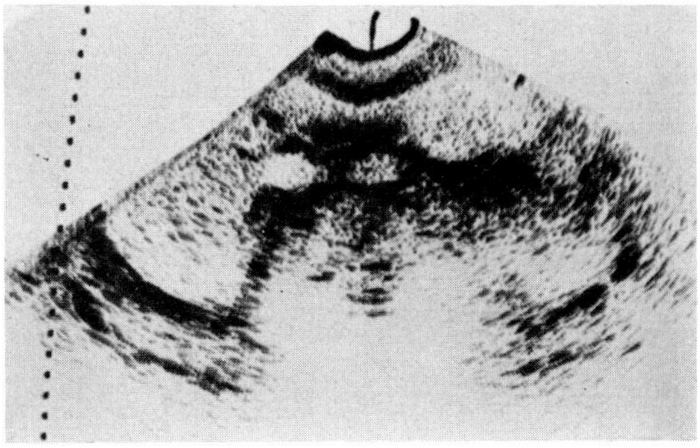

Figure 65–B. Transverse section +9. Normal caliber aorta.

REFERENCE

Filly R A, Goldberg B B: Abnormal Vessels, Chapter 4 in *Abdominal Gray Scale Ultrasonography*, B B Goldberg, editor. John Wiley & Sons, New York (1977)

Case 66

CLINICAL INFORMATION

This 28-year-old woman has no symptoms, but during a physical examination (for life insurance) her uterus is found to be enlarged and firm. The ovaries and adnexal regions are normal. She has had two uncomplicated pregnancies and has been on birth control pills for the past 18 months. She has never had any gynecologic problems. All laboratory data are normal. A plain abdominal radiograph shows a soft tissue mass in the pelvis, but the radiologist is uncertain whether it contains any calcifications.

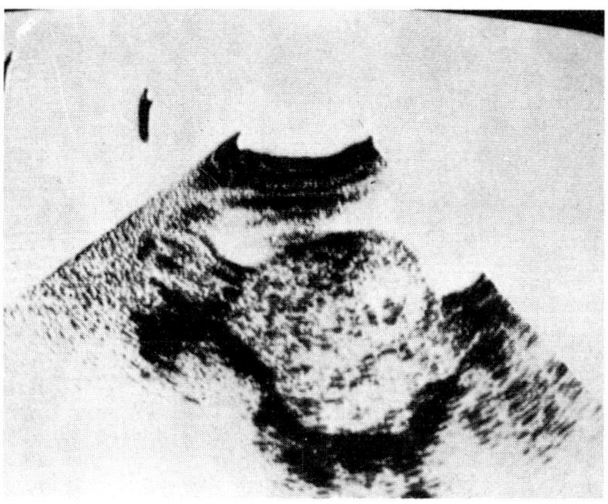

Figure 66–A. Longitudinal section 0L.

DIAGNOSIS

Uterine leiomyoma (fibroid).

ULTRASOUND FINDINGS

There is a 9-centimeter mass in the body of the uterus. The internal texture is heterogeneous, with scattered high-intensity echoes interspersed with nearly echoless areas. This pattern is quite compatible with a uterine leiomyoma or "fibroid," as it is more commonly known. Fibroids have many different appearances, depending upon their internal composition. They can be quite homogeneous, with a uniform, low-level echo pattern (Figure 66–B), or they may be bizarre with loud echoes and shadowing when they contain calcifications (Figure 66–C). Anything in between is also possible. These benign tumors grow under estrogen stimulation and frequently enlarge dramatically during pregnancy (Figure 66–D).

We do not think there is any way to differentiate fibroids from malignant uterine tumors on the basis of the *ultrasound scan appearance alone.* In a young patient with the proper clinical setting we can be reasonably certain that we are dealing with a benign tumor, but in an older patient further evaluation is required to exclude a malignancy. Figures 66–E and 66–F show endometrial carcinoma that is nearly identical in appearance to the fibroid in Figure 66–B. Figures 66–G and 66–H show a large leiomyosarcoma of the uterus (the bladder is empty in this patient). In this case the irregular margins of the tumor and its extreme heterogeneity do suggest malignant tumor.

Aside from confirming the presence of a pelvic tumor discovered or suspected by some other means, ultrasound can usually localize the tumor, define which pelvic organs are involved, and separate uterine from ovarian or adnexal masses.

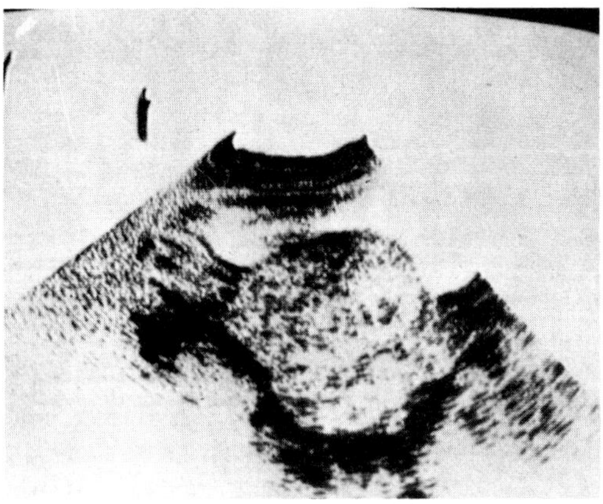

Figure 66–A. **Longitudinal section 0L.** Uterine fibroid.

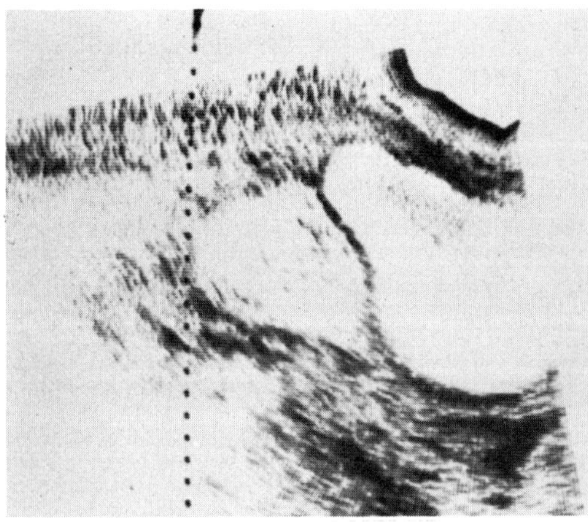

Figure 66–B. **Longitudinal section 0L.** Uterine fibroid.

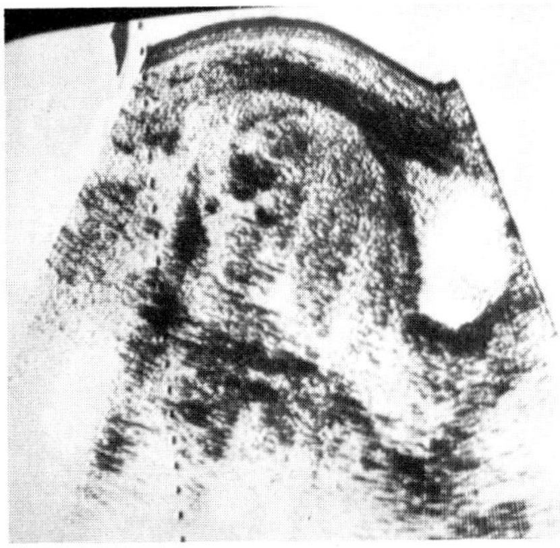

Figure 66–C. Longitudinal section 0L. Uterine fibroid.

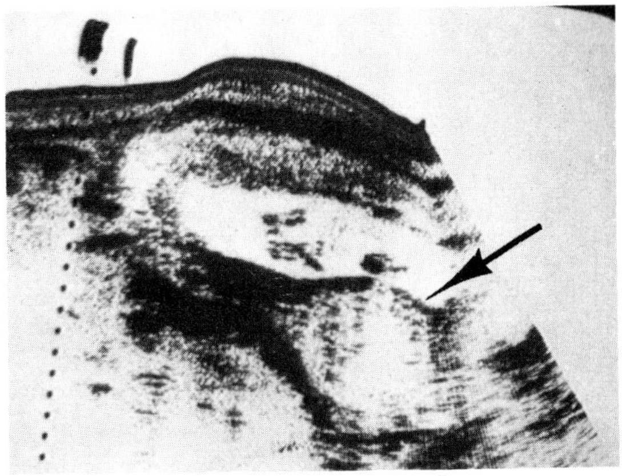

Figure 66–D. Longitudinal section 1R. Uterine fibroid in pregnancy.

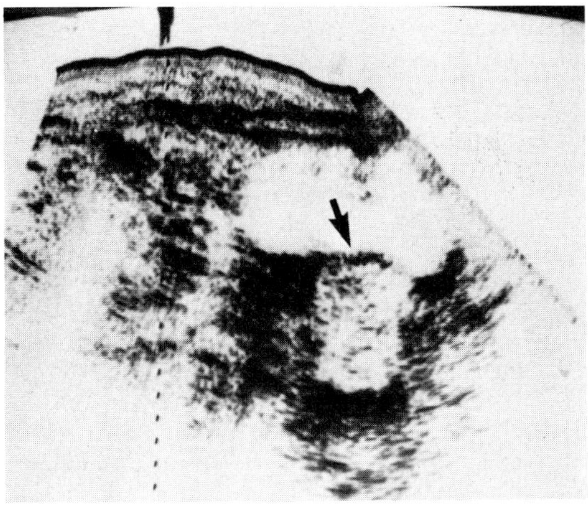

Figure 66–E. Longitudinal section 2R. Endometrial carcinoma.

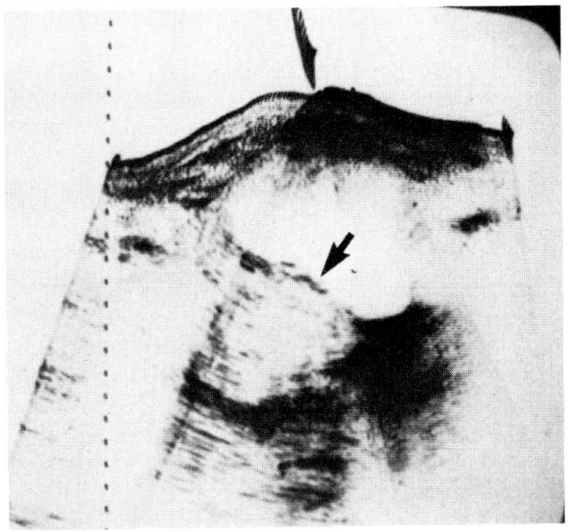

Figure 66–F. Transverse section −12. Endometrial carcinoma.

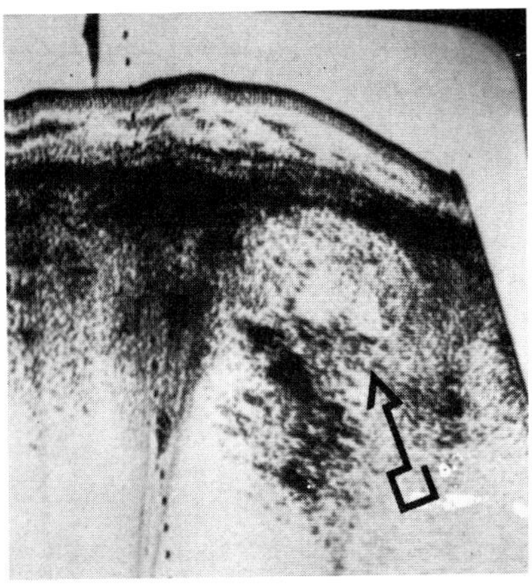

Figure 66–G. Longitudinal section 2R. Leiomyosarcoma of uterus.

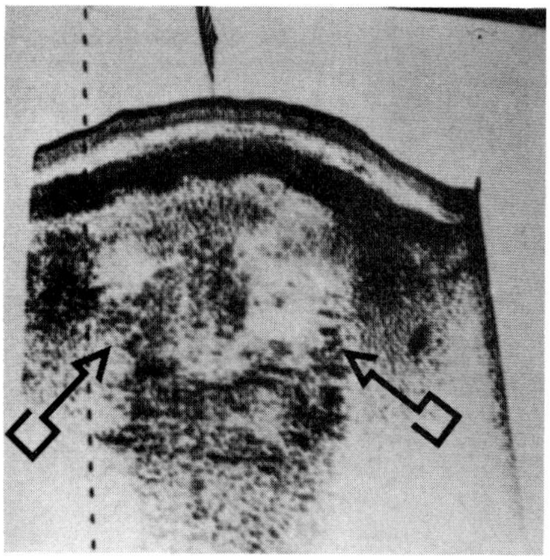

Figure 66–H. Transverse section −10. Leiomyosarcoma of uterus.

REFERENCES

Bowie JD: Ultrasound of gynecologic pelvic masses: The indefinite uterus and other patterns associated with diagnostic error. J Clin Ultra 5:323 (1977)

von Micsky LI: Sonographic study of uterine fibromyomata. In *Ultrasonography in Obstetrics and Gynecology,* Sanders and James, editors. Appleton-Century-Crofts, New York (1977)

Case 67

CLINICAL INFORMATION

This 76-year-old man has been treated with both radiotherapy and chemotherapy for a small cell carcinoma of the lung. The original diagnosis was made 16 months ago and a metastatic work-up at that time was negative. The pulmonary lesion seems to have stabilized — the chest film is unchanged over the past six months — but on his most recent physical examination a mass was felt in the left upper quadrant.

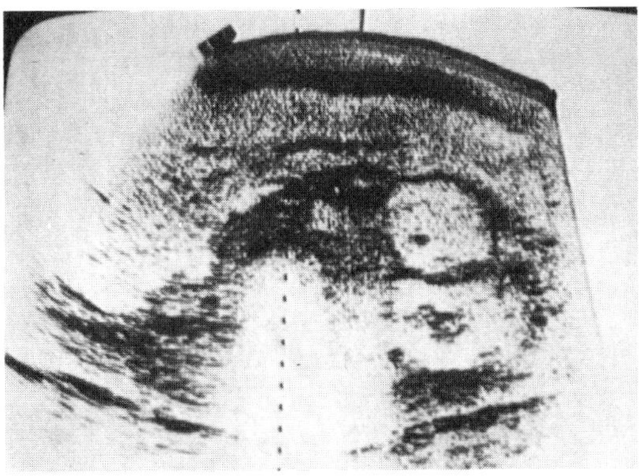

Figure 67–A. Transverse section +6.

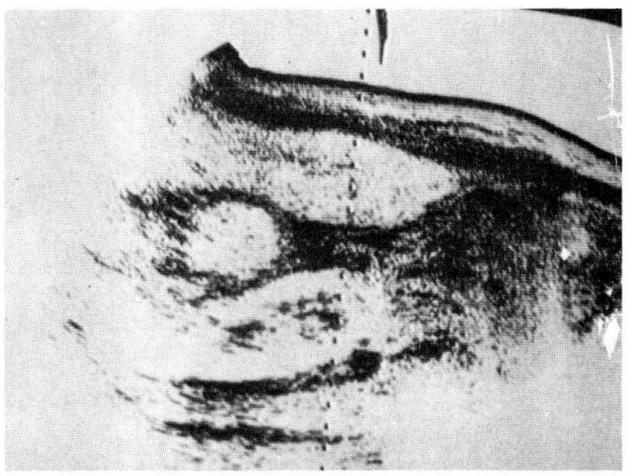

Figure 67–B. Longitudinal section 6L.

DIAGNOSIS

Metastatic tumor in retroperitoneal lymph nodes.

ULTRASOUND FINDINGS

There is a round, well-circumscribed mass lying between the left kidney and the left lobe of the liver. The mass is separate from these organs and not related to the spleen. The presence of internal echoes tells us that we are dealing with a solid tumor. Our first thought is to be certain we are not seeing part of the GI tract. The patient says he has not had anything to eat in the past 12 hours so we can eliminate food in the stomach. This could be a gastric tumor, of course, and to exclude that possibility we recommend an upper GI series. (This was performed the next day and showed that the stomach was displaced but not involved by the mass.) What other tissue is in this area? A tumor in the tail of the pancreas is one consideration. The other differential diagnostic possibility is tumor in the retroperitoneal lymph nodes; the patient has a known primary tumor and this could well represent a metastasis. On the basis of the scan picture alone we would lean toward a pancreatic tumor since the mass is so well localized: tumor in lymph nodes, either primary or metastatic, is generally more widespread, involving both sides of the aorta and infiltrating among the retroperitoneal structures. A biopsy is required to make the final diagnosis, and in this case yielded small anaplastic tumor cells consistent with small cell tumor of the lung.

This patient illustrates an uncommon presentation of metastatic tumor to the lymph nodes. Although most lymph node disease is diffuse, it can present as a focal mass. Figure 67–C is another example of this phenomenon. The mass in the left retroperitoneum is a metastasis from endometrial carcinoma.

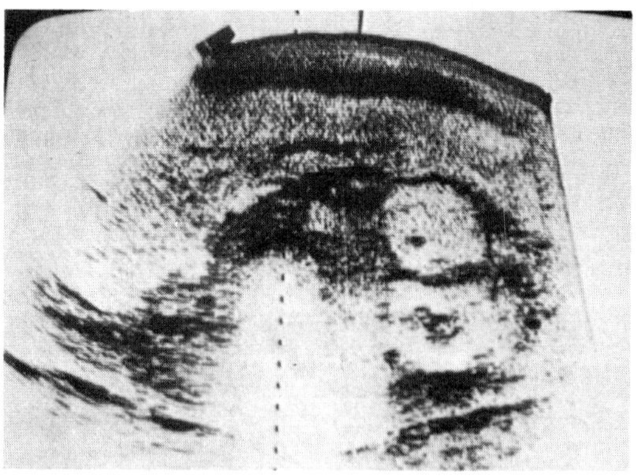

Figure 67–A. Transverse section +6. Lymph node metastases (oat cell carcinoma).

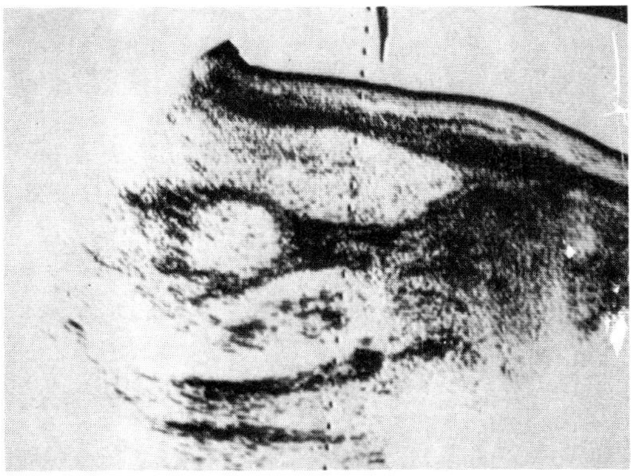

Figure 67–B. Longitudinal section 6L. Lymph node metastasis (oat cell carcinoma).

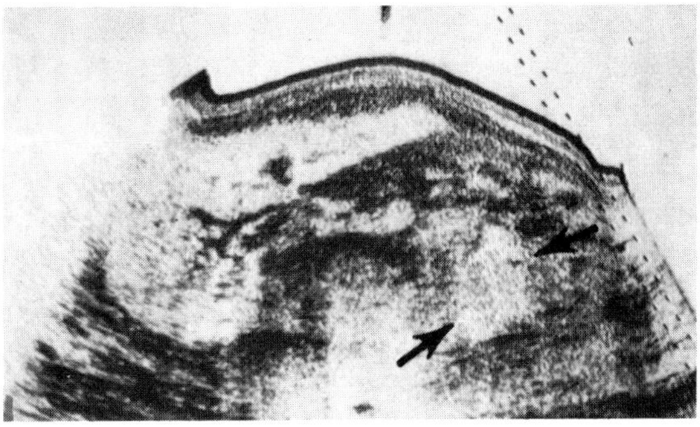

Figure 67–C. Transverse section +6. Lymph node metastasis (endometrial carcinoma).

REFERENCE

Taylor KJW: Reticuloendothelial system, Chapter 8 in *Abdominal Gray Scale Ultrsonography*, BB Goldberg, editor. John Wiley & Sons, New York, 1977

Case 68

CLINICAL INFORMATION

This 47-year-old woman has passed several renal stones in the past. Three weeks ago she had a brief episode which she thought was another stone on the left: the initial symptoms were typical but lasted only three hours so she did not seek medical attention. In the past two weeks she has had increasing flank pain and tenderness, and in the past two days she has had a fever and some nausea. Physical examination is unremarkable except for left flank tenderness. The white cell count is 13,000 with a shift to the left: the creatinine is normal and the urinalysis shows many white cells but no blood. An IVP shows the lower pole of the left kidney to be normal but the upper pole collecting system is never visualized. The ureter and renal pelvis are normal in size and there is no delay in contrast excretion. This is a cross-section through the upper pole of the left kidney.

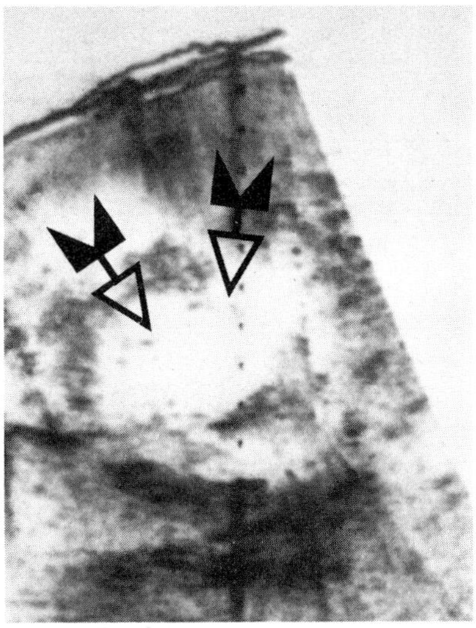

Figure 68–A. Transverse section +5. Prone scan of left kidney.

DIAGNOSIS

Renal abscess (renal carbuncle).

ULTRASOUND FINDINGS

There is an echo-free mass in the upper pole of the right kidney (arrows). The mass does not have a smooth back wall and we are therefore hesitant to call it a simple renal cyst. Although this may be just a cyst we must seriously consider other lesions such as an infected cyst, necrotic tumor, or abscess. The clinical setting in this case points strongly to abscess, and this diagnosis was readily confirmed by percutaneous aspiration.

The standard treatment for a renal abscess is surgical drainage. Recently, however, there have been reports of treatment with closed drainage. Using ultrasonic guidance, a catheter is placed in the abscess cavity to provide drainage and a means of instilling antibiotics locally.

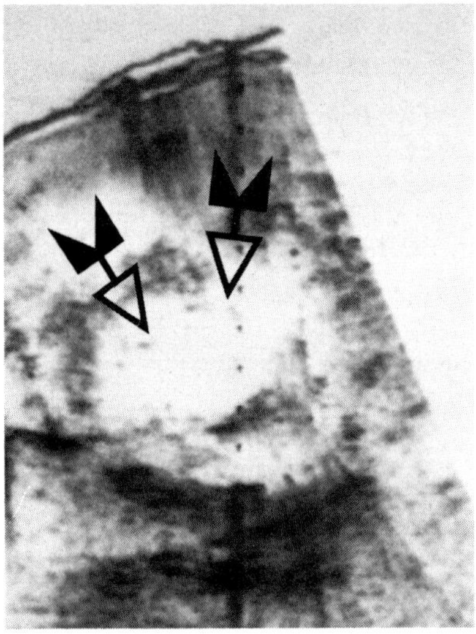

Figure 68–A. Transverse section +5. Prone scan of left kidney. Renal abscess.

REFERENCES

Goldman SM, Minkin SD, Noraval DC, et al: Renal carbuncle: use of ultrasound in its diagnosis and treatment. J Urol 118:525 (1977)

Schneider M, Becker JA, Staiano S, et al: Sonographic-radiographic correlations of renal and perirenal infections. Am J Radiol 127:1007 (1976)

Case 69

CLINICAL INFORMATION

This 58-year-old woman has been feeling weak and has had abdominal pains of increasing frequency and severity over the past two months. Although her appetite has been poor, she has not lost any weight and in fact has gained about 5 pounds. Her general health is good. The physical examination reveals a normal chest and an enlarged abdomen with a suggestion of a fluid wave. There is mild, diffuse abdominal tenderness but no organs or masses are palpable. A pelvic examination has not been performed. Routine laboratory data are all within normal limits except for the sedimentation rate, which is mildly elevated. Figure 69–A is a cross-section through the upper abdomen, and Figure 69–B and Figure 69–C are cross- and longitudinal sections through the pelvis. The urinary bladder is empty.

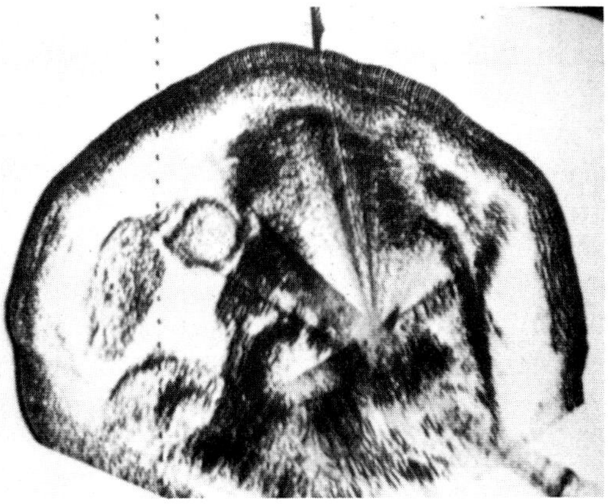

Figure 69–A. Transverse section +5.

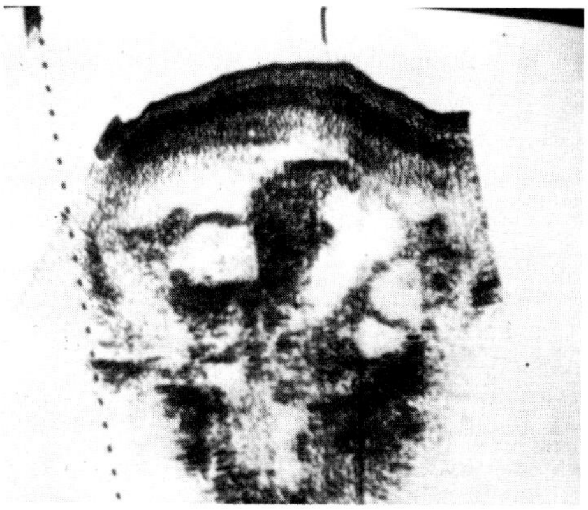

Figure 69–B. Transverse section −13.

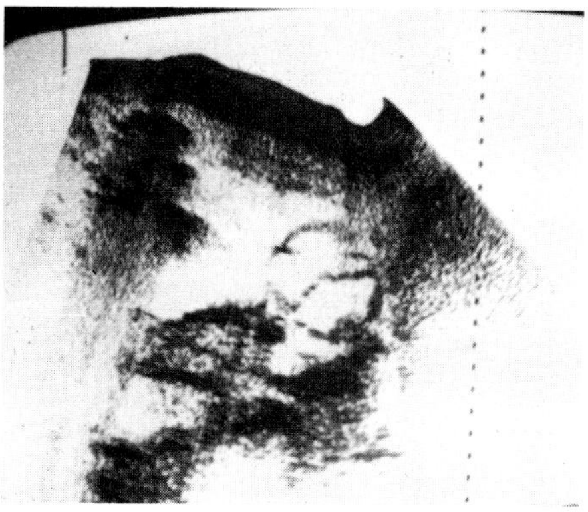

Figure 69–C. Longitudinal section 3L.

DIAGNOSIS

Malignant ascites (ovarian carcinoma).

ULTRASOUND FINDINGS

We see quickly that the patient has ascites — the abdominal organs are surrounded by echo-free areas of fluid. It is easy to make this diagnosis but not so easy to say anything about the cause. Ascites can be classed into two large general categories: benign and malignant. The former is usually secondary to heart, liver, or kidney failure (and a few rare conditions such as malabsorption hypoproteinemic states and others). Malignant ascites is obviously due to tumor involving the peritoneum. We do not think there is any reliable way to make the distinction between benign and malignant ascites *based on the ultrasonic appearance alone;* we must utilize the clinical information at our disposal. The fact that this patient does not have any evidence of the common benign causes of ascites leads us to suspect that we are dealing with a tumor. Although any intraperitoneal tumor can yield a malignant effusion, in our experience by far the most common is ovarian carcinoma. Can we find a tumor in the patient's pelvis? The complex, multiloculated mass seen there certainly makes one suspicious. This appearance is unusual for bowel loops and is strongly suggestive of ovarian tumor. A subsequent pelvic examination demonstrated a large pelvic mass, which was indeed an ovarian carcinoma.

We arrived at the correct diagnosis in this case because all the signs pointed in the same direction. It is not always so easy, however, and we must be ready to admit that in many cases we cannot tell the origin of the ascites. Tumor masses and normal bowel loops can have identical appearances when surrounded by ascitic fluid, and it is often impossible to tell them apart. If there is any uncertainty in our mind we do not make an etiologic diagnosis, but recommend a diagnostic paracentesis. If we call all such effusions in women ovarian cancer, we will be right more than we are wrong, but we are only playing the odds. Just to keep you honest we include the scan in another woman with a malignant effusion (Figure 59–D), but her underlying tumor is a metastatic carcinoma of the breast.

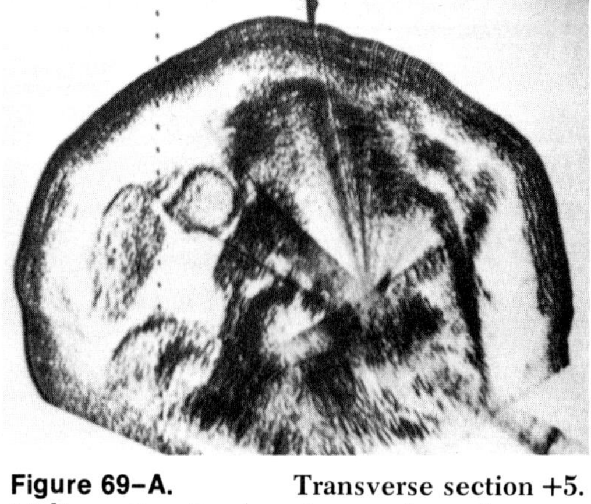

Figure 69–A. Transverse section +5. Malignant ascites (ovarian carcinoma).

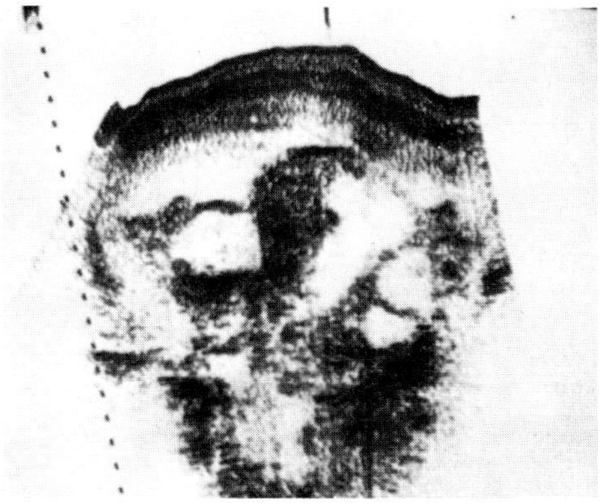

Figure 69–B. Transverse section −13. Malignant ascites (ovarian carcinoma).

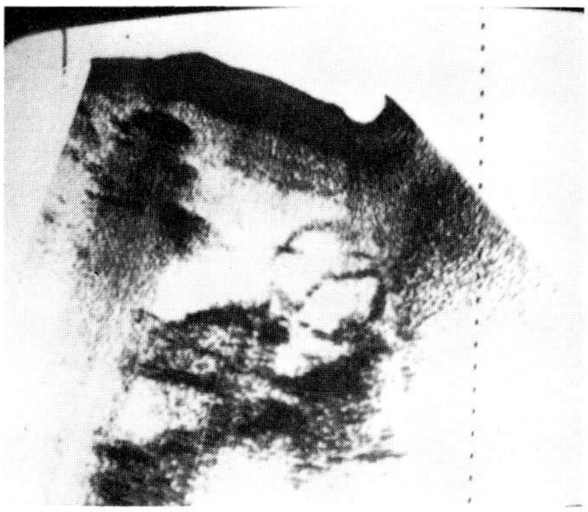

Figure 69–C. Longitudinal section 3L. Malignant ascites (ovarian carcinoma).

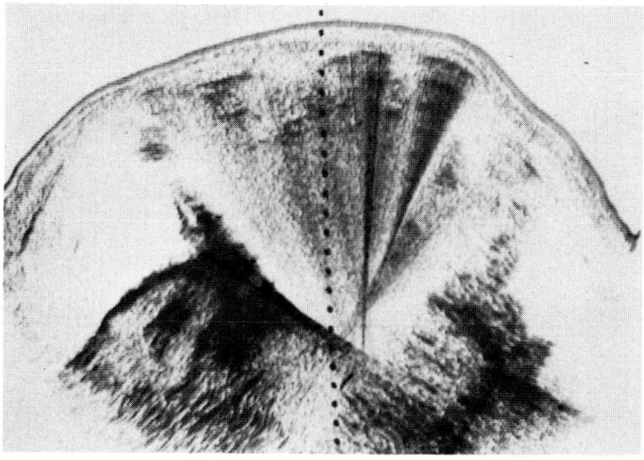

Figure 69–D. Transverse section +8. Malignant ascites (breast carcinoma).

Case 70

CLINICAL INFORMATION

This 62-year-old man has increasing nocturia and frequency associated with a decrease in the size of his urine stream. He is otherwise in good health, and his physical examination is not remarkable except for slight obesity and an enlarged prostate. His urinalysis reveals microscopic hematuria. He was scheduled for prostate surgery, and a preoperative IVP was obtained. On this study a parapelvic mass was discovered in the right kidney (Figure 70–A). An ultrasound scan through the long axis of the right kidney is shown in Figure 70–B.

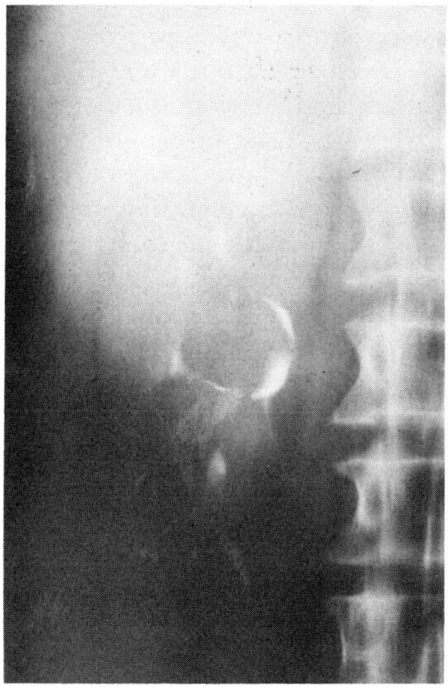

Figure 70–A. IVP, 8-minute compression tomogram of right kidney.

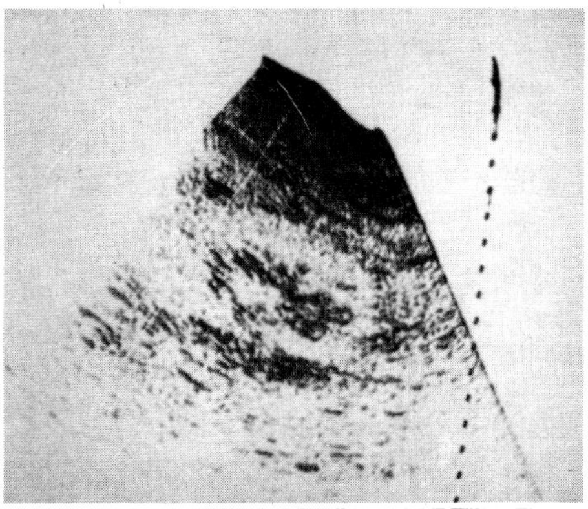

Figure 70–B. Longitudinal section 1R. Prone scan of right kidney.

DIAGNOSIS

Transitional cell carcinoma of the renal pelvis.

ULTRASOUND FINDINGS

The ultrasound scan clearly shows the parapelvic mass seen on the IVP. We can see that it contains low-level echoes, and we know, therefore, that we are dealing with a solid rather than a fluid-filled lesion. This eliminates parapelvic cyst, calyceal diverticulum, and focal hydronephrosis or extrarenal pelvis from consideration. (The IVP appearance is not consistent with the latter two possibilities anyway.) Furthermore, the lesion is somewhat lumpy or lobulated and does not merge smoothly with the adjacent renal parenchyma; there seems to be a band of high-intensity echoes separating the two.

We must now decide whether we are dealing with a benign lesion (infolding of the renal cortex, the so-called "column of Bertin") or a tumor (trasitional cell, hypernephroma, adenocarcinoma, adenoma, or metastases). The lobulated appearance and lack of continuity with the adjacent parenchyma make it unlikely that we are dealing with a column of Bertin. (Figure 70–C is a scan of a column of Bertin.) Thus, we are left with the diagnosis of renal tumor. We canot be specific as to type of tumor on the basis of the ultrasound picture *alone*. Nevertheless, by combining the ultrasound image with the IVP and a general knowledge of renal tumors, we can make some educated guesses as to tissue type. Renal adenomas are quite rare and can be considered unlikely on that basis alone. Renal metastases are very commonly found at autopsy, but infrequently manifest themselves while the patient is alive; they usually occur as multiple small lesions rather than a single dominant mass, and the patient usually is symptomatic from his primary cancer. Hypernephroma is the most common renal tumor, but is much more commonly seen in the peripheral cortex than in the parapelvic region. It must, however, still be considered as a possibility in this case. The location of this tumor is good for either a mucinous adenocarcinoma or a transitional cell carcinoma. The latter tumor is much more common, but in our case the IVP does not show any definite tumor within the renal collecting system. For this reason we opted initially for a mucinous adenocarcinoma. As always seems to be the case, when we bet on a rare condition, we lose; this patient had a transitional cell carcinoma removed at surgery. Transitional cell carcinomas are not always small parapelvic lesions, by the way. Figure 70–D is a scan of a massive transitional cell carcinoma (in a different patient), which has replaced nearly the entire kidney and is invading the pararenal structures.

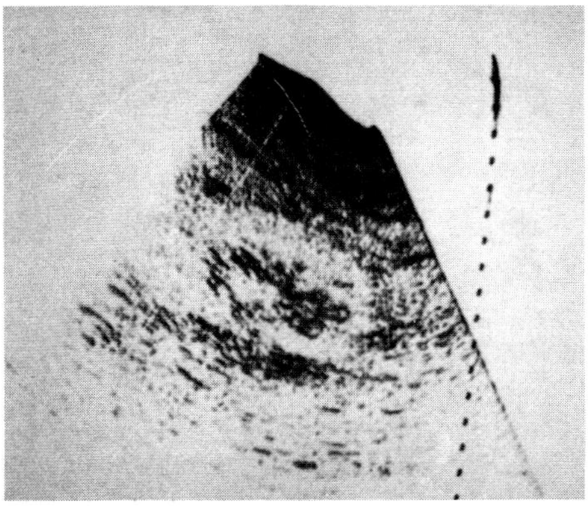

Figure 70–B. Longitudinal section 1R. Prone scan of right kidney. Transitional cell carcinoma.

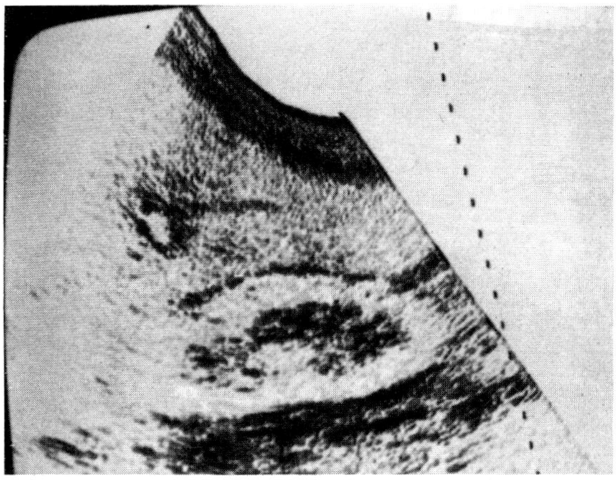

Figure 70–C. Longitudinal section 7R. Supine scan. Column of Bertin.

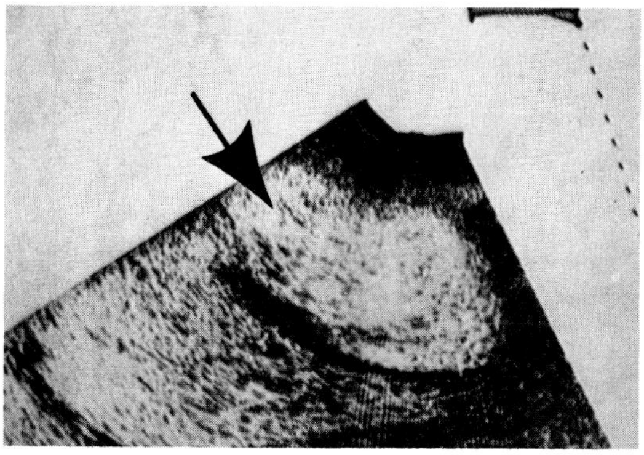

Figure 70–D. Longitudinal section 0L. Prone scan of right kidney. Massive transitional cell carcinoma (arrow).

REFERENCES

Maklad NF, Chuang VP, Doust BD, et al: Ultrasonic characterization of solid renal lesions. Radiology 123:733 (1977)

Davidson AJ: *Radiologic Diagnosis of Renal Parenchymal Disease*. Chapter 11. WB Saunders Co, Philadelphia (1977)

Case 71

CLINICAL INFORMATION

This 54-year-old soldier-of-fortune has lost his appetite and his taste for cigarettes. Moreover, some of his friends have told him that he looks yellow, and this has made him aware of some stomach pains, which he had not previously noticed. His past health has been good except for a few broken bones. Though he once had the DT's, he says he is not really a heavy drinker. On physical examination his liver edge is palpable, but there are no other abdominal masses, and the medical observers do not think he is jaundiced. The liver function tests are all mildly abnormal (bilirubin 2 mg.); the serum alkaline phosphatase and amylase are not elevated. He is mildly anemic with a normal white blood cell count. The radiology resident brings this scan to you and is concerned about a mass in the porta hepatis (arrows).

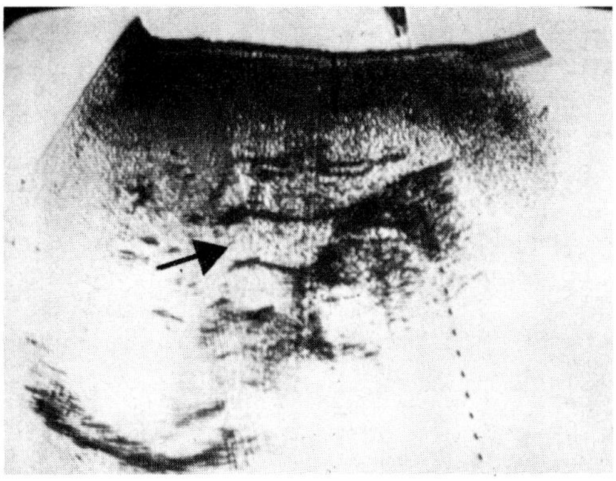

Figure 71–A. Transverse section +12.

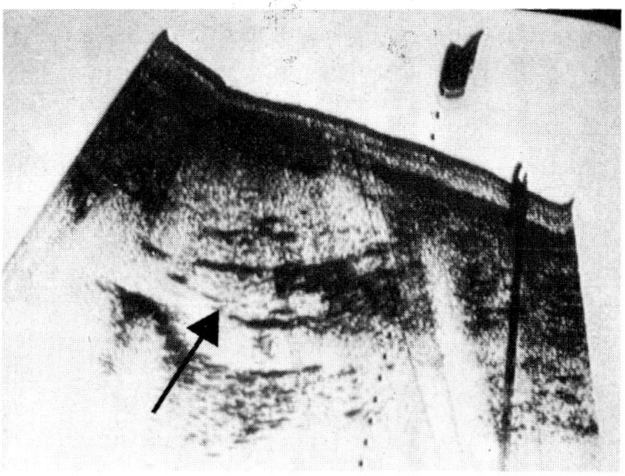

Figure 71–B. Longitudinal section 3R.

DIAGNOSIS

Prominent caudate lobe of the liver (normal variant).

ULTRASOUND FINDINGS

The arrows point to a prominent caudate lobe of the liver. If one is not familiar with this appearance, one can easily be misled into thinking there is a pathologic mass. (Most commonly this is falsely labeled as cancer of the pancreas.) The trick is to notice that the right lateral and cranial edges of the "mass" merge imperceptibly into the right lobe of the liver. On the longitudinal section we can see that the caudate lobe is located more cranially than the portal vein, whereas the pancreas normally lies caudal to the splenic and portal veins.

The liver in this patient is somewhat more echogenic than normal, and it is difficult to get good penetration of the sound beam, particularly in the right lobe. This suggests that there is diffuse liver disease, most likely early cirrhosis. We do not, however, have pathologic proof of this diagnosis in this patient. He would not permit a needle biopsy.

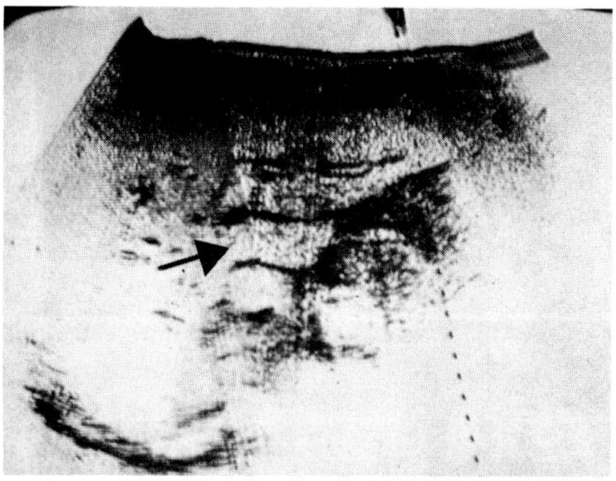

Figure 71–A. Transverse section +12. Caudate lobe of liver.

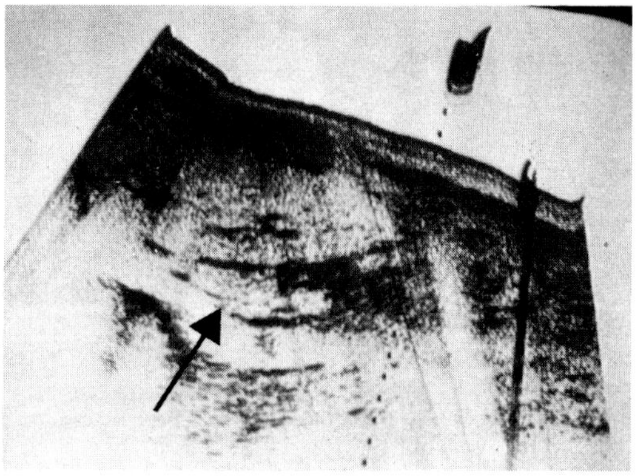

Figure 71–B. Longitudinal section 3R. Caudate lobe of liver.

Case 72

CLINICAL INFORMATION

The patient is a two-day-old boy with a large easily palpable mass in the left abdomen. He is a full-term baby, whose gestation and delivery were completely normal and who had no perinatal problems. He has no obvious congenital abnormalities (other than the abdominal mass) and is taking feedings well and producing both stool and urine. Before he was transferred to our hospital an IVP was performed; we do not have the films available, but the report of the radiologist reads: "There is a large soft-tissue mass in the left abdomen which displaces bowel and obscures the retroperitoneal structures. The right kidney is normal. There is a a suggestion of a left kidney, displaced into the pelvis by the mass, but this is equivocal. The findings suggest either massive hydronephrosis or congenital multicystic kidney."

Figure 72–A is a cross-sectional ultrasound scan through the midabdomen and Figure 72–B is a detailed scan of the left upper quadrant with a baby supine. Can we differentiate between hydronephrosis and multicystic kidney? If not, what else might we do?

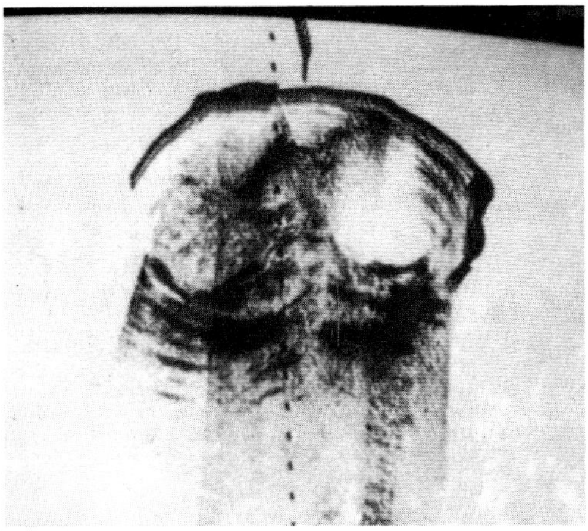

Figure 72–A. Transverse section +3.

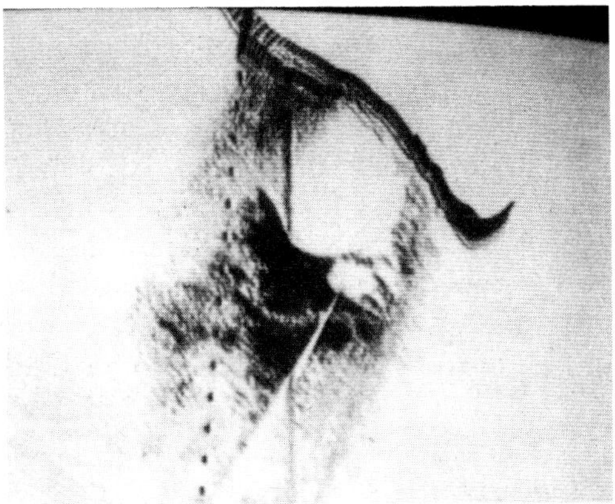

Figure 72–B. Transverse section +1.

DIAGNOSIS

Congenital multicystic dysplastic kidney.

ULTRASOUND FINDINGS

As we have said, scanning tiny babies is not easy. First, they are usually quite active and "squirmy," and this causes difficulties in defining even normal anatomy, let alone small lesions. We try to use the trick of scanning them immediately after a feeding when they are most likely to be sleeping. (If absolutely necessary, they can be sedated with small amounts of chloral hydrate or whiskey.) Second, they usually have tremendous amounts of bowel gas, which leaves no acoustic window. We do not know of any remedy for this. Fortunately, the most common indication for a scan is the evaluation of a mass; the mass itself usually displaces bowel out of the way. Finally, the small size of most infants allows the use of a high frequency (5 MHz) transducer, which is focused within 3 to 4 cm. of the transducer face. The focal length of the standard adult transducer is too great for these small patients. (Even a high-frequency adult transducer will make poor pictures because of poor near-field focus.) If your laboratory does not have a pediatric transducer, you can improvise by filling a surgical glove with water which has been boiled and cooled; the water-filled glove is placed on the baby and the scan performed through it. This moves the transducer back away from the skin and allows the area of interest to fall within the zone of optimal focus.

In this case, we see that the bulk of the abdominal mass is a large, round, fluid-filled lesion. A limited scan (Figure 72–B) also demonstrates a second, smaller fluid-filled area. Nowhere is there any structure resembling normal renal parenchyma. While the absence of identifiable renal tissue seems to suggest cystic dysplastic kidney, we cannot exclude massive congenital hydronephrosis because, in severe forms, this disease also causes extensive distortion and destruction of the renal parenchyma. (A very rare tumor, cystic hamartoma, can also look like this.) We have found that the only way to make this differentiation reliably is to perform an antegrade pyelogram (either with ultrasonic or fluoroscopic guidance). Since the child was to be operated on regardless of the diagnosis, pyelography was not performed in this case. At surgery a multicystic dysplastic kidney was removed; there was only minimal solid tissue present and no identifiable ureter.

What did ultrasound contribute to the evaluation of this patient? Although we were unable to differentiate between the two working diagnoses, we were able to reaffirm the likelihood that one or the other was correct. Another common mass in this age group, neuroblastoma, was excluded by demonstrating that the mass was fluid-filled and not solid.

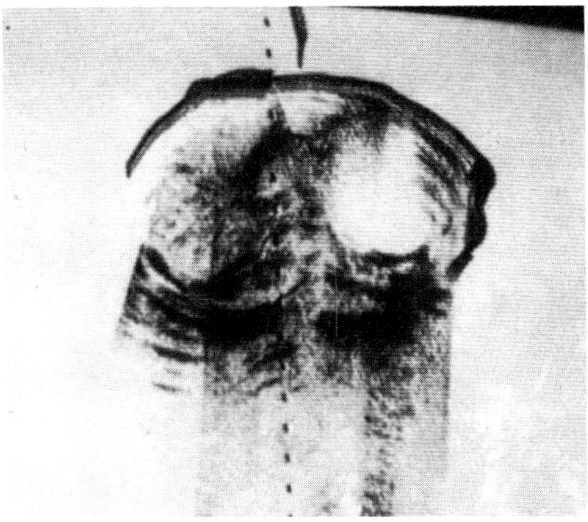

Figure 72–A. Transverse section +3. Multicystic dysplastic kidney.

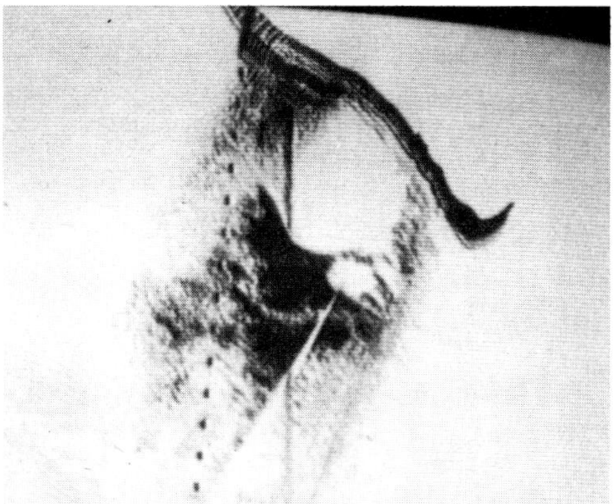

Figure 72–B. Transverse section +1. Multicystic dysplastic kidney.

REFERENCES

Bearman SB, Hene PL, Sanders RC: Multicystic kidney: a sonographic approach. Radiology 118:685 (1976)

Sanders RC: Renal ultrasound. Rad Clin N Am 13:417 (1975)

Case 73

CLINICAL INFORMATION

This 22-year-old woman suffers from severe menstrual cramps. Her menarche occurred at age 14 and her periods were always regular and without significant pain until 18 months ago. During the past year and a half she has had increasingly severe pelvic pain and has been bedridden during her last two menstrual periods. She has never been pregnant and does not use birth control pills or an IUD. Pelvic examination reveals a soft, tender 10-centimeter mass in the left adnexal region. The ovary cannot be felt. Laboratory studies, including pregnancy test and cervical culture, are normal.

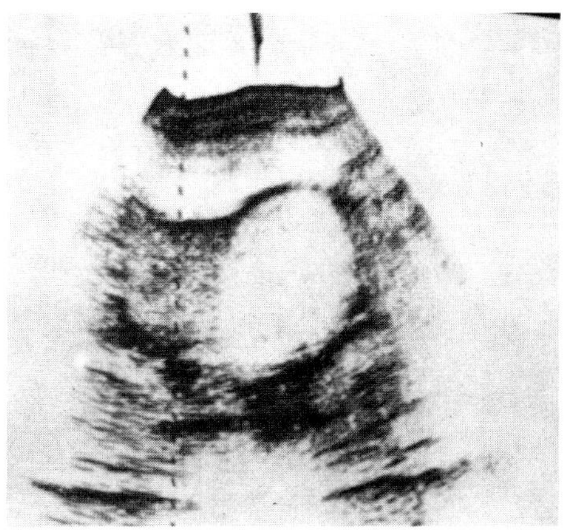

Figure 73–A. Transverse section −14.

DIAGNOSIS

Endometrioma (chocolate cyst).

ULTRASOUND FINDINGS

There is a large, echo-free mass in the pelvis, just to the left of the uterus. This corresponds in size and location to the mass palpated on physical examination, and we seem to be dealing with a cyst in the region of the left ovary. There are many different pathologic types of ovarian cysts, including follicle, corpus luteum, and paraovarian cysts. The clinical history in this patient, however, is most suggestive of a "choclate cyst" or endometrioma. There are two other conditions we should consider in our differential diagnosis: tubo-ovarian abscess and ectopic pregnancy. The first is unlikely in view of the normal laboratory data, and the second because of the normal pregnancy test. Also, the clinical setting is somewhat atypical for these conditions.

Figures 73–B and 73–C show another endometrioma (arrows), which contains solid elements as well as a multiocular cyst.

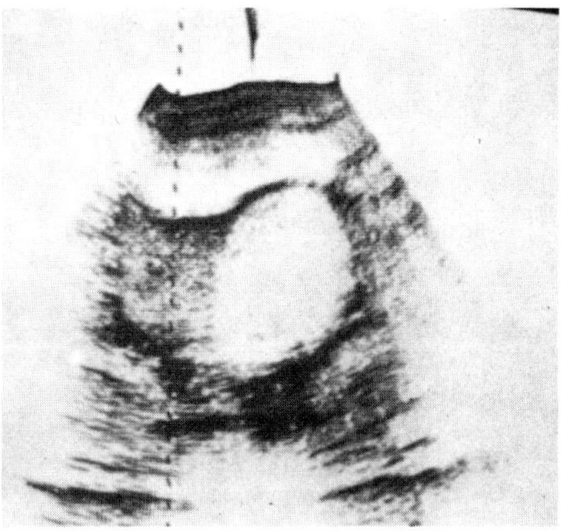

Figure 73–A. Transverse section −14. Endometrioma.

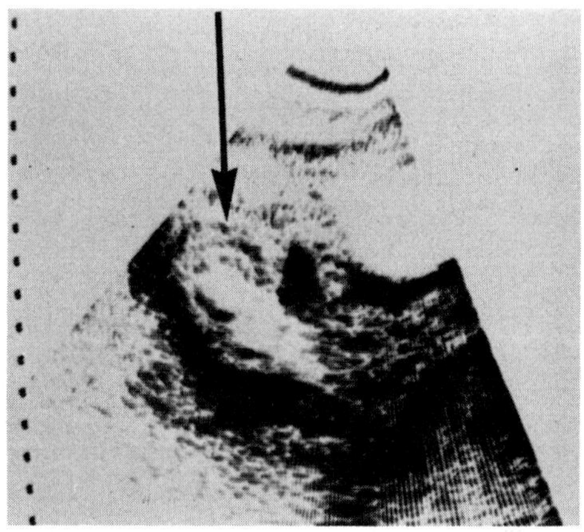

Figure 73–B. Longitudinal section 3R. Endometrioma.

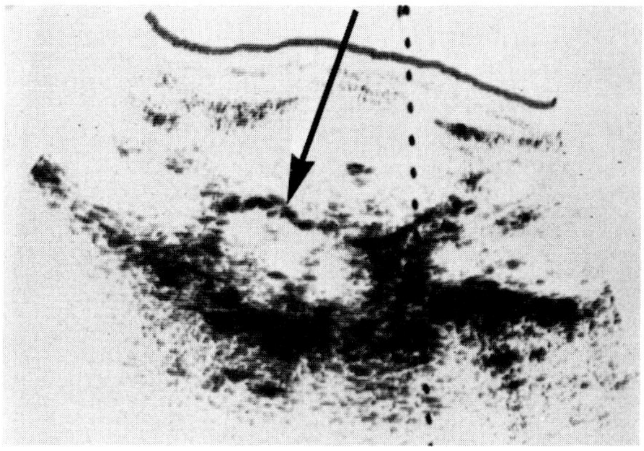

Figure 73–C. Transverse section—12. Endometrioma.

REFERENCE

Sandler MA, Karo JJ: The spectrum of ultrasonic findings in endometriosis. Radiology 127:229 (1978)

Case 74

CLINICAL INFORMATION

This 53-year-old alcoholic is a familiar fixture on the hospital wards. He has had uncountable admissions for the various complications of his alcoholism, including GI bleeding, pancreatitis, pneumonia, and slowly increasing liver failure. His present admission is for abdominal pain and fever. The physical examination revealed an enlarged, firm liver, a very tender and distended abdomen, and mild jaundice. Virtually every laboratory test is positive, including the serum amylase, which is 470, the total bilirubin, which is 3.8, and the white cell count, which is 15,250 with a shift to the left.

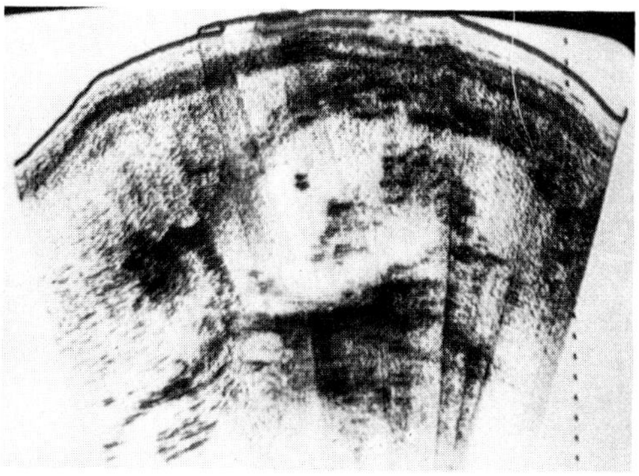

Figure 74–A. Transverse section +6.

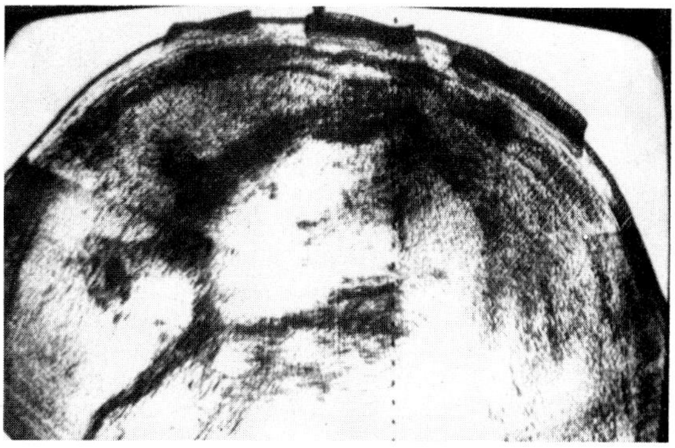

Figure 74–B. Transverse section +3.

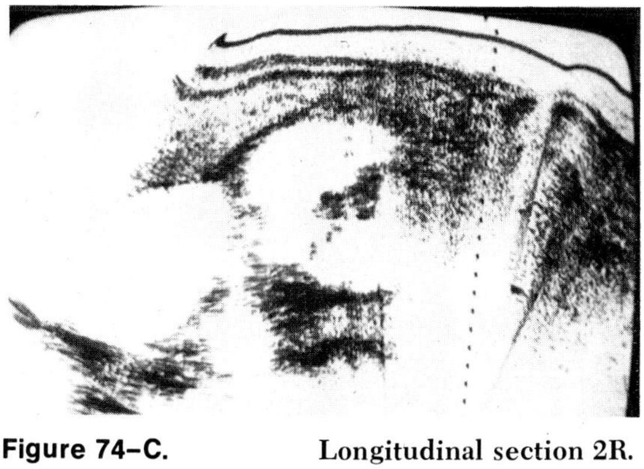

Figure 74–C. Longitudinal section 2R.

DIAGNOSIS

Dissecting pancreatic pseudocyst.

ULTRASOUND FINDINGS

This confusing scan shows an abdomen in disarray. First, we notice a large, mostly echo-free mass in the central abdomen, but there are many other echo-free areas. One large collection is seen beneath the right lobe of the liver (Figure 74–C), another in the right paravertebral gutter (Figure 74–B), and a third extending down toward the patient's pelvis. In an alcoholic with cirrhosis we must always consider the problem of ascites. This is a rather unusual appearance for free intraperitoneal fluid, however; instead of flowing passively around the liver and floating the bowel to the central abdomen, this fluid is actively dissecting into some spaces while ignoring others and is actually elevating and compressing the right lobe of the liver (Figure 74–C). Such behavior is not characteristic of simple ascites and suggests some type of active inflammatory process. A malignant peritoneal effusion can do this, but despite his many other problems there is no reason to suspect a tumor in this patient. A dissecting pseudocyst is a much more likely possibility. It is well known that pancreatic pseudocysts do not always stay confined to the pancreatic area but frequently dissect into surrounding tissue. Most commonly the spread is to adjacent areas of the abdomen and, in our experience, usually the right paravertebral gutters. They can even go to the left side or up the mediastinum into the neck and chest! Not infrequently these pseudocysts become secondarily infected, and this is certainly a distinct possibility in our patient. The best way to exclude this complication is to aspirate a specimen of cyst fluid percutaneously through a small (21 gauge) needle for laboratory analysis. As long as sterile technique is maintained, this aspiration is a risk-free procedure. The aspirate in this patient was sterile but had an amylase of 12,000.

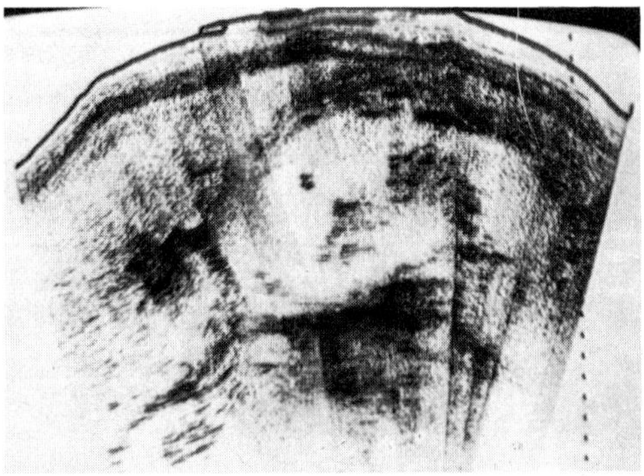

Figure 74–A. Transverse section +6. Dissecting pseudocyst.

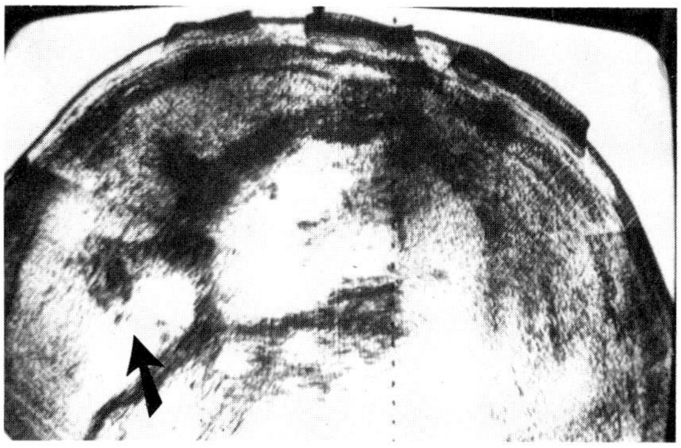

Figure 74–B. Transverse section +3. Dissecting pseudocyst.

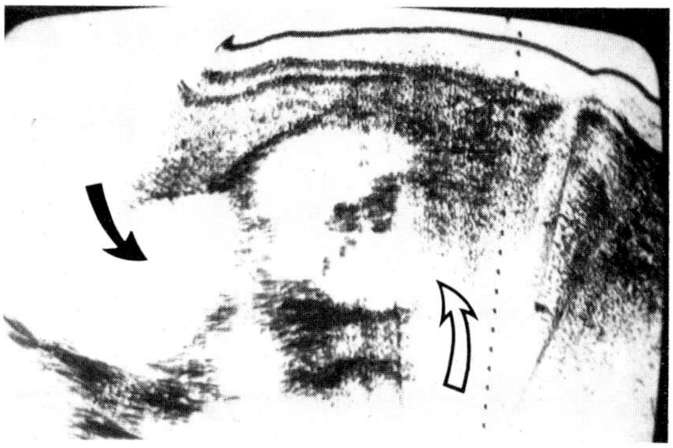

Figure 74–C. Longitudinal section 2R. Dissecting pseudocyst.

REFERENCES

Conrad MR, Landay MJ, Khoury M: Pancreatic pseudocysts: unusual ultrasound features. Am J Radiol 130:265 (1978)

Hancke S, Pedersen JF: Percutaneous puncture of pancreatic cysts guided by ultrasound. Surg Gynecol Obstet 142:551 (1976)

Case 75

CLINICAL INFORMATION

This 15-year-old tackle passed out during a high-school football game and had to be carried from the field on a stretcher. Although he was involved in several rough tackles during the game he had not been "hit" on the two plays preceding his collapse. In the Emergency Ward he is conscious and slightly disoriented; his blood pressure is 100/40, his pulse 96, and he is pale. The abdomen shows mild diffuse tenderness but no signs of peritoneal irritation. The hematocrit is 39 per cent and the white blood count 13,000. An emergency ultrasound examination is requested to evaluate the spleen for possible rupture.

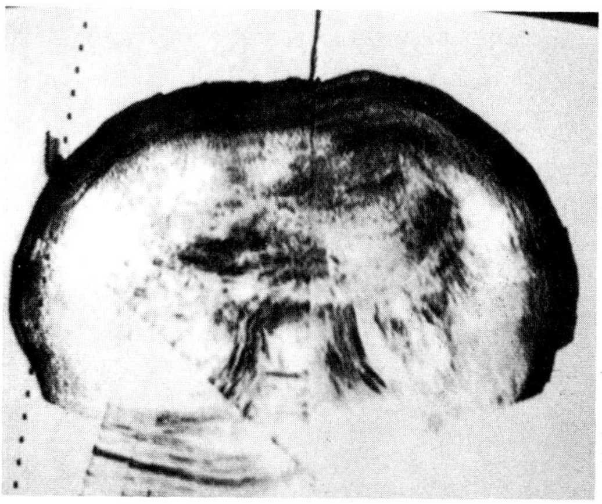

Figure 75–A. Transverse section +8.

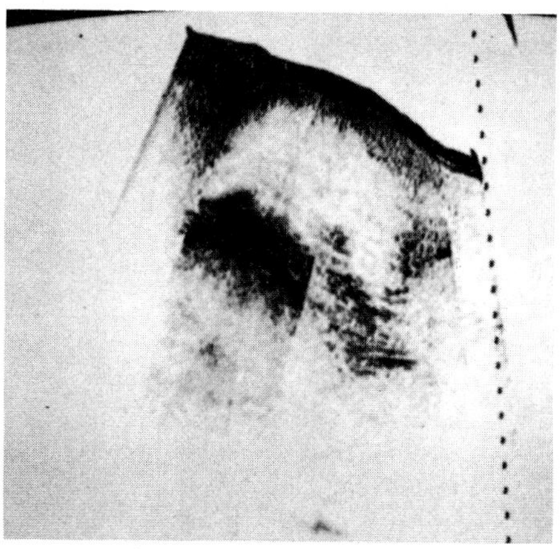

Figure 75–B. Right lateral decubitus scan; longitudinal section of spleen performed between ribs.

DIAGNOSIS

Normal spleen.

ULTRASOUND FINDINGS

The spleen is seen unusually well on the cross-section (Figure 75–A), and this always raises the possibility of splenomegaly. However, careful longitudinal scans (Figure 75–B) show that the overall length of the spleen is less than 14 centimeters and it is therefore not enlarged but is simply positioned more laterally than the average spleen. The echoes are homogeneous and low level, which is normal: there are no areas of heterogeneity to suggest splenic disruption. Likewise there are no parasplenic masses, which would raise the possibility of a hematoma. There is nothing on the ultrasound scan to suggest splenic rupture. (An abdominal paracentesis on the patient was normal; he was observed for 24 hours during which his clinical condition returned to normal, and his discharge diagnosis was "transient cardiovascular instability"—a somewhat unsatisfactory term meaning "acute exhaustion.")

We have had very few requests for this type of ultrasound examination and suspect that it is not a particularly valuable procedure for evaluating splenic trauma. We have questioned other ultrasonographers and surgeons from different parts of the country and most agree with us, preferring clinical evaluation with paracentesis, isotope spleen scanning, and arteriography as diagnostic tools. There is some support for ultrasound, however, as the reference below will attest. Our only quibble with these authors is that most of their patients had a positive paracentesis before being referred to ultrasound and therefore represented a very selected population.

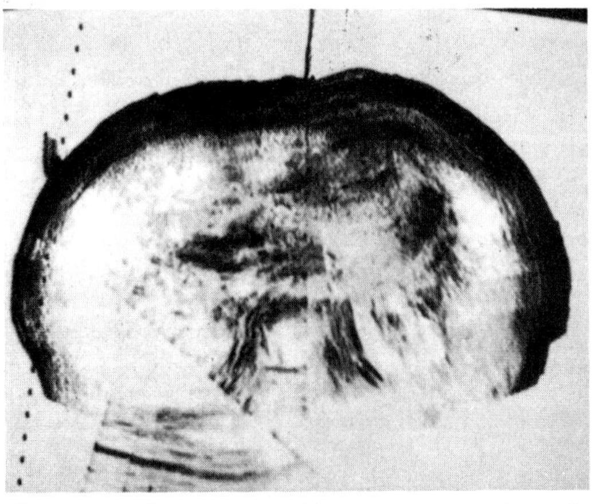

Figure 75–A. Transverse section +8. Normal spleen.

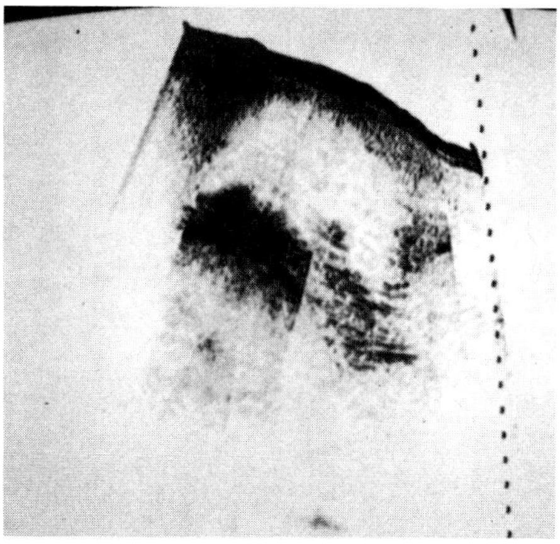

Figure 75–B.
Right lateral decubitus scan; longitudinal section of spleen performed between ribs. Normal spleen.

REFERENCE

Asher WM, Porvin S, Virgilio RW, et al: Echographic evaluation of splenic injury after blunt trauma. Radiology 118:411 (1976)

Case 76

CLINICAL INFORMATION

This 54-year-old woman was undergoing hospital evaluation for an ill-defined neurological problem and the intern noted that her thyroid was diffusely enlarged and nodular. A radioisotope scan showed patchy decrease in activity without a solitary cold area. This ultrasound scan was requested to assess the gland further. The technologist is a bit restive about the scan; she is frustrated because she finds many areas in which the echoes are decreased in intensity; there are, however, no clearly echo-free areas. What is your interpretation of the scan?

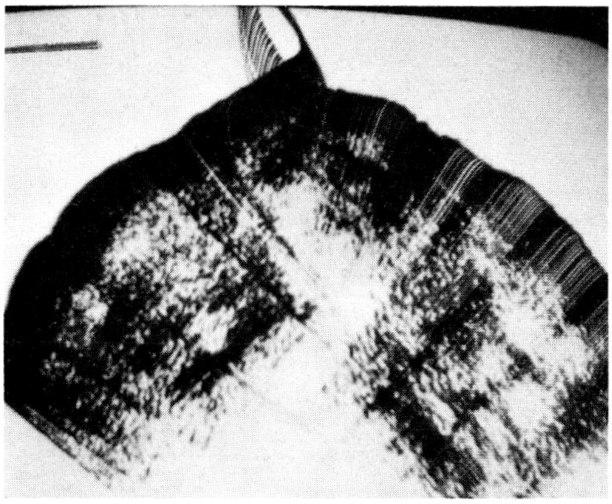

Figure 76–A. Transverse section +6. Thyroid.

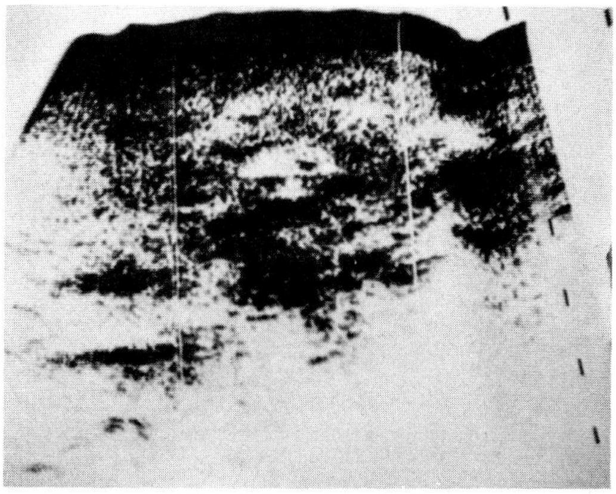

Figure 76–B. Longitudinal section 2L. Thyroid.

DIAGNOSIS

Nontoxic nodular goiter.

ULTRASOUND FINDINGS

We are not proud of the quality of the scan; it was done in the early days of our experience with the small 5 MHz transducer used in direct contact with the skin, and it is not an elegant example of what we have been able to achieve with more practice. Nonetheless, it is a rather typical picture of nodular goiter. There are numerous areas of decreased echo production where nodules have developed in the gland. The abnormality is uneven but almost universal, and this allows us to suggest the diagnosis. This is one instance in which biopsy would not be helpful; the endocrinologist who saw the patient felt it was safe to follow the gland's response to suppressive therapy. Although we do not see many of these cases — most patients sent for ultrasound scanning of the thyroid come because they have discrete cold nodules — the appearance is strongly suggestive of the diagnosis, and we offer it so that you will recognize it when it comes your way. Unfortunately, the demonstration of nodular goiter does not exclude the presence of cancer. Physical examination and response to therapy have to be relied upon to help make the differentiation. Also, thyroiditis has been reported to cause a decrease in echogenicity. In this case nonultrasonic laboratory tests are necessary to make the diagnosis.

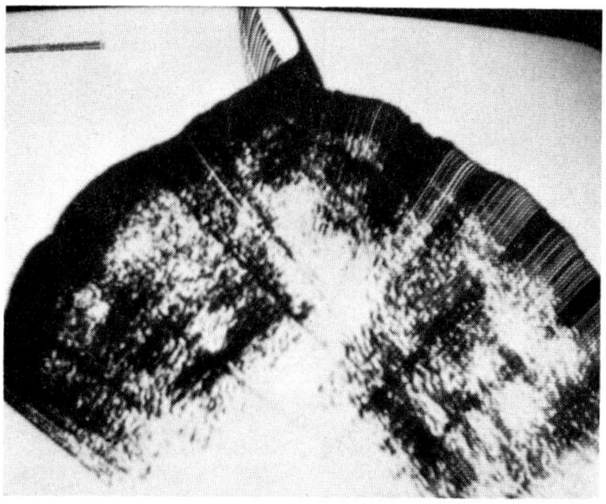

Figure 76–A. Transverse section +6. Nodular goiter.

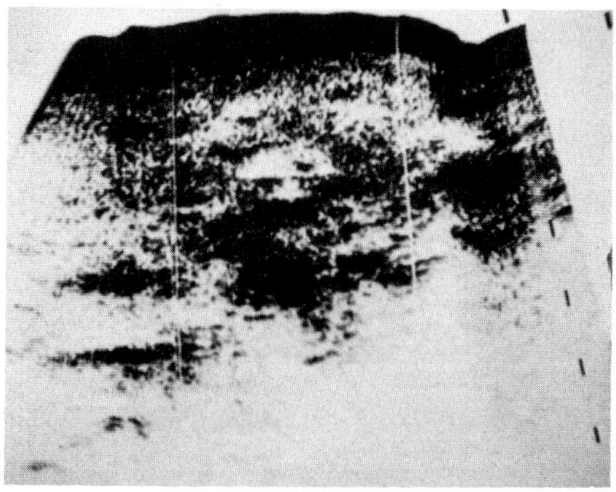

Figure 76–B. Longitudinal section 2L. Nodular goiter.

REFERENCES

Sackler JP, Passalaqua AM, Blum M, et al: A spectrum of diseases of the thyroid gland as imaged by gray scale water bath sonography. Radiology 125:467 (1977)

Blum M, Goldman AB, Herskovic A, et al: Clinical applications of thyroid echography. N Engl J Med 287:1164 (1972)

Miskin M, Rosen IB, Walfish PG: B-mode ultrasonographic assessment of thyroid gland lesions. Ann Int Med 79:505 (1973)

Case 77

CLINICAL INFORMATION

This 47-year-old woman had her first attack of pancreatitis six and a half years ago. At that time she suffered fairly typical symptoms which resolved on medical therapy. An oral cholecystogram was normal, and the basis of the pancreatitis was felt to be alcoholism, despite the patient's denial of excess consumption. Twenty-two months later she had another acute attack, and a repeat oral cholecystogram showed several 1 cm. gallstones. A cholecystectomy was performed, but there were no stones in the common duct, and the stones in the gallbladder were larger than would be expected to enter the ductal system. The etiology of the pancreatitis was, therefore, unclear. She has had three subsequent attacks of pancreatitis in the intervening years and in the last three or four months has had nearly continual abdominal pain. Her laboratory studies, including serum and urinary amylase, are normal. The upper GI series shows only mild edema of gastric rugae.

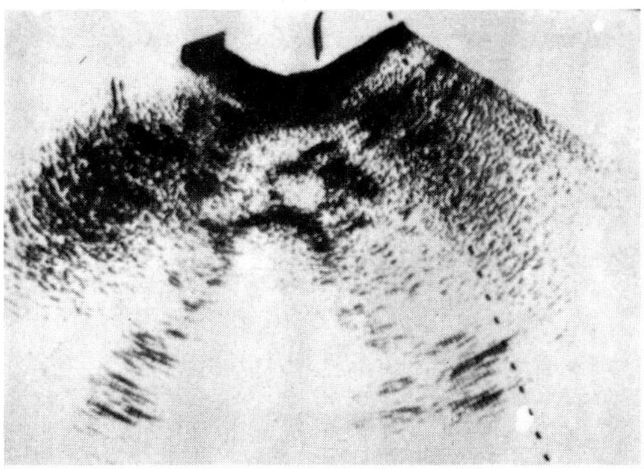

Figure 77–A. Oblique section +9, 80°.

DIAGNOSIS

Chronic pancreatitis.

ULTRASOUND FINDINGS

This is an oblique scan made through the long axis of the pancreas. Although not a part of our routine scanning procedure, these oblique scans can be useful when we are particularly interested in the pancreas. The goal is to orient the scanning plane parallel to the long axis of the pancreas, which lies roughly along a line connecting the hilum of the right kidney with the hilum of the spleen. With the scanner aligned in this fashion, we make a series of quick scans to locate the splenic vein, making minor adjustments in the scanning plane until the entire vein is displayed. Because the splenic vein usually demarcates the cranial border of the pancreas, all we need do is make a series of scans progressing caudally, and we will be scanning through the pancreas.

The pancreas in this case is less echogenic than normal; in fact, it has an appearance suggesting diffuse edema from acute pancreatitis. Yet, this patient does not have acute pancreatitis; rather, she has chronic relapsing pancreatitis. In theory, the pancreas in this situation is not edematous and should generate higher-level echoes than this. So much for theory. Chronic pancreatitis can often be indistinguishable from acute disease on the basis of the ultrasound scan alone, and we cannot make this distinction in most instances. To complicate matters even further, a "normal" pancreas sometimes appears relatively more prominent and echo-free than usual and can simulate either acute or chronic pancreatitis (Figure 77–B). Fortunately, in our experience the "normal" pancreas is usually a very echogenic organ and does not often present a diagnostic dilemma.

What we wish to emphasize is the fact that the ultrasound picture cannot stand alone; it must always be interpreted with full knowledge of the clinical setting. It is fine to make ultrasound scanning a game, but use all the clues you can get.

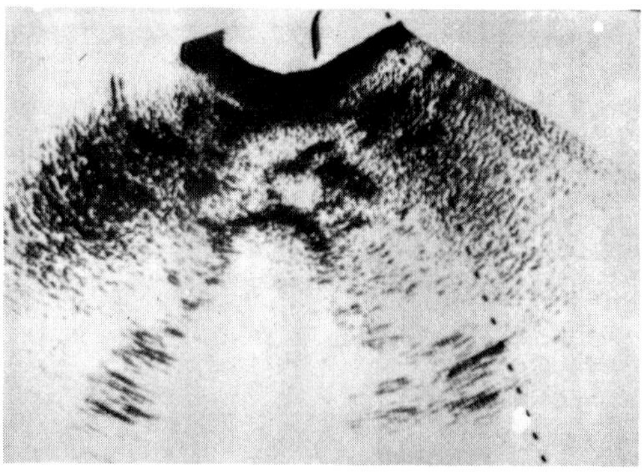

Figure 77–A. **Oblique section +9, 80°,** Chronic pancreatitis.

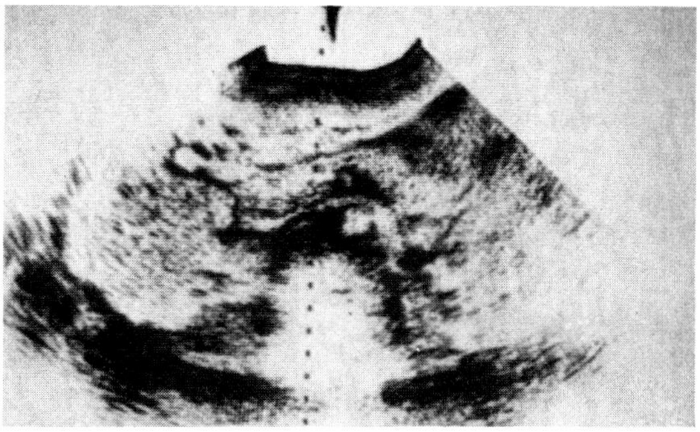

Figure 77–B. Oblique section +9, 80°. Normal pancreas.

REFERENCES

Ghorashi B, Rector WR: Gray scale sonographic anatomy of the pancreas. J Clin Ultra 5:25 (1977)

Sample WF, Po JB, Gray RK, et al: Gray scale ultrasonography: Techniques in pancreatic scanning. App Radiol 4:63 (1975)

Case 78

CLINICAL INFORMATION

This 42-year-old man consulted his doctor because of increasing headaches. His blood pressure was 114/92. Routine laboratory data were normal except for a serum creatinine of 2.7. There was no past history of kidney disease. On an IVP performed at an outside hospital the kidneys were large but the collecting systems were very poorly seen and could not be evaluated. Figures 78–A and 78–B are longitudinal and cross-sectional scans of the left kidney and Figures 78–C and 78–D are similar scans of the right kidney.

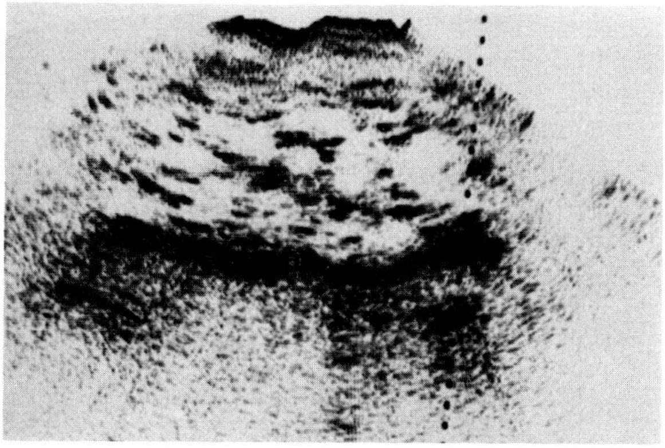

Figure 78–A. Longitudinal section 0L. Prone scan of left kidney.

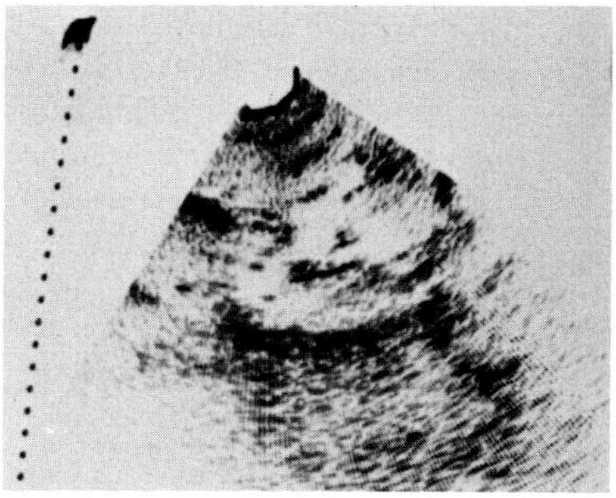

Figure 78–B. Transverse section +3. Prone scan of left kidney.

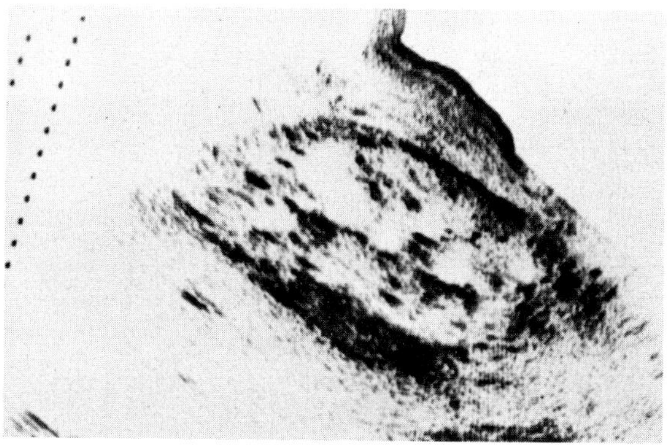

Figure 78–C. Longitudinal section 0L. Prone scan of right kidney.

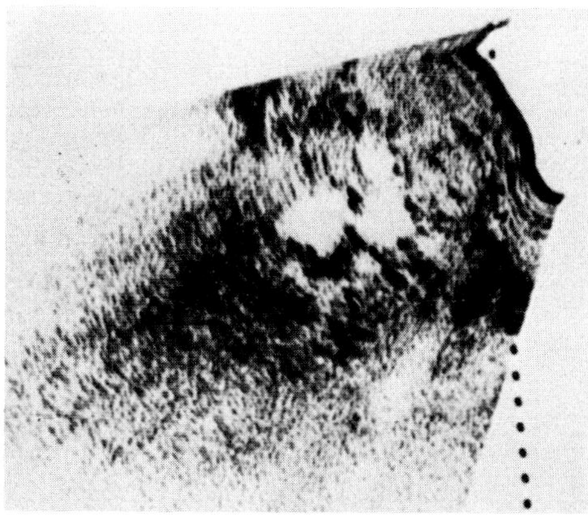

Figure 78–D. Transverse section +3. Prone scan of right kidney.

DIAGNOSIS

Polycystic kidneys.

ULTRASOUND FINDINGS

Both kidneys contain multiple echo-free areas: they look like pieces of Swiss cheese. Our differential diagnosis is limited to two conditions: polycystic kidneys and hydronephrosis. The scan findings in this case all suggest polycystic kidneys. First, the condition is bilateral. Polycystic disease of the kidneys is a bilateral condition; if only one kidney is involved this diagnosis can be excluded. Hydronephrosis can occur bilaterally, of course, but does not usually involve the kidneys to the same degree. Moreover, hydronephrosis must have an underlying cause, such as chronic infection or obstruction, and there is nothing to suggest this in our patient. Second, there is no evidence of a dilated renal pelvis on these scans. Hydronephrosis severe enough to produce this many cystic areas should have a large, easily identifiable renal pelvis: cross-sections through the center of the kidneys (Figures 78–B and 78–D) fail to demonstrate such a pelvis.

Polycystic renal disease is a congenital disease but patients do not usually come for examination until they begin to develop renal failure, most often in early middle age. Children can be affected, however, and the disease must be considered whenever the scan shows the "Swiss cheese" pattern bilaterally. Figure 78–E shows a polycystic kidney in a four-year-old girl. Polycystic disease is multisystem and frequently involves the liver as well as the kidneys. Figure 78–F shows multiple liver cysts in another patient with polycystic kidneys.

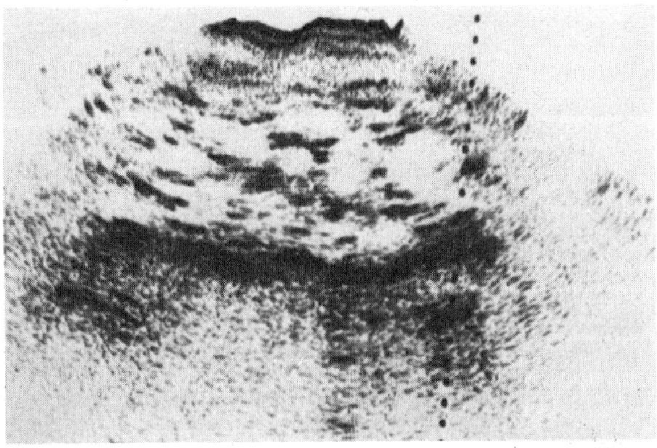

Figure 78–A. **Longitudinal section 0L.** Prone scan of left kidney. Polycystic kidney.

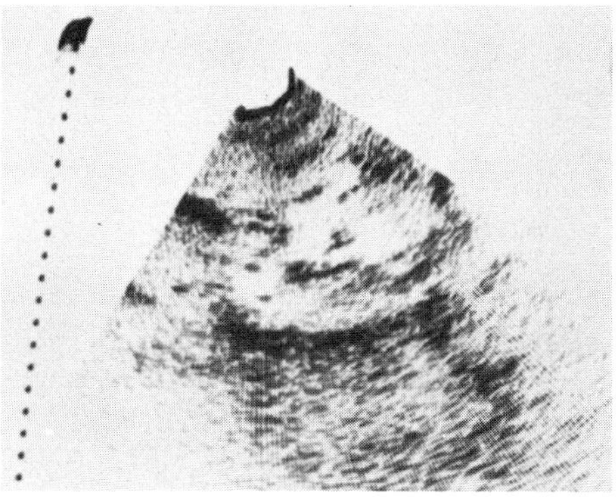

Figure 78–B. Transverse section +3. Prone scan of left kidney. Polycystic kidney.

Figure 78–C. Longitudinal section 0L. Prone scan of right kidney. Polycystic kidney.

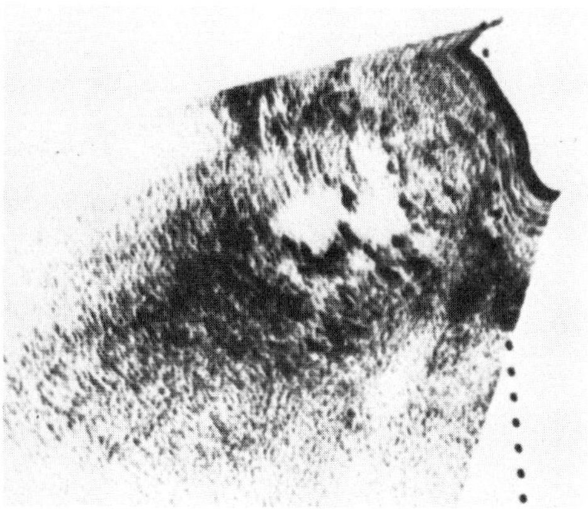

Figure 78–D. Transverse section +3. Prone scan of right kidney. Polycystic kidney.

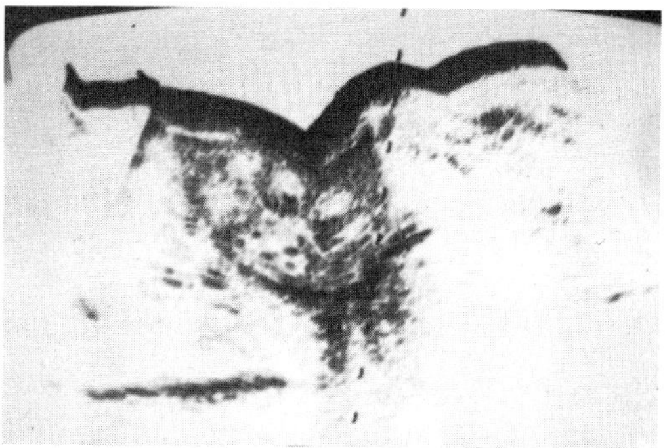

Figure 78–E. Longitudinal section 0L. Prone scan of right kidney. Polycystic kidney in a child.

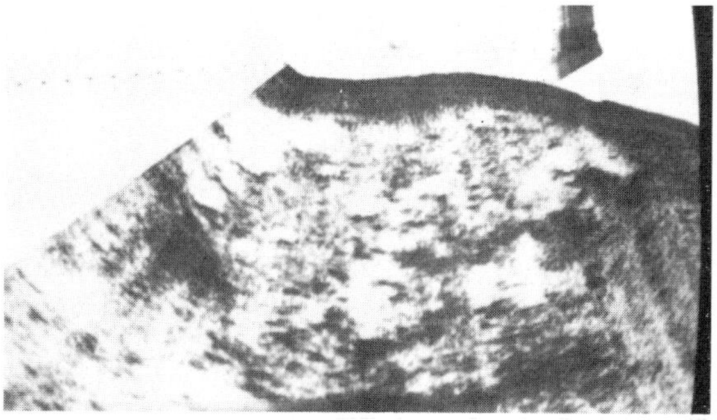

Figure 78–F. Longitudinal section 4R. Polycystic disease of the liver.

REFERENCES

Kelsey JA, Bowie JD: Gray scale ultrasonography in the diagnosis of polycystic kidney disease. Radiology 122:791 (1977)

Case 79

CLINICAL INFORMATION

This six-month-old boy has a large mass in the right upper abdomen that was discovered by his father, a physician. The child had an uncomplicated gestation, delivery, and perinatal period. A "well-baby" examination at age three months was normal. Aside from the mass, there are no abnormal findings — the baby is robust, is eating and growing normally, and has normal routine laboratory studies. An IVP shows that the right kidney is displaced into the pelvis by a large soft-tissue mass, which does not contain any calcifications. The radiologist feels quite certain that the kidney is *not* involved by the mass and favors a diagnosis of neuroblastoma. There are no other radiologic abnormalities, and, in particular, no bony lesions can be found.

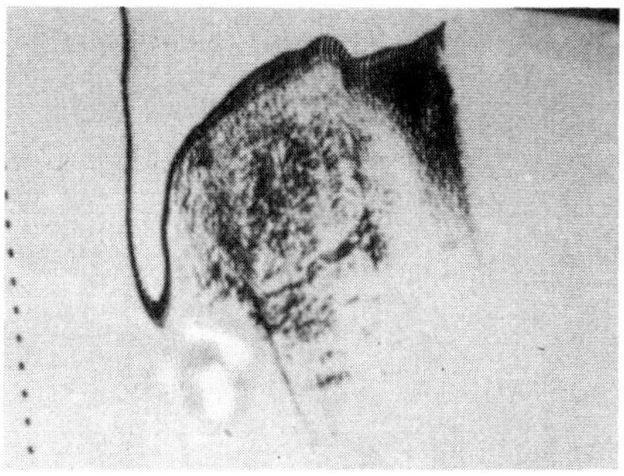

Figure 79–A. Transverse section +3.

Figure 79–B. Longitudinal section 3R.

DIAGNOSIS

Hepatoblastoma.

ULTRASOUND FINDINGS

If you got this correct you are better than we because we also believed this to be a neuroblastoma. Maybe we should have been tipped off by the intimate association between the tumor and the liver and the similarity of the echo texture? We do not think so. Because solid tumors in this region in children are usually neuroblastomas and liver invasion occurs frequently, if you make the diagnosis of neuroblastoma, you will be right more often than wrong. Of course, in doing this we are only playing the odds, and we should really give a differential diagnosis that includes hepatoblastoma, benign and malignant adrenocortical tumors (non-neurogenic), and Wilms' tumor. In most cases we can distinguish a Wilms' tumor (or "renal infantile hamartoma," as some pathologists label this tumor in children less than two) from the other lesions by demonstrating renal involvement. This may be easier with an IVP than with ultrasound. In all patients a biopsy is required for definitive diagnosis. We have included two other cases to illustrate the difficulty in making a histologic diagnosis based on the ultrasound criteria noted above. Figures 79–C and 79–D are cross-sectional and longitudinal scans from a three-and-a-half-year-old boy with an asymptomatic abdominal mass, which proved to be a neuroblastoma. The mass is shown by the arrows, while the triangular arrowhead points to the right kidney (which was displaced but not invaded). Figure 79–E is a scan of a four-year-old girl with an asymptomatic left upper quadrant mass, which proved to be a benign adrenocortical tumor of non-neurogenic origin.

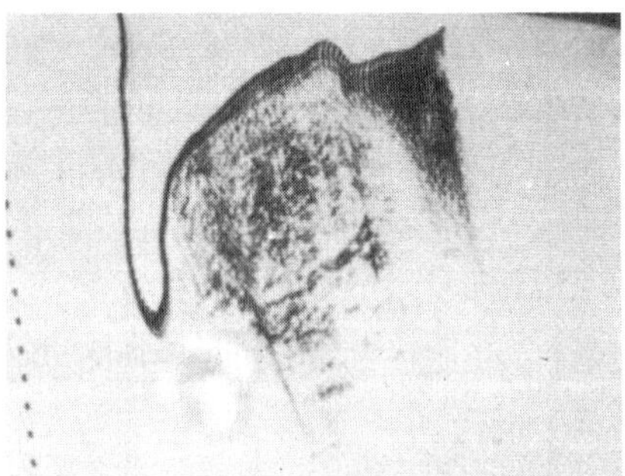

Figure 79–A. Transverse section +3. Hepatoblastoma.

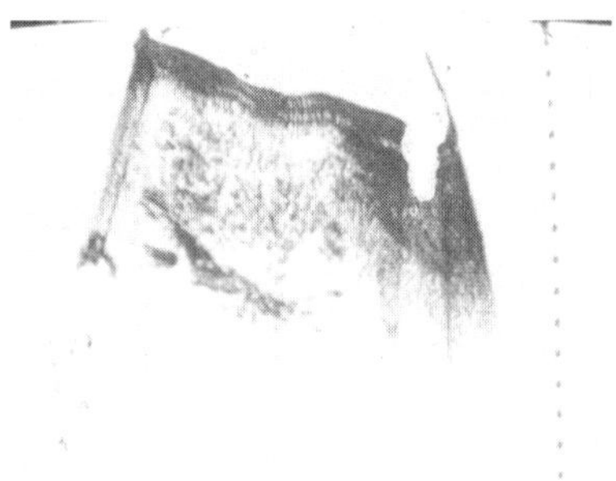

Figure 79–B. Longitudinal section 3R. Hepatoblastoma.

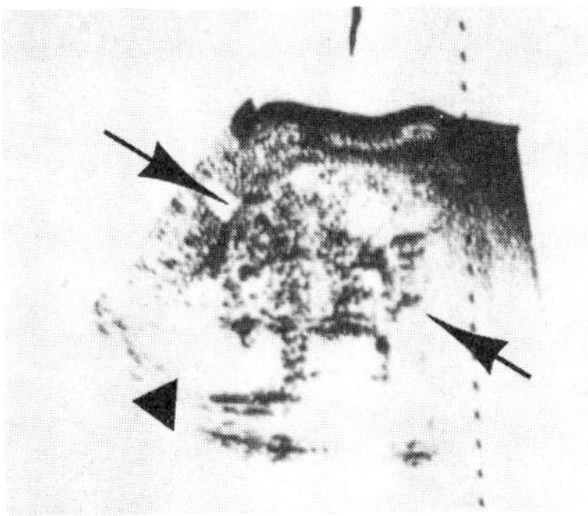

Figure 79–C. Transverse section +3. Neuroblastoma.

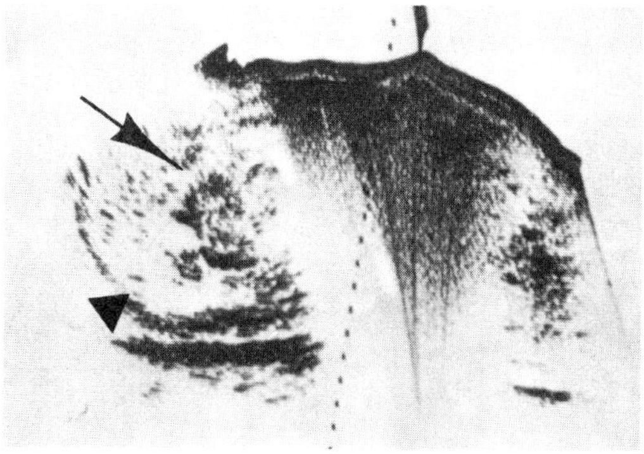

Figure 79–D. Longitudinal section 2R. Neuroblastoma.

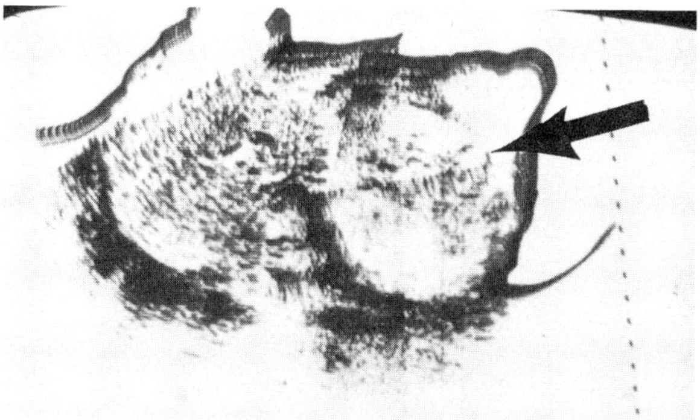

Figure 79–E. Transverse section +7. Adrenocortical adenoma.

REFERENCES

Rose JS, Becker JA, Staiano SJ, et al: B-mode sonographic evaluation of abdominal masses in the pediatric patient. Am J Radiol 120:691 (1974)

Bearman SB, Sanders RC, Oh KS: B-scan ultrasound in evaluation of pediatric abdominal masses. Radiology 108:111 (1973)

VonMicsky LI, Radkowski MA, Hecker J, et al: Optimal diagnosis of renal masses in children (with) sonography and radiography. Am J Radiol 120:438 (1974)

Case 80

CLINICAL INFORMATION

This 17-year-old high school junior has a painless scrotal enlargement. He has known of it for some time, but just recently decided that he would like to have it taken care of. He is asymptomatic and has played varsity football and basketball for his high school; his principal concern is cosmetic inasmuch as his athletic activities now make it difficult for him to hide the fact that he is "different" from his peers. Although the diagnosis is obvious from the clinical examination, the urologist to whom the boy was referred by his team physician has sent him to you for study. Of what help can an ultrasound scan be? Figure 80–A is a cross-section through the scrotum. (The straight arrows point to the testicles and the median raphe is indicated by the letter *R;* the curved arrow points to the epididymis of the right testicle.) Figure 80–B is a longitudinal scan of the left testicle.

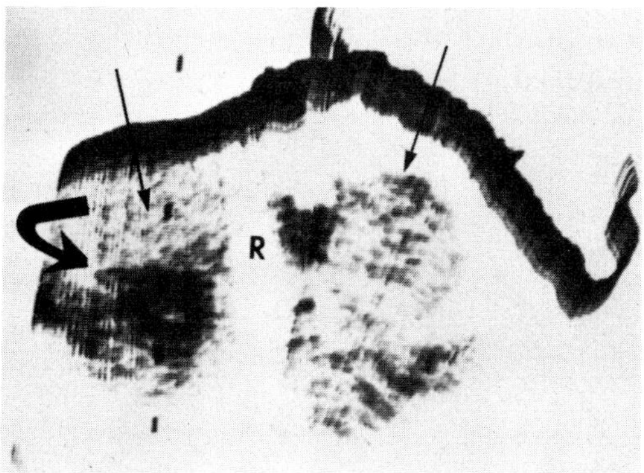

Figure 80–A.
Transverse scan of scrotum.

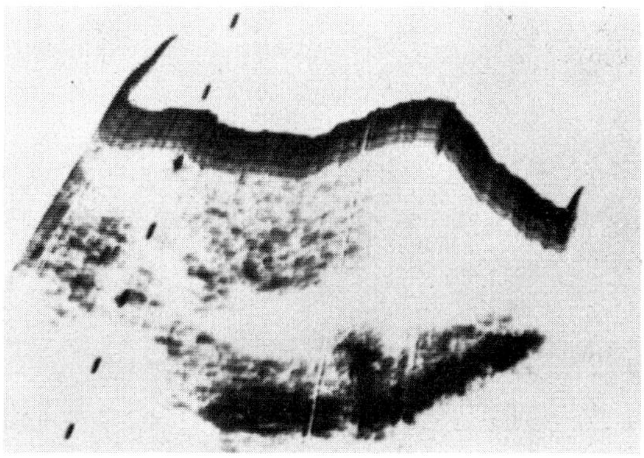

Figure 80–B.
Longitudinal scan through left testicle.

DIAGNOSIS

Normal testicles, left hydrocele.

ULTRASOUND FINDINGS

This young man typifies both the problems and the opportunities which examination of the scrotum presents to the ultrasonographer. First a word about technique. We use a 5 MHz, 6 mm. diameter transducer with a short internal focus, the same as we employ for thyroids. Furthermore, we use direct contact rather than a water bath. The scan is displayed on the 1:1 scale because the structures we are viewing are small and the detail obtained in this manner is important in interpreting the findings. The testicles must be immobilized with one hand while passing the transducer with the other. We have found that both our female technologists and the patients are uneasy about this examination; therefore, we have male personnel perform the scans. (This does not mean that we are sexist or unprofessional, but the study requires patience and perservance. If the technologist is uncomfortable doing the study, she probably will not do a thorough, careful job. Such are the facts of life.) As the present case illustrates, in spite of the technical difficulties, ultrasound does provide considerable information about scrotal disease.

The diagnosis of hydrocele is easy enough: in Figure 80–A both testicles are seen in cross-section, the left one surrounded by a prominent echo-free zone. The urologist knew that much from his physical examination. Why the scan? Hydroceles are reported to be associated with testicular tumors, and it was for this reason that the ultrasound was requested. We were able to tell the referring physician that this was an uncomplicated hydrocele and that both testicles were normal.

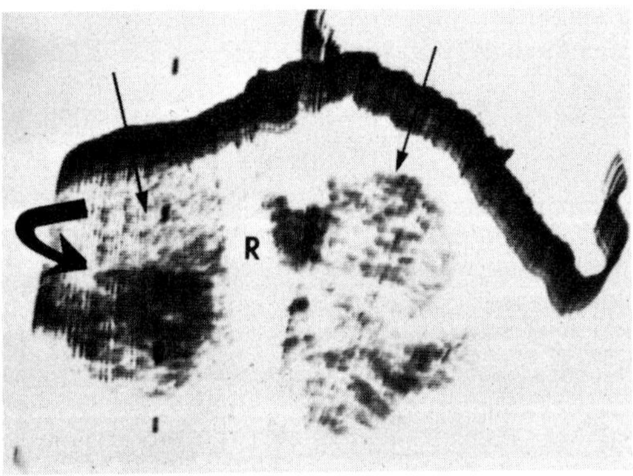

Figure 80–A. Transverse scan of scrotum. Left hydrocele.

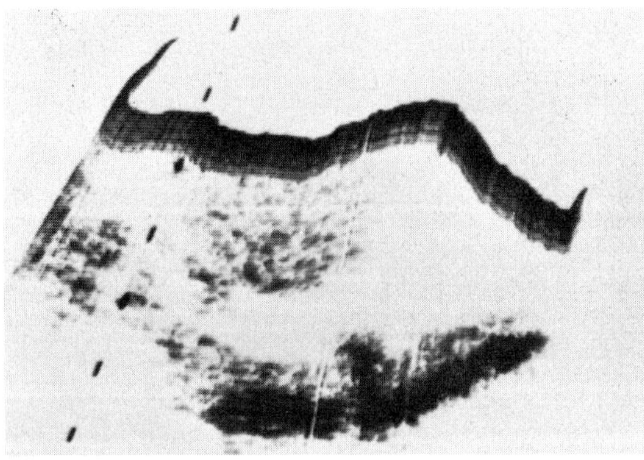

Figure 80–B. Longitudinal scan through left testicle. Hydrocele.

REFERENCES

Sample WF, Gottesman JE, Skinner DG, et al: Gray scale ultrasound of the scrotum. Radiology 127:225 (1978)

Miskin M, Busckspan M, Bain J: Ultrasonographic examination of scrotal masses. J Urol 117:185 (1977)

Case 81

CLINICAL INFORMATION

This 40-year-old woman has symptoms suggesting peptic ulcer disease but her upper GI series shows only mild hypersecretion with some edema of the duodenal folds; no definite ulcer crater is identified. Her doctor wonders if her pain could be due to gallbladder disease and orders an ultrasonic cholecystogram. Figure 81–A was obtained with the patient supine, Figure 81–B is a longitudinal section made with the patient upright, and Figure 81–C is a longitudinal scan obtained after the patient had returned to the supine position.

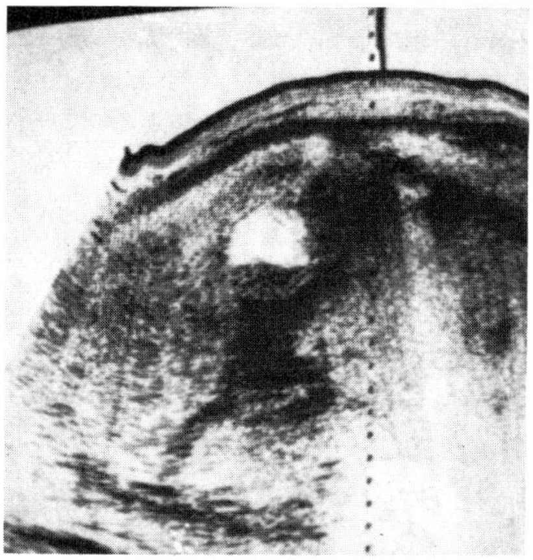

Figure 81–A. Transverse section +8.5.

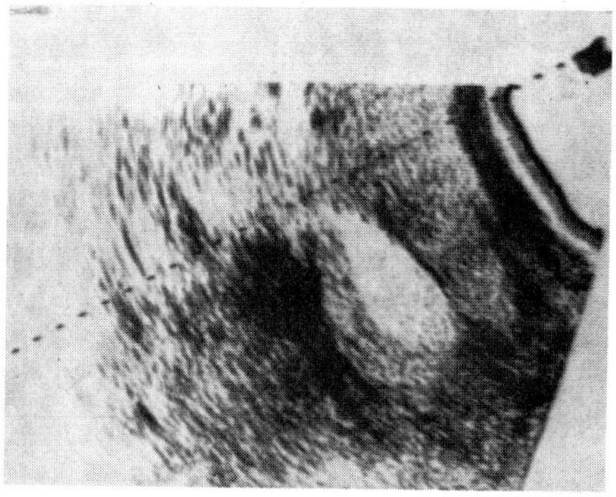

Figure 81–B. Longitudinal section 4R. Upright scan.

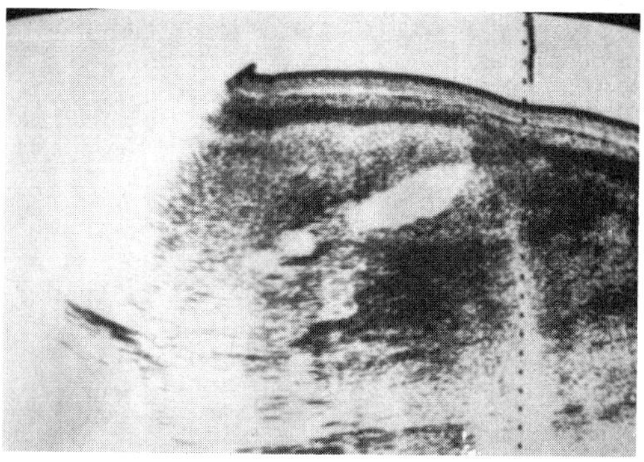

Figure 81–C. Longitudinal section 4R. Supine scan.

DIAGNOSIS

Echogenic bile (no gallstones).

ULTRASOUND FINDINGS

There is a thick layer of echoes in the posterior half of the gallbladder; are these due to stones or to echogenic bile? Before considering this question we should define what we mean by "echogenic bile." Bile is a complex fluid which tends to stratify into layers of differing densities (which explains why some gallstones float). These small density differences constitute acoustical mismatches, which can generate low-level echoes. Fortunately, the density differences are usually small and the echoes too weak to be recorded. On occasion, however, these echoes are loud enough to be registered by the scanner (particularly with the highly sensitive late-model machines) and will appear in the gallbladder lumen to complicate the life of the ultrasonographer. When this occurs, we say the bile is "echogenic." Echogenic bile *does not* contain particulate matter, such as small stones or sand, but it is thick and tenacious and often referred to as gallbladder "sludge" by surgeons.

There is no sure-fire method to distinguish between small stones, gravel, and echogenic bile, but there are a few helpful indications we can use. First, bile without stones does not cast an acoustic shadow; the echoes are too weak. Second, owing to its consistency — it resembles maple syrup at 35 degrees — bile flows slowly and does not shift position readily when the patient is put in the upright position. Figures 81–A, 81–B, and 81–C illustrate this phenomenon. Figure 81–A was obtained after the patient had been lying supine for several minutes; Figure 81–B was made just after the patient sat upright — the bile can be seen flowing slowly down the back wall; Figure 81–C was made immediately after the patient returned to the supine position and again demonstrates the flowing of the bile. Theoretically, if the bile could be well mixed by having the gallbladder shaken up, the densities would be dispersed and the echoes would vanish. Unfortunately, the usual patient is not athletic enough to accomplish this maneuver.

Figures 81–D through 81–H are other examples of echogenic bile (arrowheads). The patient in Figure 81–G has cirrhosis of the liver and marked ascites; the gallbladder is well seen in the ascitic pool (arrow) and is filled with echogenic bile. The patient in Figure 81–H has both echogenic bile and a gallstone (arrow).

The clinical significance of echogenic bile is controversial. If these gallbladders are removed, the pathologic diagnosis will always be "chronic cholecystitis" or "acalculous cholecystitis." Whether these are real diseases depends, we feel, on the philosophy of the individual gastroenterologist and surgeon.

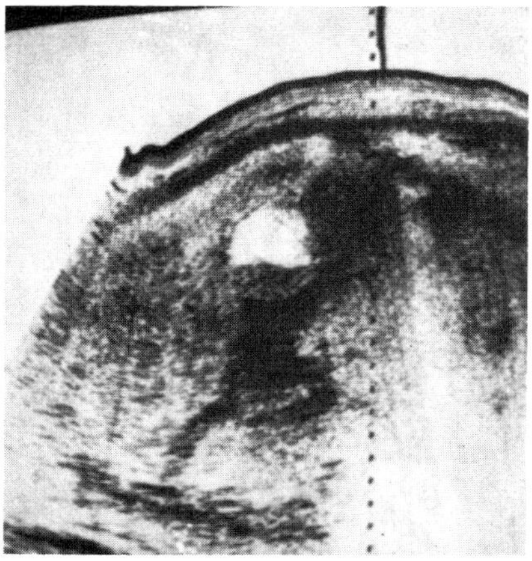

Figure 81–A. Transverse section +8.5. Echogenic bile.

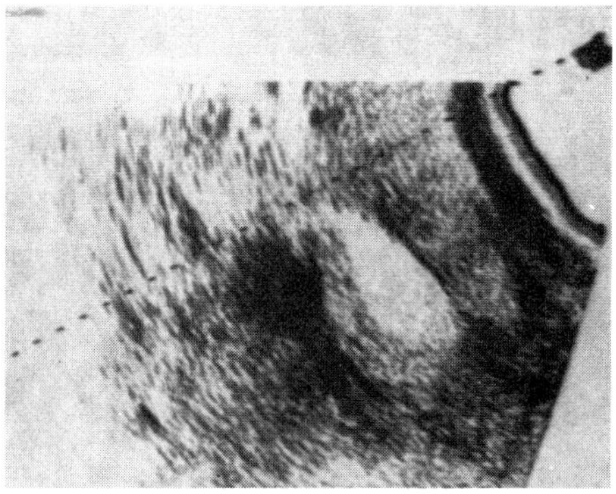

Figure 81–B. Longitudinal section 4R. Upright scan. Echogenic bile.

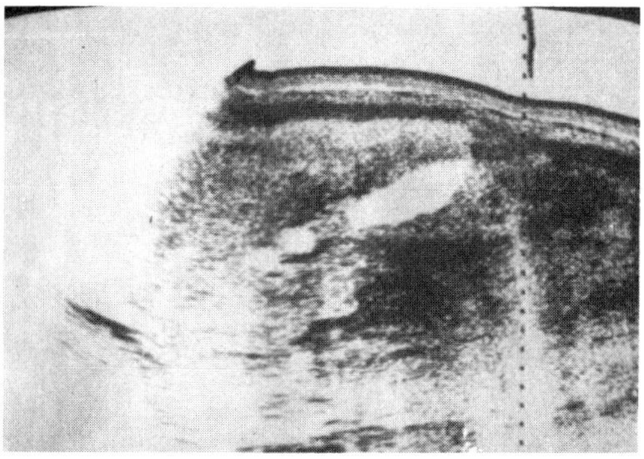

Figure 81–C. Longitudinal section 4R. Supine scan. Echogenic bile.

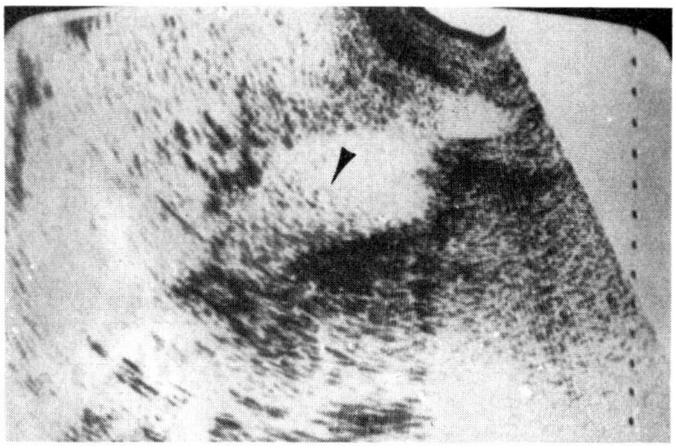

Figure 81–D. Longitudinal section 4R. Echogenic bile.

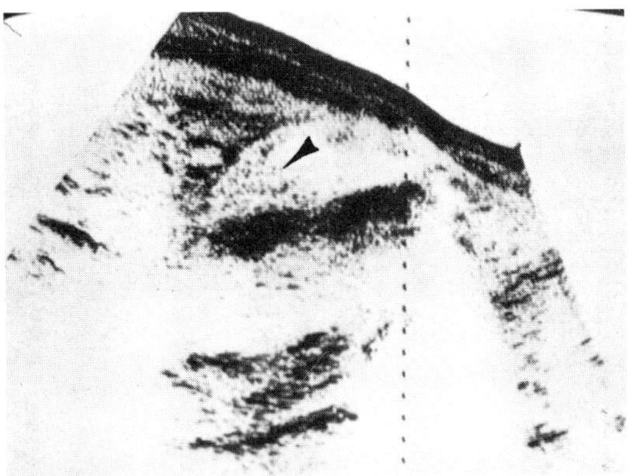

Figure 81–E. Longitudinal section through gallbladder. Echogenic bile.

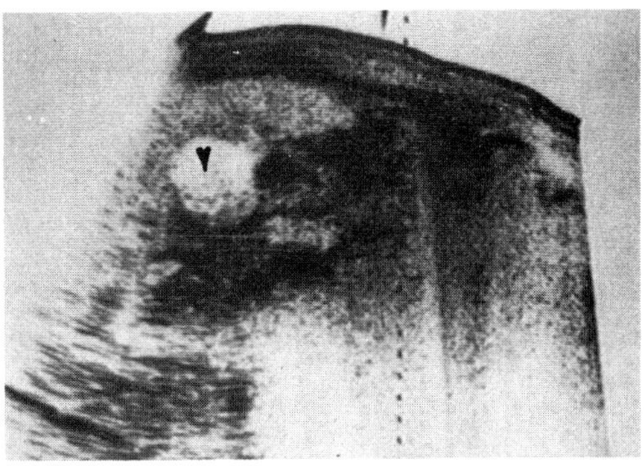

Figure 81–F. Transverse section +7. Echogenic bile.

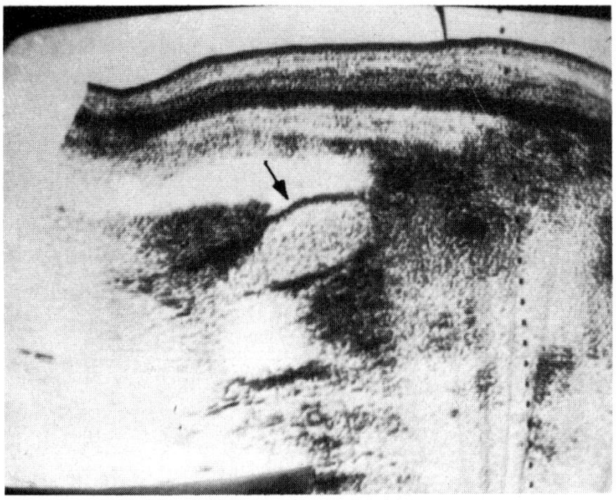

Figure 81–G. Longitudinal section 4R. Echogenic bile and ascites.

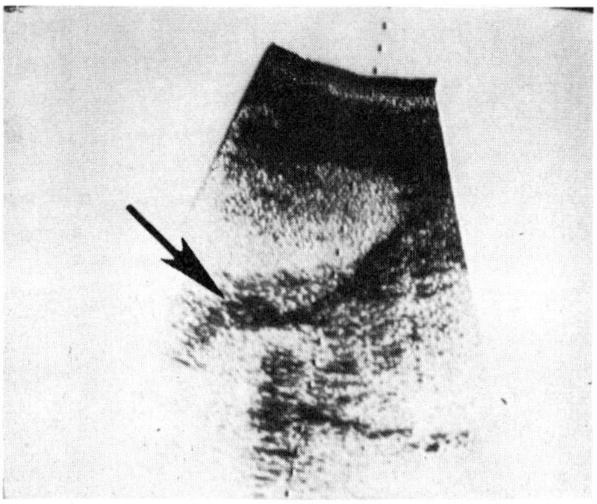

Figure 81–H. Longitudinal section 3R. Echogenic bile and gallstone.

REFERENCE

Goldberg BB: Gallbladder and bile ducts. Chapter 6 in *Abdominal Gray Scale Ultrasonography.* BB Goldberg, editor. John Wiley & Sons, New York (1977)

Case 82

CLINICAL INFORMATION

This 47-year-old woman has had trouble with her kidneys for several years. About 15 years ago she began having episodes of "cramps and bad back pain" on the left. These cramps occurred once or twice a year and usually lasted no more than six or eight hours. On one occasion she asked her doctor about them and was told that they were probably "kidney infections." About five years ago the attacks ceased, and she was well until two weeks before this scan was obtained. At that time she had another bout of pain and consulted her doctor, who ordered an IVP. The film showed a normal right kidney, but no function of the left kidney; in fact, the kidney could not even be identified with certainty.

The ultrasound technician has been having a terrible time with this scan; despite attempts in many different positions, this is the only image she can produce.

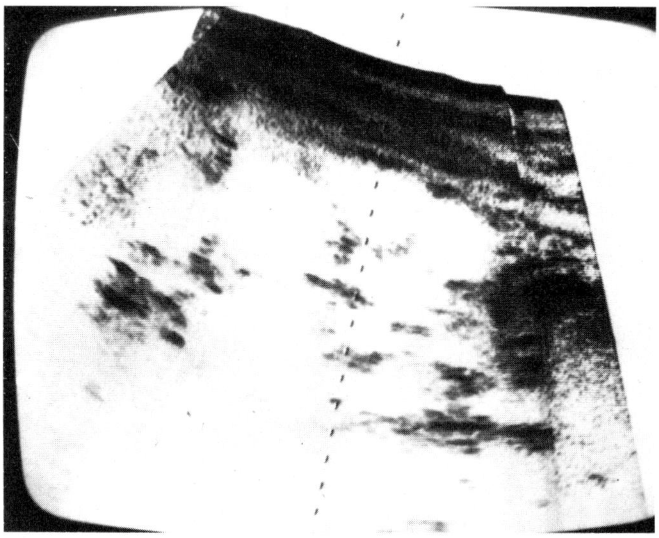

Figure 82–A. Prone scan of left kidney. **Longitudinal section 0L.**

DIAGNOSIS

Massive hydronephrosis with nearly total destruction of the kidney.

ULTRASOUND FINDINGS

The normal "doughnut or coffee bean" appearance of the kidney can never be identified in this case; instead we see multiple cystic masses in close continuity. Even careful scanning at 5 mm. intervals fails to turn up anything resembling renal cortex. There are two diagnoses to be considered in this situation — multiple renal cysts and massive hydronephrosis. Although we cannot distinguish between these two with certainty, we lean toward hydronephrosis, if for no other reason than that hydronephrosis is much more common than replacement of the entire kidney by multiple renal cysts. We have excluded polycystic disease by scanning the other kidney, which we found was normal. The left kidney was removed surgically and was just a large hydronephrotic sac. The ureter was only slightly dilated, but there was a stricture at the uretero-vesical junction. The cause of the stricture is still obscure; perhaps the previous episodes of cramps were related to the passage of renal stones, which led ultimately to the ureteral stricture.

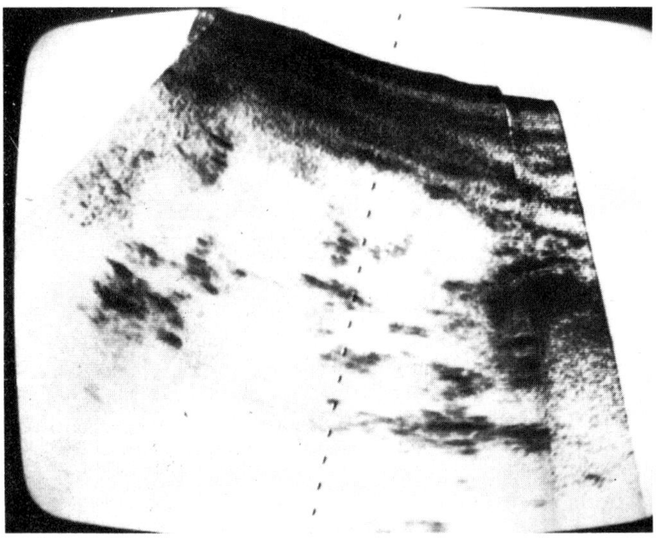

Figure 82–A. **Longitudinal section 0L.** Prone scan of left kidney. Massive hydronephrosis.

REFERENCES

Sanders RC, Bearman S: B-scan ultrasound in the diagnosis of hydronephrosis. Radiology 108:375 (1974)

Bearman SB, Hene PL, Sanders RC: Multicystic kidney: a sonographic approach. Radiology 118:685 (1976)

Case 83

CLINICAL INFORMATION

This 52-year-old man, a chronic alcoholic, was admitted to the hospital five days ago for upper GI bleeding. Endoscopy was performed and extensive bleeding gastritis was found; there were no esophageal or gastric varices nor were there any dominant ulcers. The bleeding subsided within 24 hours of conservative management (four units of blood), and his condition improved rapidly. His liver edge is just palpable and firm but the spleen is not enlarged and there are no other abnormalities on physical examination. The total bilirubin is 1.8, the SGOT and SGPT are slightly elevated, and the albumin is 2.9.

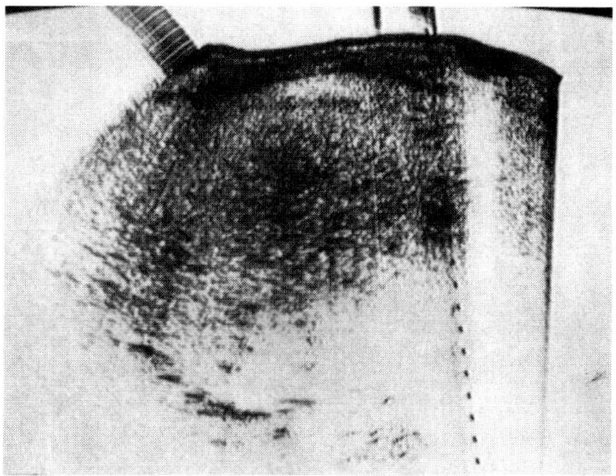

Figure 83–A. Transverse section +13.

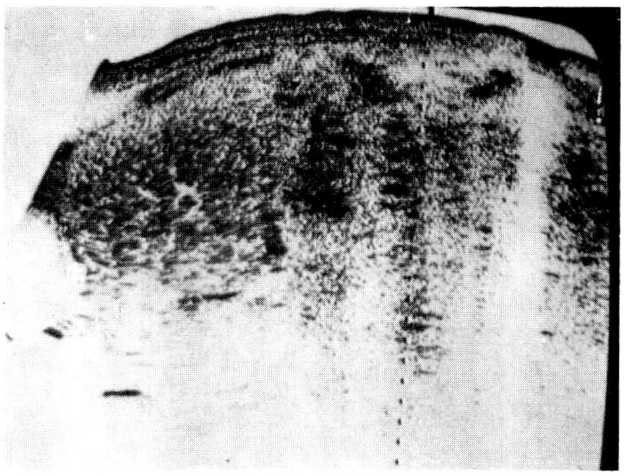

Figure 83–B. Longitudinal section 0L.

DIAGNOSIS

Cirrhosis.

ULTRASOUND FINDINGS

Figures 83–A and 83–B are cross-sectional and longitudinal scans through the right lobe of the liver — but where is it? The normal hepatic pattern of the slightly heterogeneous midlevel echoes (Figures 83–C and 83–D) has been replaced by multiple, dense, high-level echoes. In fact, this liver produces such strong echoes that we cannot get good sound beam penetration and cannot visualize the deep portions of the right lobe or the right kidney: it is difficult to separate the liver from the surrounding bowel. This echo pattern is associated with cirrhosis and represents one of the few times we can suggest a specific histologic diagnosis based on the echo pattern.

There is some disagreement as to why the cirrhotic liver produces such loud echoes. Some ultrasonographers claim it is due to increased collagen content but we have never found this explanation satisfactory from either a theoretic or an empiric point of view. Collagen has nearly the same density as intracellular fluid and should not give rise to very great acoustic impedance mismatches and should not generate unusually loud echoes. (This is verified by the normal appearance of the liver.) Moreover, as collagen content in the liver increases from cirrhosis, the density of the liver should become more homogeneous, and softer echoes should be produced. Other pathologic conditions which increase liver collagen, such as myelofibrosis, do not cause a noticeable increase in the intensity of the liver echoes. Although we have no proof, we suspect that it is the increasing *fat* content of the cirrhotic liver that causes the loud echoes. The clinical ultrasonographer does not have to be bothered by these questions of theory; he need only learn to recognize the pattern of greatly increased liver parenchymal echoes and he can confidently suggest the diagnosis of hepatic cirrhosis. (It is theoretically possible that other liver parenchymal diseases such as hepatitis and fatty infiltration could also produce this appearance. We have no experience with these conditions and cannot make further comment; it would seem, however, that the clinical setting would be sufficiently different to preclude confusion.)

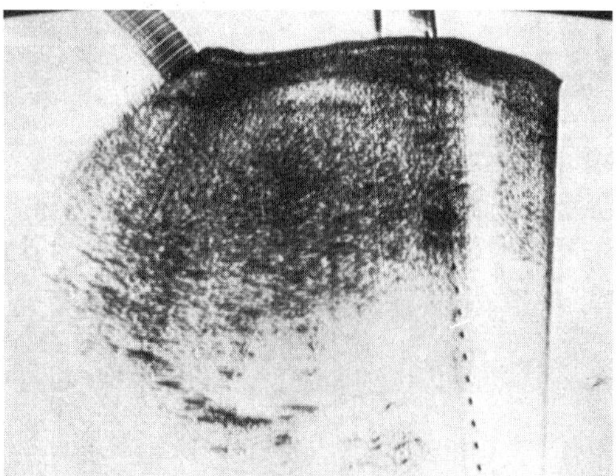

Figure 83–A. Transverse section +13. Cirrhosis.

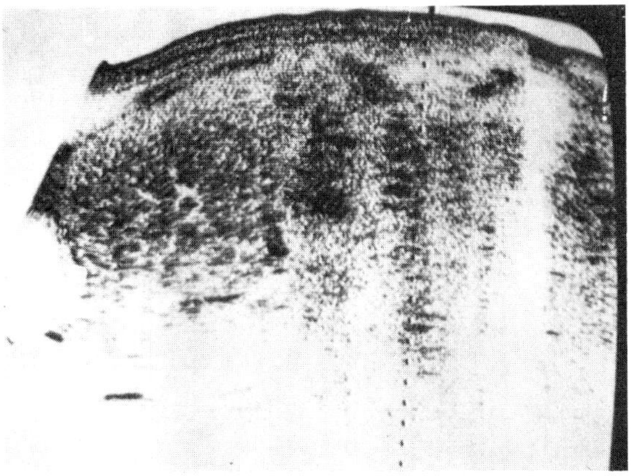

Figure 83–B. Longitudinal section 0L. Cirrhosis.

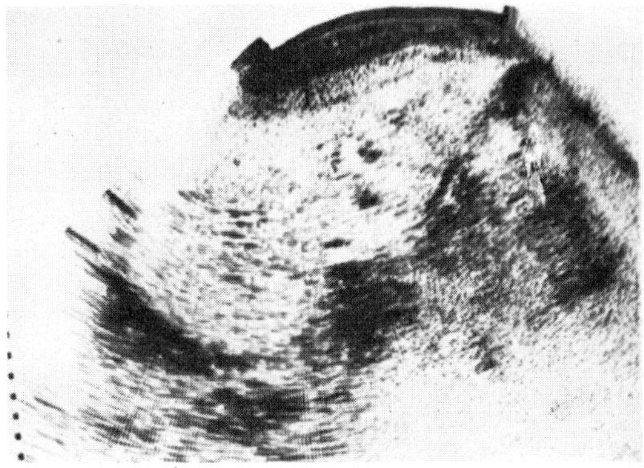

Figure 83–C. Transverse section +16. Normal liver.

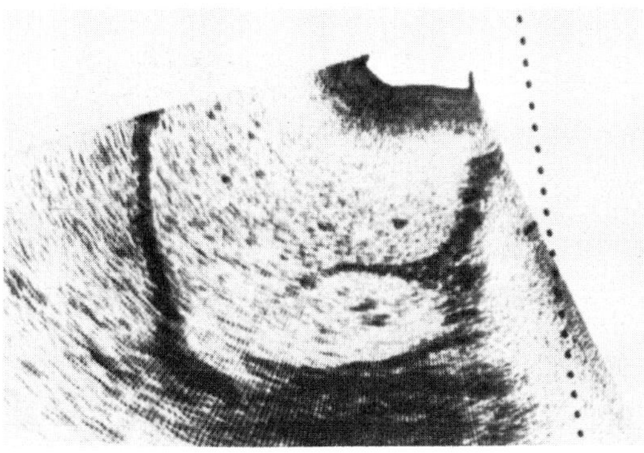

Figure 83–D. Longitudinal section 8R. Normal liver.

REFERENCE

Albertson KW, Leopold GR: Liver, Chapter 5, in *Abdominal Gray Scale Ultrasonography*, BB Goldberg, editor. John Wiley & Sons, New York (1977)

Case 84

CLINICAL INFORMATION

This 30-year-old man complains of having difficulty with urination. He describes hesitancy and often feels as though he has not been able to empty his bladder completely. Examination of his urine reveals several RBC's; an IVP is reported to be normal. Because complaints of obstructive uropathy are unusual for a patient of this age, the urologist has requested this ultrasound scan.

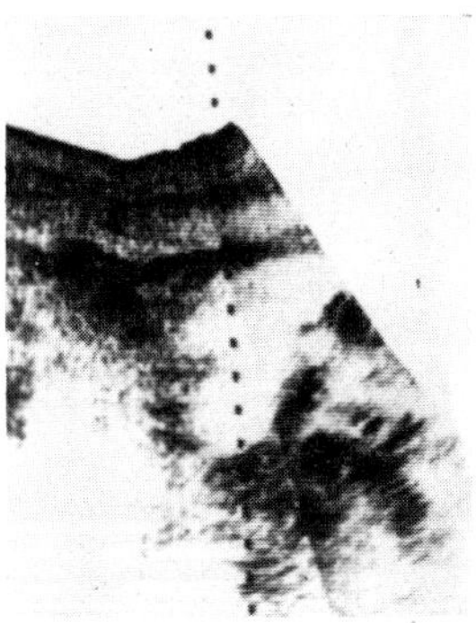

Figure 84–A. Longitudinal section 0L.

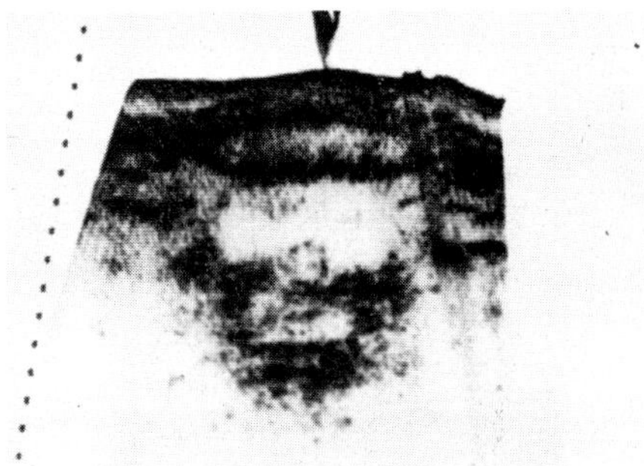

Figure 84–B. Transverse section −12.

DIAGNOSIS

Rhabdomyosarcoma of the prostate.

ULTRASOUND FINDINGS

If you correctly diagnosed this case, go to the head of the class. This is a rare tumor and there are no characteristic findings on the scan. We show it as an example of how well the prostate can be seen with a conventional B-mode scanner (without resorting to any specialized equipment). In both longitudinal and transverse scans a normal-sized prostate can be seen in Figures 84–A and 84–B (fat arrow); projecting upward from the gland there is a small nodule of tissue, which proved to be the neoplasm (small arrow). The body of the prostate and its lateral limits are well defined, particularly in the transverse scan. The referring physician was concerned about the possibility of spread of the malignancy to the pelvic wall, and, based on our scan, we can say that as yet there is none. The patient would not allow radical removal of his prostate. Both local recurrence and distant metastasis eventually did occur, and he died of widespread disease.

Figure 84–C is a longitudinal scan of the pelvis of a young man. The normal prostate is visible (arrow) at the base of the urinary bladder. Contrast this with the greatly enlarged prostate and dilated bladder in an elderly man with benign prostatic hypertrophy and bladder outlet obstruction (Figure 84–D). The urine-filled bladder provides an excellent acoustic window for demonstrating the prostate and allows us to estimate its size. The texture of the gland is variable, so that it is not usually possible to distinguish tumor from benign hypertrophy. In cases of known cancer, however, it is frequently possible to demonstrate local spread beyond the capsule.

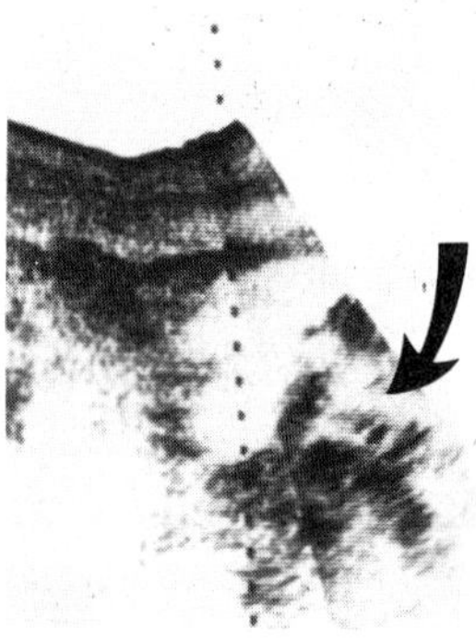

Figure 84–A. **Longitudinal section 0L.** Rhabdomyosarcoma of prostate.

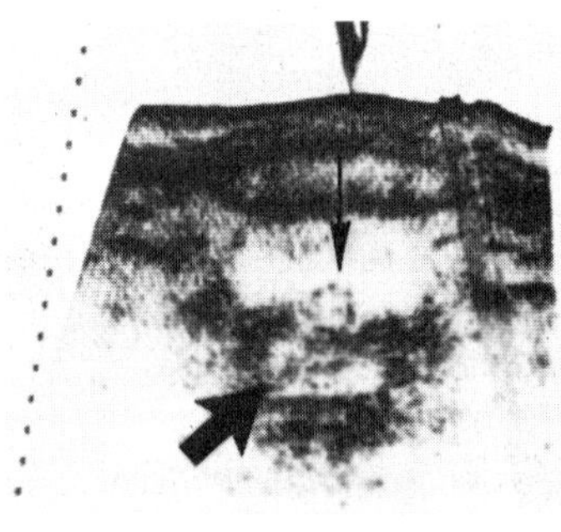

Figure 84–B. **Transverse section −12.** Rhabdomyosarcoma of prostate.

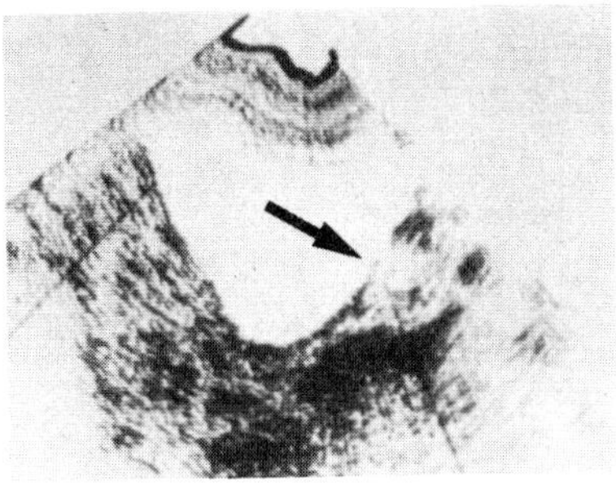

Figure 84–C.
Longitudinal section 0L.
Normal prostate.

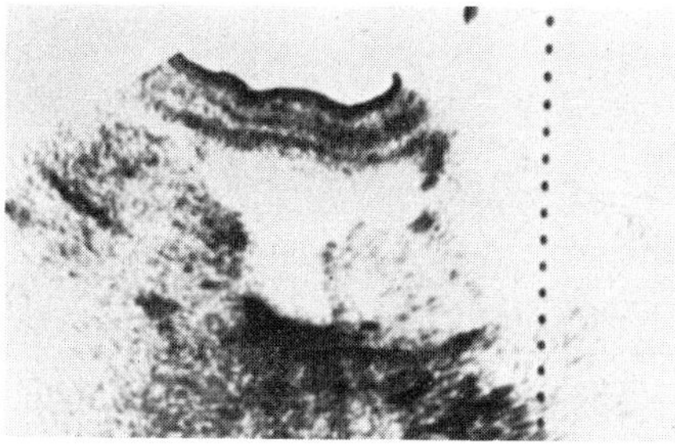

Figure 84–D.
Longitudinal section 0L.
Benign prostatic hypertrophy.

REFERENCE

Mittelstaedt CA, Gosink BB, Leopold GR: Gray scale patterns of pelvic disease in the male. Radiology 123:727 (1977)

Case 85

CLINICAL INFORMATION

This 72-year-old woman is under treatment for a chronic illness. (We will tell you what it is in a moment.) In the past few weeks she had had increasing abdominal aching and some nausea. Her doctor palpated a pulsatile mass in the midabdomen and suspects she may have an aortic aneurysm.

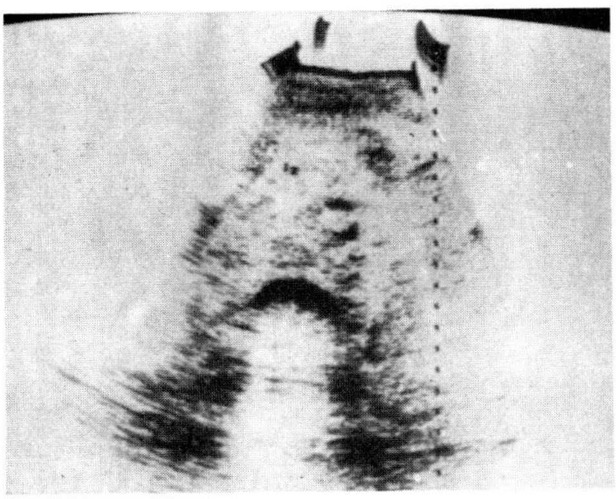

Figure 85–A. Transverse section +14.

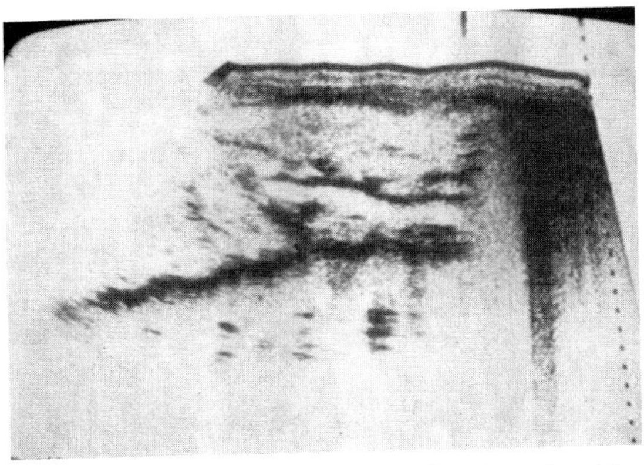

Figure 85–B. Longitudinal section 1L.

DIAGNOSIS

Retroperitoneal lymphadenopathy due to metastases from carcinoma of the breast.

ULTRASOUND FINDINGS

Figures 85–A and 85–B are scans through the aorta. The vessel is intrinsically normal but is encased in an echogenic mass (arrows). The cross-section has the appearance of a "doughnut," which is almost pathognomonic of enlarged retroperitoneal and para-aortic lymph nodes. Para-aortic hematoma can also cause a "doughnut," but this is a very uncommon lesion which has a unique clinical setting. For all practical purposes two conditions give this picture: lymphoma and metastatic disease.

Usually the clinical information will allow us to make this differential. Our patient's "chronic illness" was carcinoma of the breast; but any metastatic tumor can give an identical appearance. Figure 85–C is from a patient with carcinoma of the lung; Figures 85–D and 85–E are teratocarcinomas of the testicle; and Figure 85–F is endometrial carcinoma.

Figure 85–E does not have the characteristic "doughnut" appearance, since the tumor in this patient did not infiltrate between the aorta and spine. At one time we thought this sparing of the retroaortic space was characteristic of lymphoma but we now know that both metastases and lymphoma can give this appearance, even though the retroaortic "sparing" is more common in lymphoma.

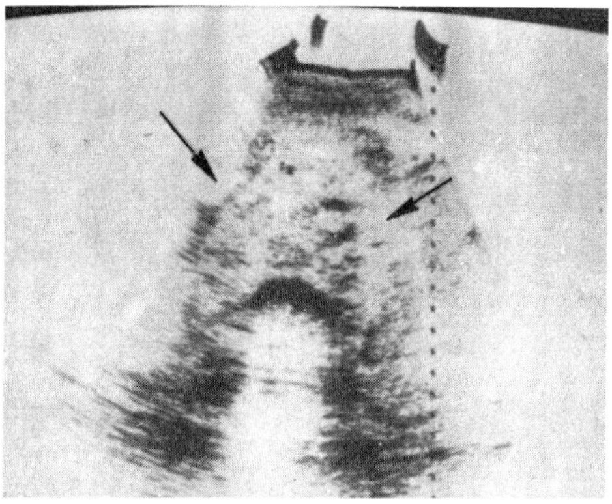

Figure 85–A. Transverse section +14. Lymph node metastases.

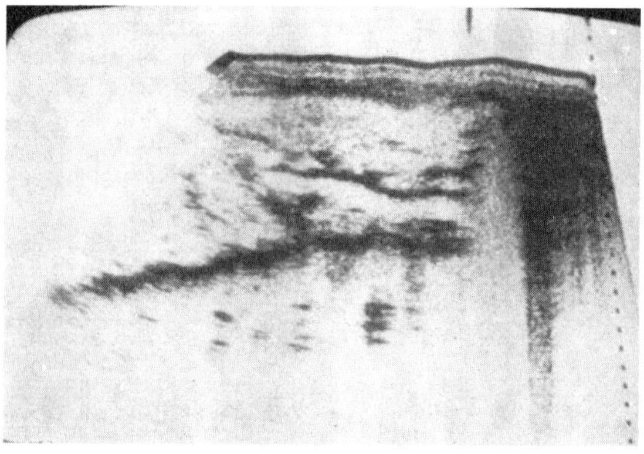

Figure 85–B. Longitudinal section 1L. Lymph node metastases.

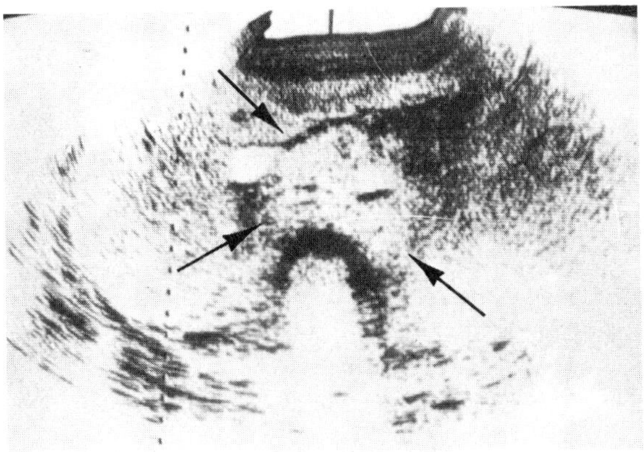

Figure 85–C. Transverse section +13. Lymph node metastases (lung).

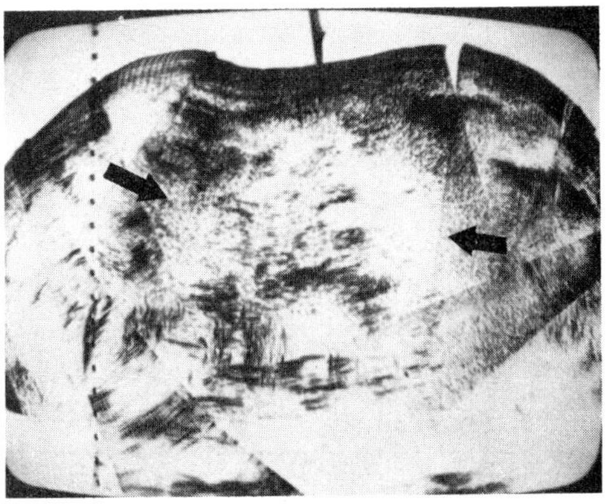

Figure 85–D. Transverse section +12. Lymph node metastases (testicle).

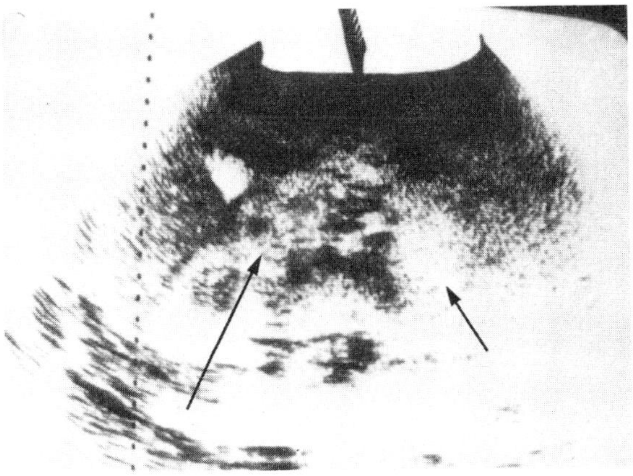

Figure 85–E. Transverse section +10. Lymph node metastases (testicle).

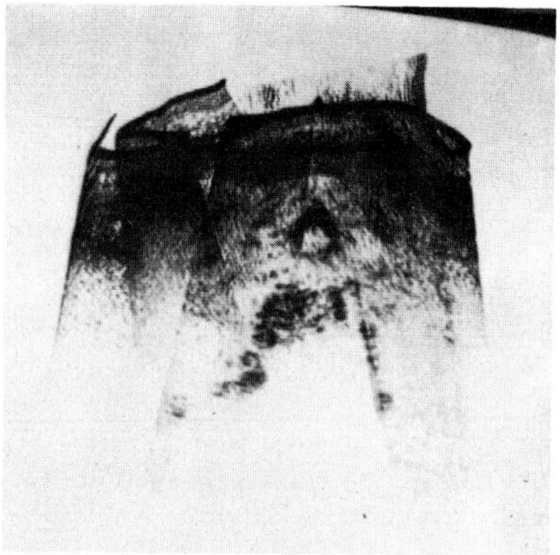

Figure 85–F. Transverse section +3. Lymph node metastases (endometrium).

REFERENCE

Taylor KJW: Reticuloendothelial system. Chapter 8, in *Abdominal Gray Scale Ultrasonography*. BB Goldberg, editor. John Wiley & Sons, New York (1977)

Case 86

CLINICAL INFORMATION

This 24-year-old woman is 18 weeks pregnant by dates and is having a routine ultrasound examination for gestational dating, placental localization, and so on. There is a strange lump on the anterior wall of the uterus (arrow).

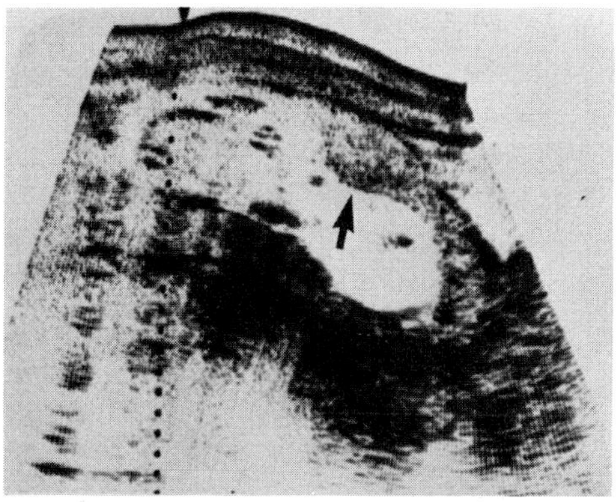

Figure 86–A. Longitudinal section 2R.

DIAGNOSIS

Unsoftened portion of the uterus (focal uterine contraction).

ULTRASOUND FINDINGS

Focal thickenings of the myometrium are frequently seen in the early stages of pregnancy. They are almost universal in the first trimester and decrease in frequency until they are only rarely seen after 20 weeks. No one is quite sure what causes these areas: some ultrasonographers have suggested they represent focal areas of contraction, while others believe they are portions of the uterus that have not yet stretched to accommodate the full size of the growing fetus. There is no official name for this observation but recently the term "unsoftened portion of the uterus" (UPU) has been coined. This is as good a description as any.

What is the significance of a UPU? None, clinically. It is a normal finding in the first 20 weeks of gestation and is more of a nuisance than anything else, because it can be confused with the placenta or a uterine fibroid. Figure 86–B shows a UPU (curved arrow) and a posterior placenta (open arrow). Notice that the UPU has exactly the same echo texture as the adjacent myometrium, while the placenta produces louder echoes. Figure 86–C shows a leiomyoma (fibroid) of the uterus. We can distinguish this from a UPU by observing the following: (1) it has a heterogeneous echo texture that is not the same as the adjacent uterine wall; (2) it is too large for a UPU at this stage of gestation, and (3) it has not decreased in size as the pregnancy progressed (in fact, it has been growing).

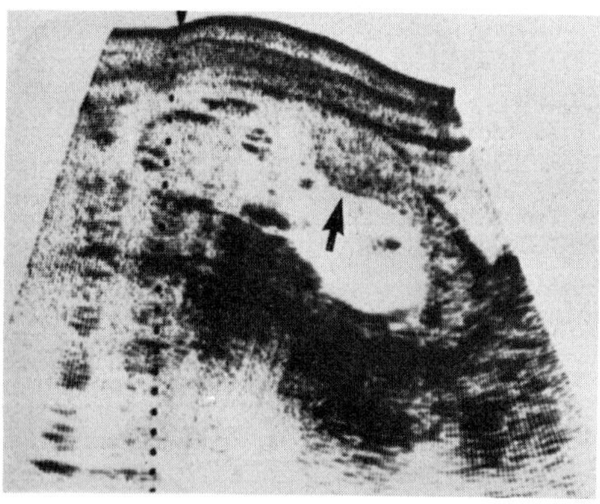

Figure 86–A. Longitudinal section 2R. Unsoftened portion of uterus.

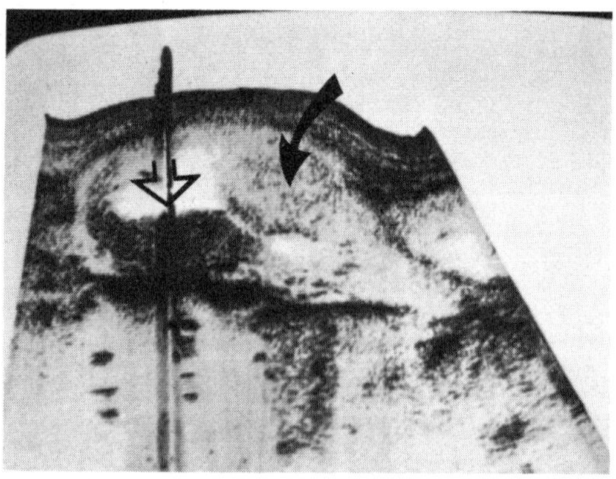

Figure 86–B. Longitudinal section 1L. Unsoftened portion of uterus with posterior placenta.

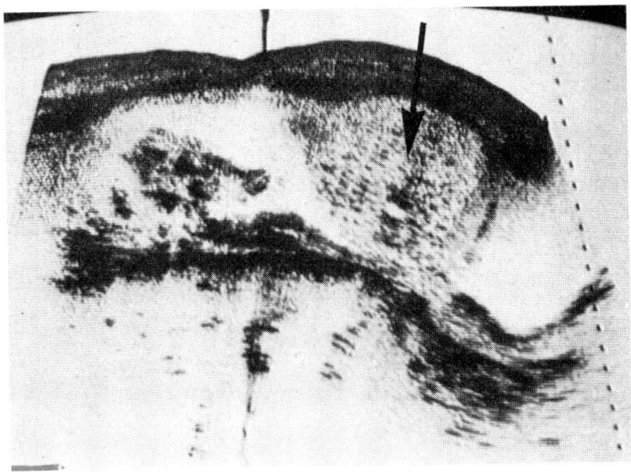

Figure 86–C. Longitudinal section 0L. Leiomyoma.

REFERENCE

Sample WF: The unsoftened portion of the uterus. Radiology 126:227 (1978)

Case 87

CLINICAL INFORMATION

This 81-year-old woman complains of "stomach" pains. Over the years she has had "stomach troubles," which various doctors have told her were due to ulcers, hiatus hernia, peptic stomach, excess acid, and gallstones. Some or all of these diagnoses may have been correct, inasmuch as many of the treatment recommendations she has received over the years have given her relief. Twelve years ago she had her gallbladder removed, and since that time she has been feeling well with only infrequent bouts of "indigestion." Over the past couple of months her stomach has been acting up again and "aching." The pain has dulled her appetite, and she has lost a little weight and is beginning to feel "weak." For the past week she has been nauseated and is unable to keep food down. She has not been able to take her "heartpills" (Digoxin) either and is now slightly short of breath and unable to lie in bed without several pillows for support. Her physical examination reveals (in addition to findings of chronic congestive heart failure) mild scleral icterus. No abdominal masses are palpable. Her hematocrit is 29, and the conjugated bilirubin 3.6; but all other laboratory tests are normal and a single stool specimen tested for occult blood is negative.

The ultrasound technician thinks this scan is abnormal: There is a strange "mass" in the left upper quadrant. Figure 87–A is a cross-section through the "mass" (arrow) and Figure 87–B is a longitudinal scan. Your diagnosis?

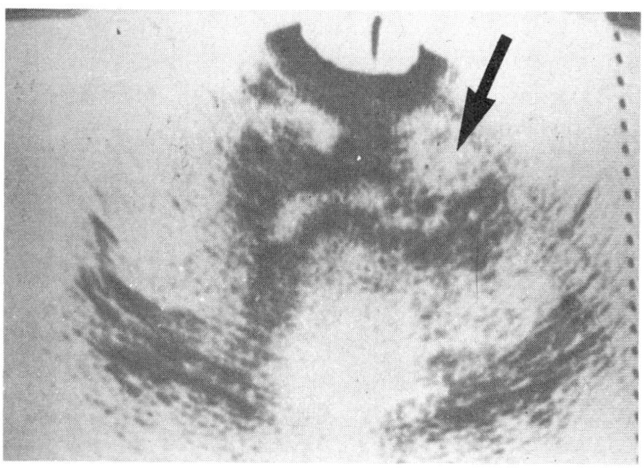

Figure 87–A. Transverse section +10.

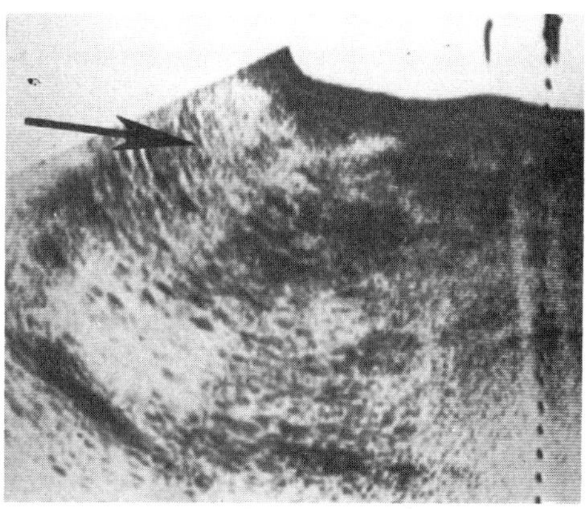

Figure 87–B. Longitudinal section 4L.

DIAGNOSIS

Adenocarcinoma of the stomach.

ULTRASOUND FINDINGS

Here is an example of a left upper quadrant echo-producing mass, this time related to disease of the stomach. On the transverse scan we can see that it contains echoes and, therefore, is a solid (more properly, not fluid-filled) lesion, which is not part of the liver. Furthermore, it is unlikely that it is associated with the pancreas, because the mass is located somewhat anteriorly. The organ we expect to find in this position is the stomach. Could this be food in an otherwise normal stomach? We asked the patient, and she says that she has had nothing to eat for three days. It seems, therefore, more likely that we are dealing with a stomach tumor. (We would like to verify our diagnosis by having the patient fill her stomach with water and then repeat the scan in the upright position; unfortunately, she is too sick for this maneuver.)

Review of the transverse scan shows a second "mass" on the right (Figure 87–A), small arrow). The longitudinal scan made slightly to the right of the midline (Figure 87–C) shows an echo-free area, which indicates that this mass contains some fluid. We know that the patient's gallbladder has been removed, so we can eliminate it as a source of the ultrasound findings. She is jaundiced; how about a dilated common bile duct? Unlikely; the position and course are wrong; there are no dilated intrahepatic ducts, and her bilirubin is not that much elevated. Could this be a pancreatic pseudocyst? Perhaps, but in that case we would expect that the amylase would be elevated and that there would be a history of pancreatic disease. We want to stress, however, that a pseudocyst cannot be excluded. How about an abscess? Possible, but again the clinical setting and the laboratory data are not suggestive. Maybe this is part of the GI tract — a portion of the stomach or the duodenum. Not a bad bet, but not a diagnosis that we can confirm by ultrasound scan alone. We cannot come to a specific diagnosis on the basis of the ultrasound scan in this case; all we can do is describe the anatomic findings and recommend further studies. Because we suspect a gastric tumor, we recommend an upper GI series.

The upper GI series (Figure 87–D) confirms our suspicion of a cancer in the body of the stomach (arrow) and shows that the fluid-filled mass on the right is probably the gastric antrum. Figures 87–E and 87–F show a different patient with the same diagnosis — adenocarcinoma of the stomach (arrows).

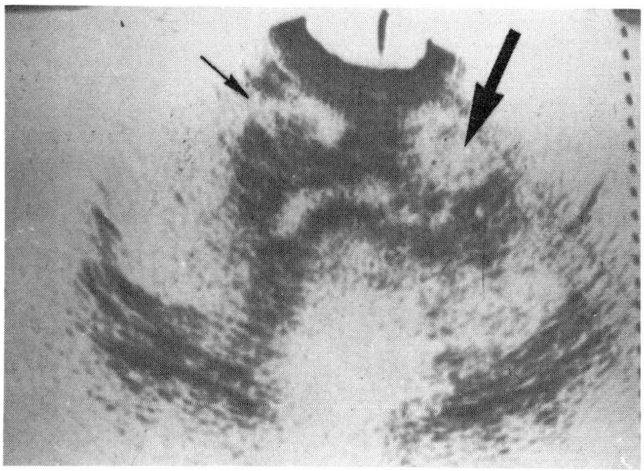

Figure 87–A. Transverse section +10. Carcinoma of the stomach.

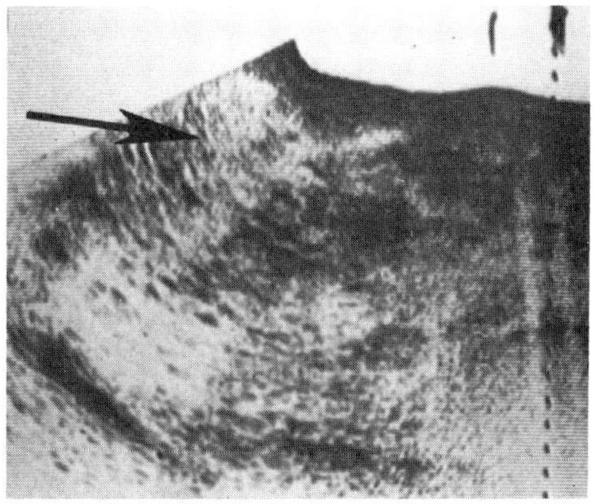

Figure 87–B. Longitudinal section 4L. Carcinoma of the stomach.

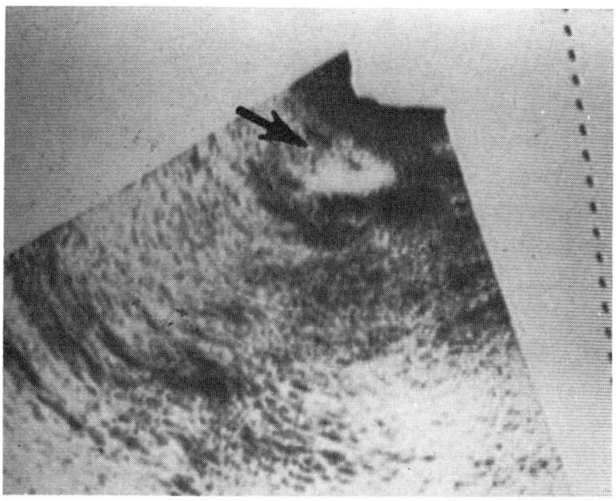

Figure 87–C. Longitudinal section 3R. Gastric antrum (fluid-filled).

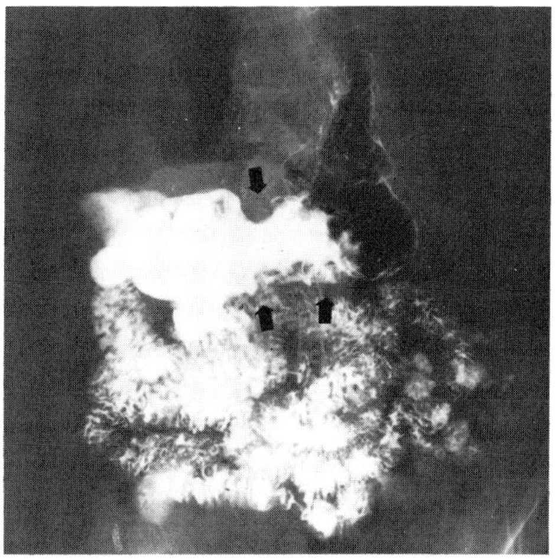

Figure 87–D. Carcinoma of the stomach. Same patient as in Figures 87–A, B, and C.

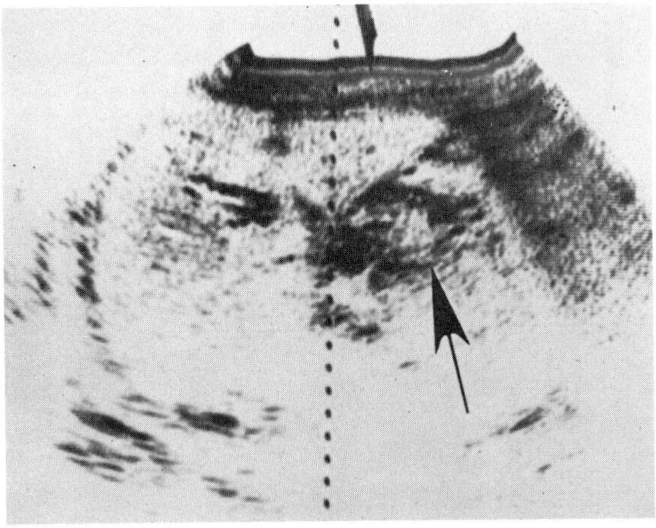

Figure 87–E.
Transverse section +12.5. Carcinoma of the stomach.

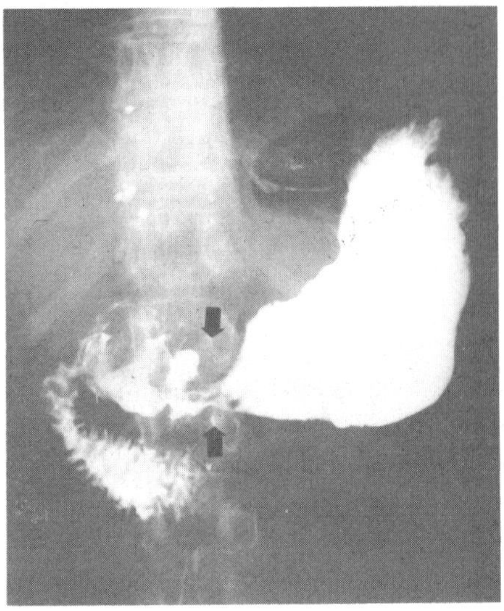

Figure 87–F.
Carcinoma of the stomach. Upper gastrointestinal film of patient in Figure 87–D.

REFERENCES

Mascattello VJ, Carrera GF, Teele RL, et al: The ultrasonic demonstration of gastric lesions. J Clin Ultra 5:383 (1977)

Walls W: The evaluation of malignant gastric neoplasm by ultrasonic B scanning. Radiology 119:159 (1976)

Case 88

CLINICAL INFORMATION

This 50-year-old man has a cough and fever. His doctor obtained a chest radiograph which revealed a partially collapsed and consolidated left upper lobe. When this area did not improve on antibiotics, tomography was performed and a deformity of the upper lobe bronchus was identified. Bronchoscopy and biopsy confirmed the diagnosis of bronchogenic carcinoma. Metastatic evaluation is underway and a liver scan is obtained (Figure 88–A); the anterior and posterior projections are normal but the right lateral view is equivocal. Multiple ultrasound sections of the liver are all similar; Figure 88–B is a typical example.

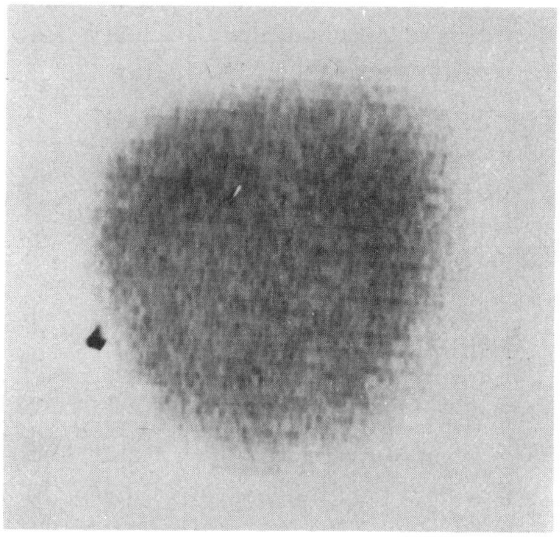

Figure 88–A.
Tc-sulfur colloid liver scan; right lateral view.

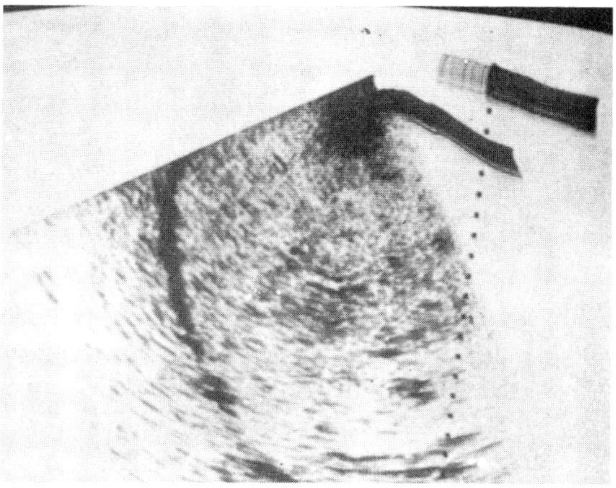

Figure 88–B. Longitudinal section 8R.

DIAGNOSIS

Diffuse liver metastases.

ULTRASOUND FINDINGS

We think that the ultrasound scan is equivocal: there may be more heterogeneity in the liver than one would normally expect, but this is a highly subjective reading. Figure 88–C is a similar section through a normal liver; is there any real difference in the liver texture? Maybe, and maybe not. This patient subsequently had a closed liver biopsy that demonstrated extensive diffuse metastatic disease in the liver.

When the liver is invaded by large focal metastases it is usually quite easy to make an accurate ultrasound diagnosis (see discussion of Case 52); but small, diffusely distributed metastases are a much different problem. The alteration of the normal liver texture may be so subtle as to defy differentiation from normal. Figures 88–D, 88–E, and 88–F are all scans of patients with subtle liver metastases. Figure 88–G is a scan of a normal liver; how much difference is there between this scan and the one in Figure 88–F? This is a tough problem, particularly as we usually do not know the final diagnosis at the time the scan is performed. (In retrospect it may be quite easy to recognize subtle metastases, but, unfortunately, retrospective diagnoses are not of great value to the patient.)

There is controversy among experienced ultrasonographers about the accuracy of ultrasound in diagnosing liver metastases. A few feel that this is the method of choice and claim a diagnostic accuracy approaching 100 per cent. The great majority of ultrasound laboratories, including ours, do not have such impressive results. There is no argument that when the liver contains multiple focal areas of echo heterogeneity there are metastases present. But what about the patient whose liver appears normal or equivocal, as in this case? Can we say that there are no metastases present? If we do, we will be wrong a significant percentage of the time! The exact figure depends upon the make-up of the population we study. If there are a large number of normal patients in the population, our accuracy rate will be better, since we will get all these diagnoses correct by default.

For example, suppose we examine 100 patients for liver metastases: assume that 98 of these are normal and two have metastases. The normal patients all have a uniform liver texture and we correctly diagnose them as negative. Let us further imagine that one of the patients with metastases has an unequivocally abnormal scan, which we correctly diagnose as positive, while the other patient has small diffuse metastases that do not alter the texture and we make an incorrect diagnosis of negative. Now let us look at our "accuracy." We correctly diagnosed 99 of our patients, so our overall accuracy is a whopping 99 per cent. Overall accuracy is not worth much to either the patient or referring doctor; what they want to know is how good we are with patients who actually have metastatic disease. Here our accuracy is only 50 per cent, hardly an inspiring figure.

The point we mean to make here is that we must interpret reported accuracy figures with care. Overall accuracy is a rather meaningless figure and is best disregarded; false positive and false negative rates are the numbers of concern. Even these are of limited value unless we know the patient population. We already know that big liver metastases are easy to diagnose, so that any study which has a high proportion of such patients will have a high accuracy. But how about patients such as the one in this case? If there are a large number of such patients in the study the accuracy will obviously be lower.

We suspect that differences in patient population, diagnostic criteria, and adequacy of follow-up account for the variability in results from different ultrasound laboratories. We think that the best way for the average clinical ultrasonographer to proceed is as follows: If the scan shows unequivocal evidence of metastases, a positive diagnosis can be made with a high degree of certainty; if the scan is equivocal or normal, it is best to proceed with caution and say that although no metastases are seen they cannot be excluded either.

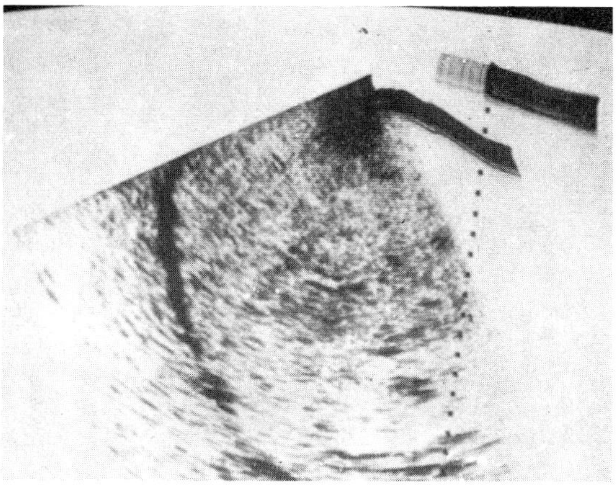

Figure 88–B. Longitudinal section 8R. Diffuse liver metastases.

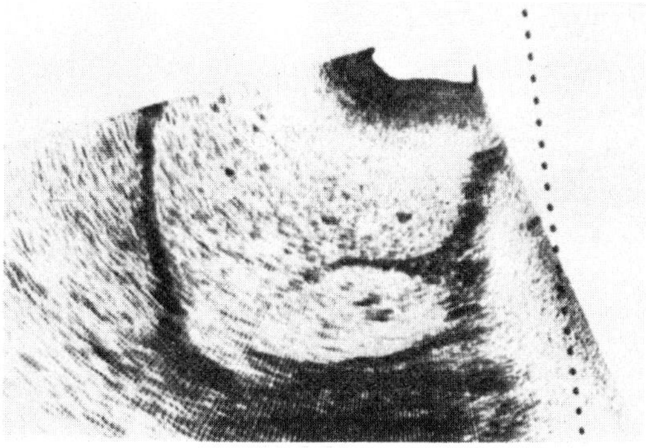

Figure 88–C. Longitudinal section 8R. Normal liver.

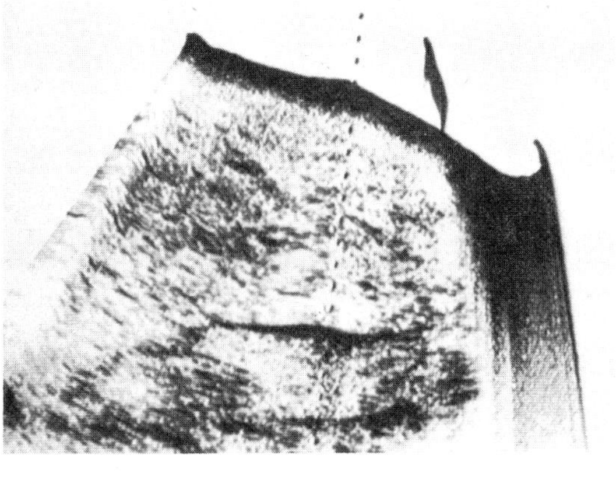

Figure 88–D. Longitudinal section 7R. Diffuse liver metastases.

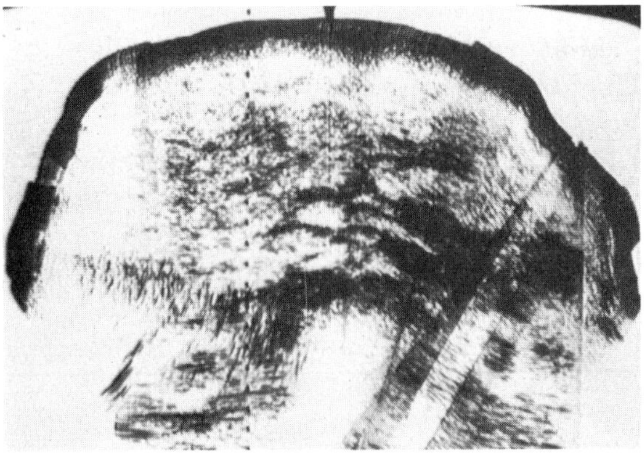

Figure 88–E. Transverse section +14. Diffuse liver metastases.

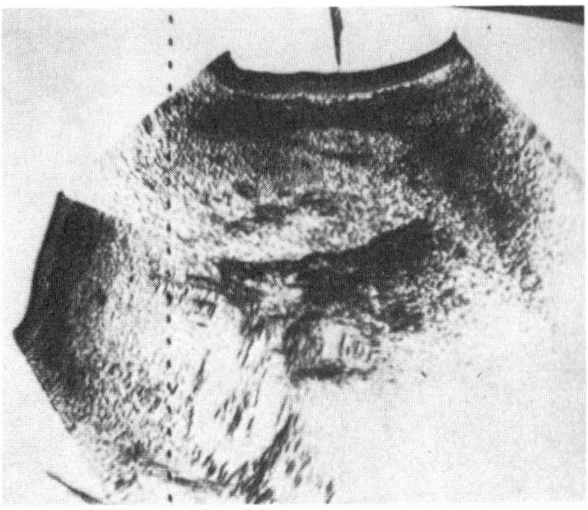

Figure 88–F. Transverse section +10. Diffuse liver metastases.

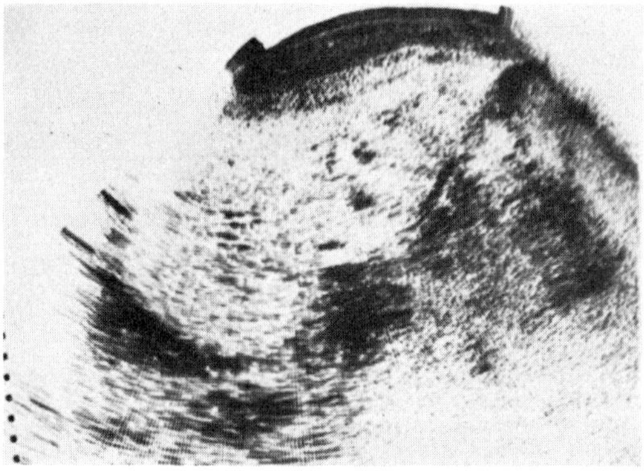

Figure 88–G. Transverse section +14. Normal liver.

REFERENCES

Scheible W, Gosink BB, Leopold GR: Gray scale echographic patterns of hepatic metastatic disease. Am J Radiol 129:983 (1977)

Green B, Bree RL, Goldstein HM, et al: Gray scale ultrasound evaluation of hepatic neoplasms. Radiology 124:203 (1977)

Zatz LM, Gouldin JA, Hanley GA: A comparison of gray scale ultrasound and radionuclide for the detection of focal hepatic lesions. J Clin Ultra 5:178 (1977)

Case 89

CLINICAL INFORMATION

This 38-year-old woman is 26 weeks pregnant by dates, but her uterus is much larger, nearer to 36 weeks in size. Figures 89–A and 89–B are longitudinal scans through the left and right halves of the uterus and Figures 89–C and 89–D are cross-sections through the fundus and just above the cervix.

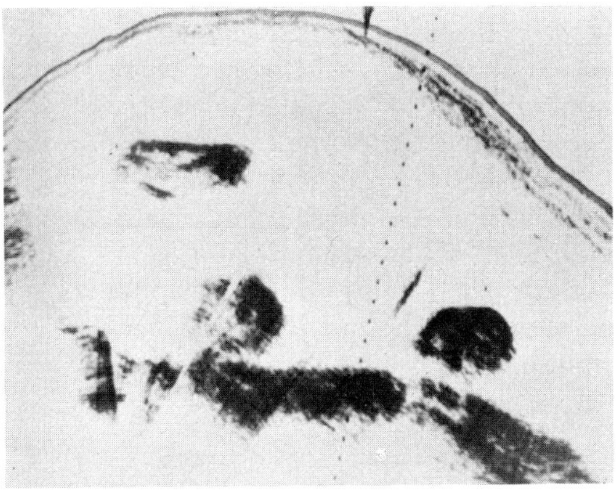

Figure 89–A. Longitudinal section 3L.

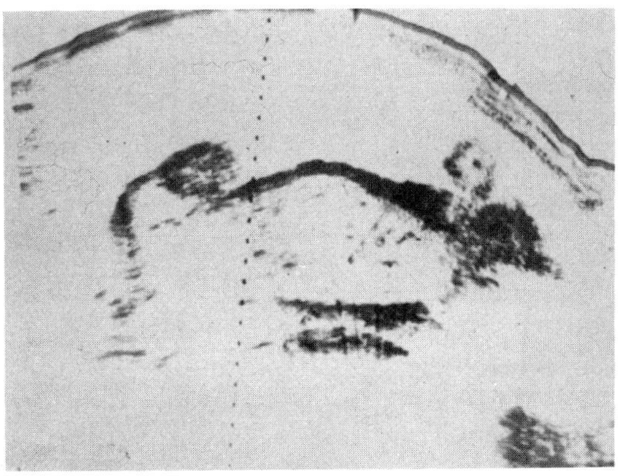

Figure 89–B. Longitudinal section 3R.

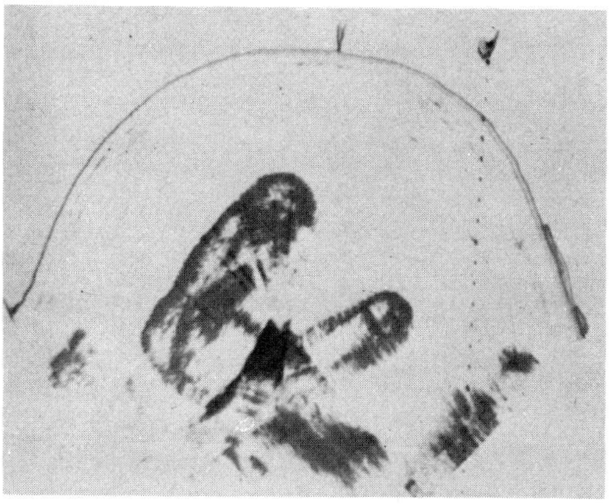

Figure 89–C. Transverse section 0.

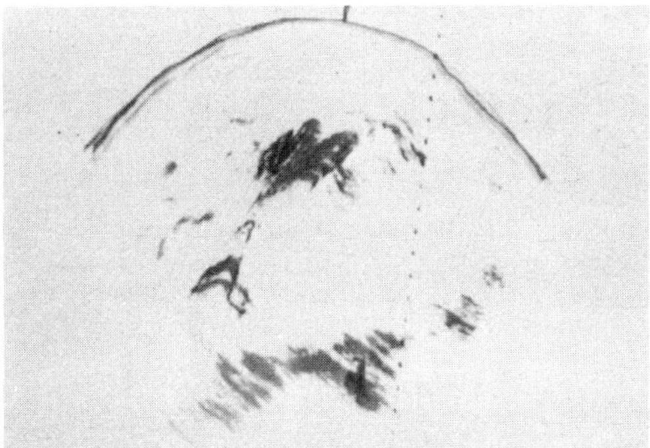

Figure 89–D. Transverse section –8.

DIAGNOSIS

Anencephaly (with associated polyhydramnios).

ULTRASOUND FINDINGS

These scans cover the entire uterus and there is no evidence of a fetal head anywhere. In addition, the ratio of amniotic fluid to fetus is much greater than normal (even allowing for the fact there is no fetal head) and hence there is polyhydramnios, a condition often associated with anencephaly. This is a simple diagnosis — just make sure you are correct before getting everyone excited. Search carefully in all parts of the uterine cavity — the head may be way down deep in the pelvis or way off to one side! A word of caution: if you suspect anencephaly from the ultrasound findings, you should suggest that this be confirmed by a radiograph of the uterus, especially if the obstetrician plans to induce labor early. This added maneuver is especially worthwhile for the neophyte ultrasonographer. As you acquire confidence, you will not feel the need of this back-up because the normal fetal head is so easily recognized by the experienced scanner.

A technical note: this scan was done at night with an analog scan converter. There was not time to let the apparatus warm up adequately, so there are very few low- and medium-level grays on the film. We do not leave our scanner turned on all night, as this is hard on the scan converter; however, it is important to turn it on at least an hour before the first case in the morning so as to avoid this problem. Also, when you are doing an emergency case, you can take advantage of the usual delays in any hospital by having the technologist turn on the instrument before the patient is called to your department — simple enough, but it does take remembering. The digital scan converters should not require the long warm-up.

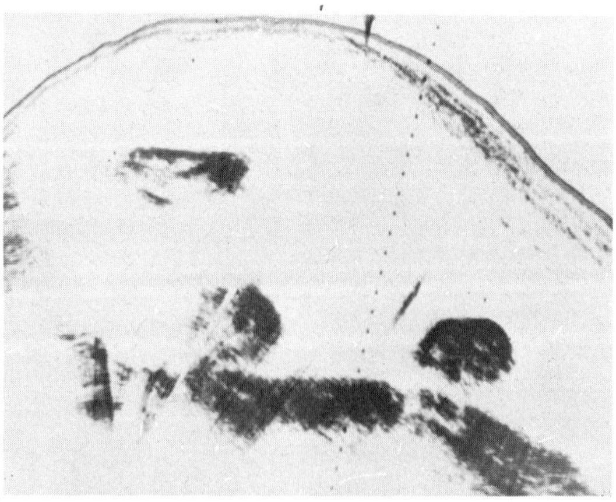

Figure 89–A. Anencephaly. **Longitudinal section 3L.**

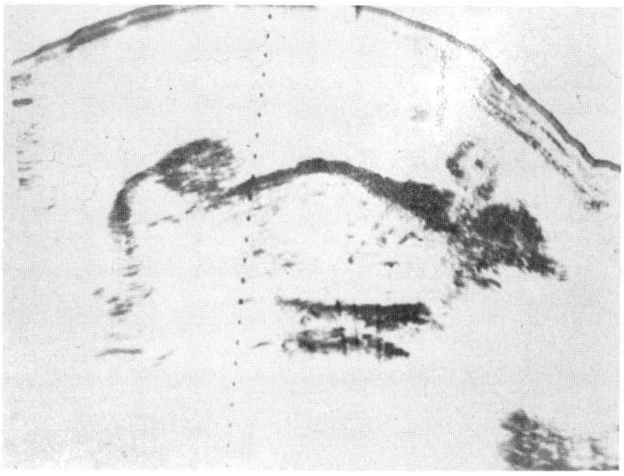

Figure 89–B. Longitudinal section 3R. Anencephaly.

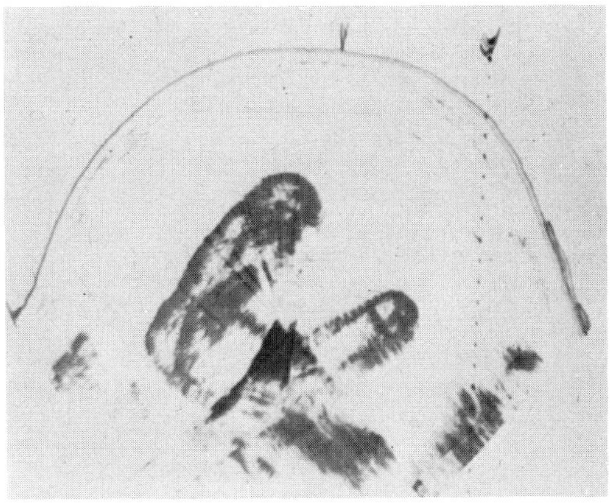

Figure 89–C. Transverse section 0. Anencephaly.

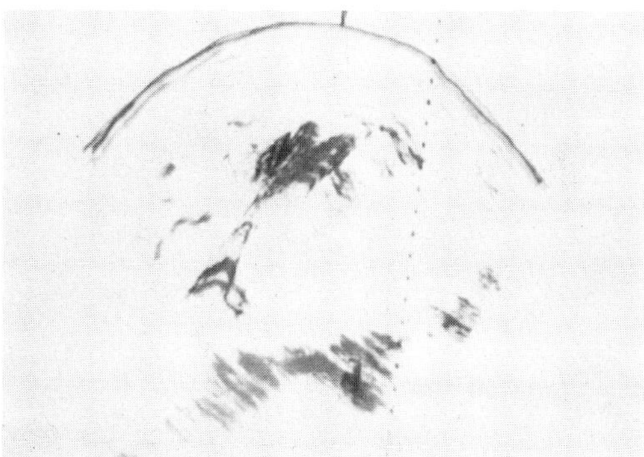

Figure 89–D. Transverse section –8. Anencephaly.

REFERENCE

Campbell S, Johnstone FD, Holt EM, et al: Anencephaly: early ultrasonic diagnosis and advice management. Lancet 2:1226 (1972)

Case 90

CLINICAL INFORMATION

This 77-year-old man has a nagging pain in his back and "bad stomach aches." His past health has not been good — he admits to drinking too much and suffering some of the consequences. In the past he has had trouble with ulcers, and two years ago he "vomited blood." He was in the hospital for three weeks on that occasion and had an extensive evaluation. Although his liver was slightly enlarged, the liver function tests were normal, and he was felt to have only mild cirrhosis. Endoscopy showed that the GI bleeding was due to gastritis; there were no varices nor was there an active duodenal ulcer. He was treated medically, the bleeding stopped, and he has been stable in the interim. For the past three days he has had a new type of abdominal pain, which has been steady and increasing in severity. This pain seems to run down his abdominal wall and into his back. He has been nauseated but has not vomited. Physical examination revealed a "6 cm. pulsatile mass in the upper abdomen, probably an aneurysm." He is, therefore, sent for ultrasound of the aorta. No laboratory data are available at the time of the examination. The ultrasound scans show a slightly ectatic aorta but no evidence of an aneurysm. There is, however, a large abdominal mass. What do you think it is?

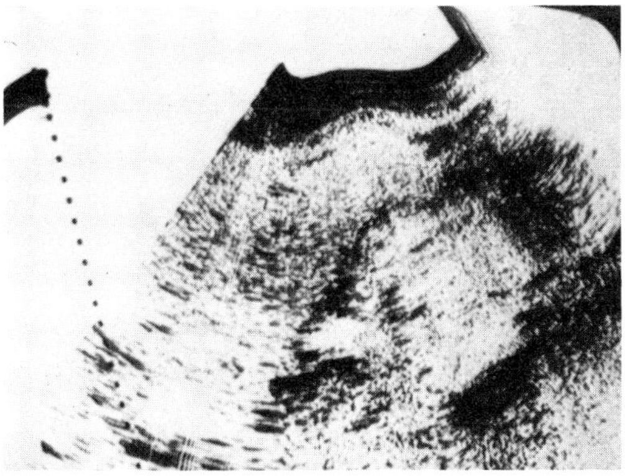

Figure 90–A. Transverse section +15.

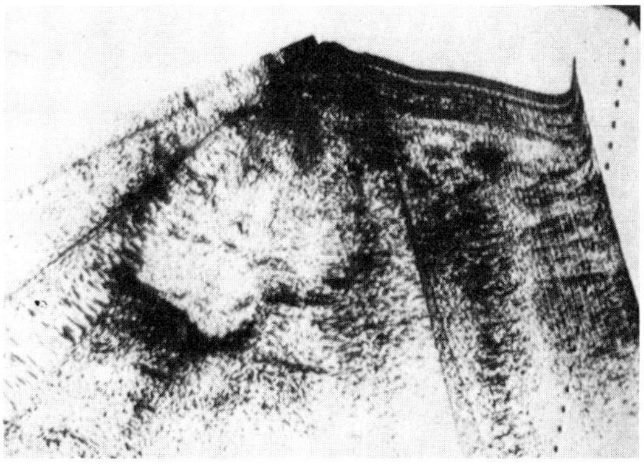

Figure 90–B. Longitudinal section 5L.

DIAGNOSIS

Massive focal pancreatitis.

ULTRASOUND FINDINGS

This is quite an unusual case — an atypical presentation of a common disease. Although the mass encountered was subsequently shown to be a massively enlarged pancreas secondary to acute pancreatitis, there is nothing about the ultrasound appearance alone that would allow us to reach this conclusion. The mass itself is quite clearly seen; its location is excellent for a lesion arising from the tail of the pancreas, but it could just as well represent a nonpancreatic tumor of the retroperitoneal area. The irregular borders and fairly high-intensity echoes suggest neoplasm rather than pancreatitis, and this should be the first diagnostic consideration. We felt this most likely represented carcinoma of the pancreas and recommended aspiration biopsy for verification.

The patient's doctor decided to defer this procedure (and all other diagnostic evaluation as well) because the patient had a very high amylase and the clinical setting was compatible with pancreatitis. He reasoned that if indeed this were a tumor it would be quite advanced and basically untreatable. The patient's subsequent clinical course proved him correct; three months after these scans were made the amylase had returned to normal and the pain and mass were gone. Nine months after the scans the patient returned with a dissecting abdominal aortic aneurysm, which was treated successfully with a graft. At the time of the surgery, the pancreas and left upper abdominal area were carefully examined, and no mass could be found.

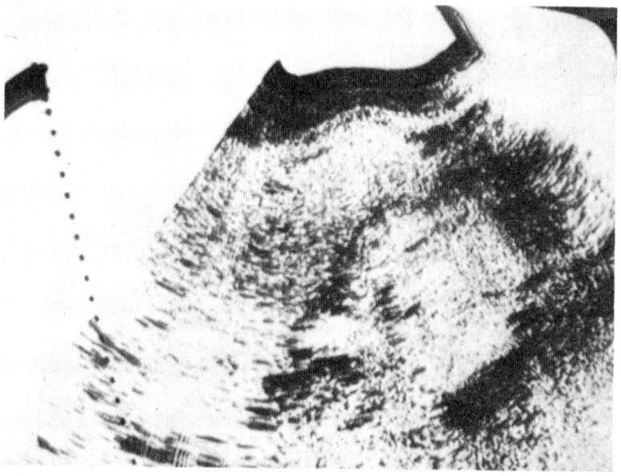

Figure 90–A. **Transverse section +15.** Focal pancreatitis.

Figure 90–B. **Longitudinal section 5L.** Focal pancreatitis.

REFERENCES

Doust BD: Ultrasonic examination of the pancreas. Rad Clin N Am 13:467 (1975)

Perlmutter GS: Pancreas, Chapter 7, in *Abdominal Gray Scale Ultrasonography*, BB Goldberg, editor. John Wiley & Sons, New York (1977)

Case 91

CLINICAL INFORMATION

This young man was first seen by a urologist because of an enlargement in the left side of his scrotum. The patient has no sexual dysfunction (he has fathered a child), but reports that he has begun to experience aching in his scrotum after sexual stimulation, especially if it is prolonged. On physical examination the scrotum is not painful, but there is a relatively pliable, ropey mass lying alongside the left testis.

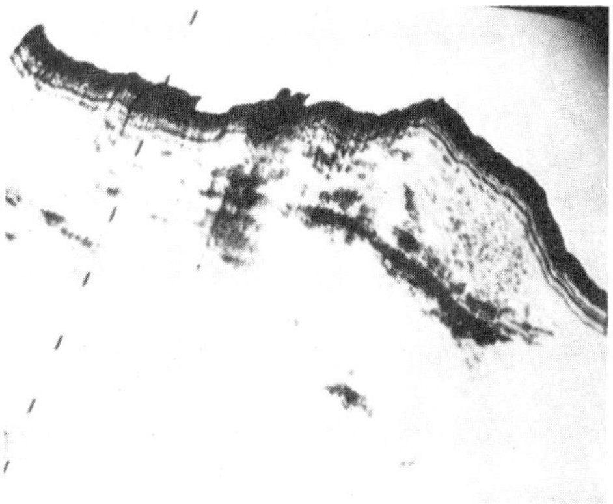

Figure 91–A.
Longitudinal scan through center of left testicle.

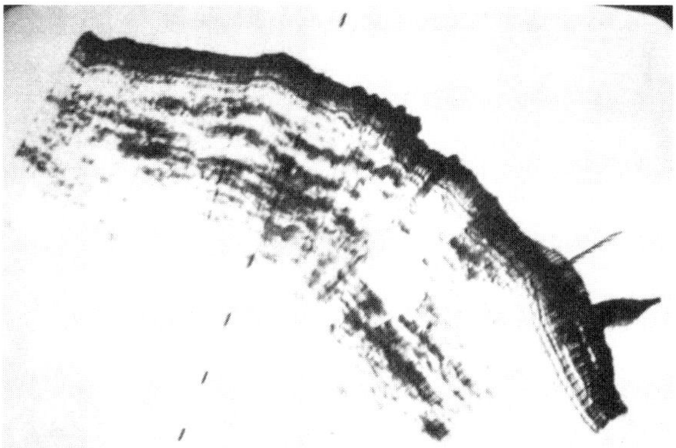

Figure 91–B.
Longitudinal scan of left testicle and spermatic cord.

DIAGNOSIS

Varicocele of left testicle.

ULTRASOUND FINDINGS

Figures 91–A and 91–B are views of the left testicle. The gland itself is easily recognizable as a region containing numerous low-level echoes, the pattern of which is reminiscent of the thyroid (arrow in Figure 91–A). Superior to this there are several linear echo-free structures, which can be followed upward toward the spermatic cord (Figure 91–B). A little imagination is all that is needed to understand that these represent somewhat enlarged blood vessels, in this instance veins within a varicocele.

There are other pathologic entities which have to be considered in the differential diagnosis; however, none is as common as varicocele. Figure 91–C is one such case. Again, there is an irregular echo-free area adjacent to the testis (arrow). We were unable to demonstrate any extension of this structure upward away from the testis itself (as in the varicocele); this patient had a lymphangioma of the scrotum. A spermatocele of the epididymis also appears as an echo-free mass lying beside the testicle. It, too, is unlikely to have any cephalad continuation toward and into the spermatic cord, as was the case with the varicocele, so that if such extension is seen we can eliminate spermatocele as a possible diagnosis.

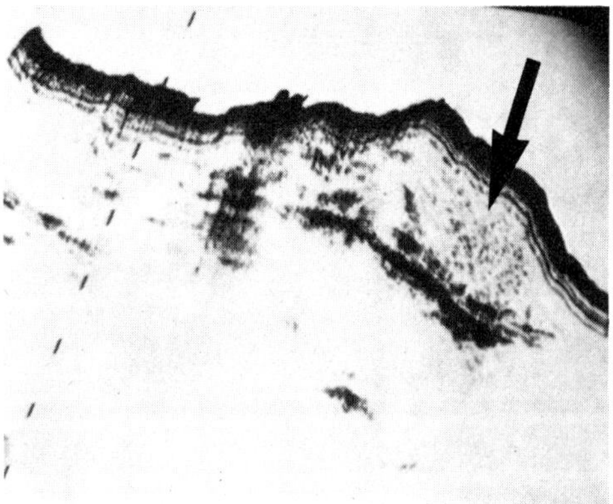

Figure 91–A.
Longitudinal scan through center of left testicle. Varicocele.

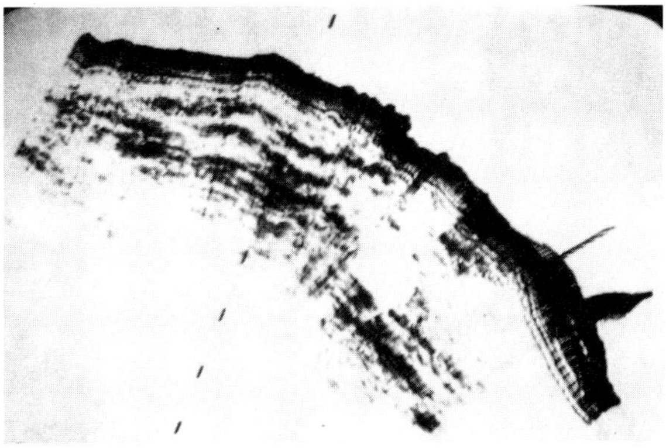

Figure 91–B.
Longitudinal scan of left testicle and spermatic cord. Varicocele.

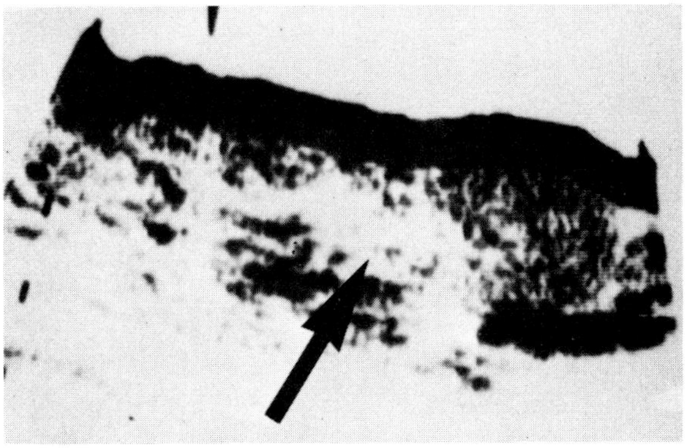

Figure 91–C.
Longitudinal scan of right testicle. Scrotal lymphangioma.

REFERENCES

Sample WF, Gottesman JE, Skinner DG, et al: Gray scale ultrasound of the scrotum. Radiology 127:225 (1978)

Miskin M, Busckspan M, Bain J: Ultrasonographic examination of scrotal masses. J Urol 117:185 (1977)

Case 92

CLINICAL INFORMATION

This 17-year-old girl has had "flu" for a week and it is getting worse. The initial problem was severe abdominal cramps and pain, associated with vomiting, mild diarrhea, and fever. The pain improved slowly over a 48-hour period but the nausea and the fever have persisted. In the last day the abdominal pain has started to increase again, and she has become much weaker. Her previous health has been good; she has had no operations and is certain that she is not pregnant. On physical examination the abdomen is tender, with mild rebound, and there is a suggestion of a mass in the right lower quadrant. The hematocrit is 44, the white cell count 13,600, and creatinine, liver functions, amylase, urinalysis, pregnancy test, and cervical smear are all normal. Plain films of the abdomen are almost totally airless and there are no unusual calcifications present. These scans were obtained in the right lower quadrant, just over the most tender area.

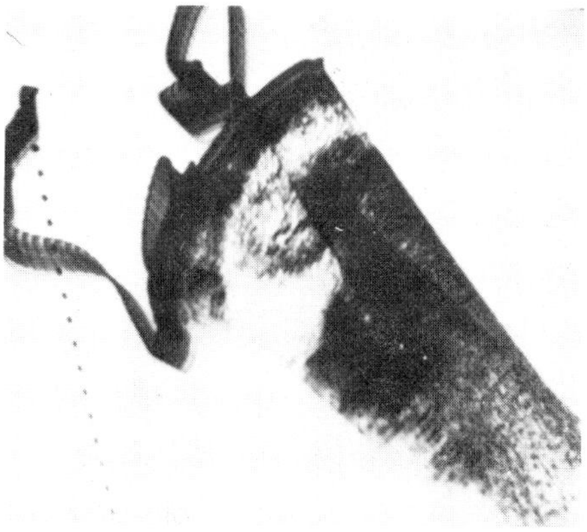

Figure 92–A. Transverse section 0.

Figure 92–B. Longitudinal section 8R.

DIAGNOSIS

Periappendicular abscess.

ULTRASOUND FINDINGS

Well, the history alone is a dead giveaway. But there is an echo-free area in the right flank (arrows on Figures 92–A and 92–B), which implies intraperitoneal fluid. A search of the remainder of the abdomen fails to turn up any other collections and so we can rule out a generalized process such as ascites. There is no communication with the region of the pancreas so that a pseudocyst also is unlikely. This seems to be a focal abnormality in the right lower quadrant and in this clinical setting points strongly to a periappendicular abscess. We verified this diagnosis with a percutaneous aspiration biopsy and the patient was taken to the operating room where a loculated abscess arising from a ruptured appendix was evacuated.

Another condition which should be considered is granulomatous disease of the bowel (Cröhn's disease), which can also produce right lower quadrant abscesses in young patients. Figure 92–C is an intra-abdominal abscess secondary to granulomatous disease in a 14-year-old boy.

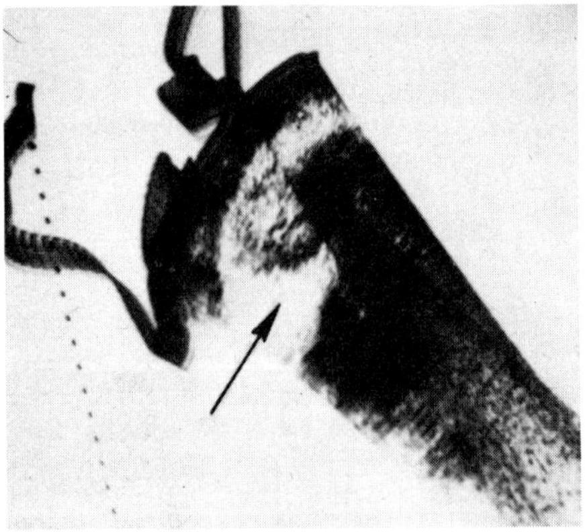

Figure 92–A. Transverse section 0. Periappendicular abscess.

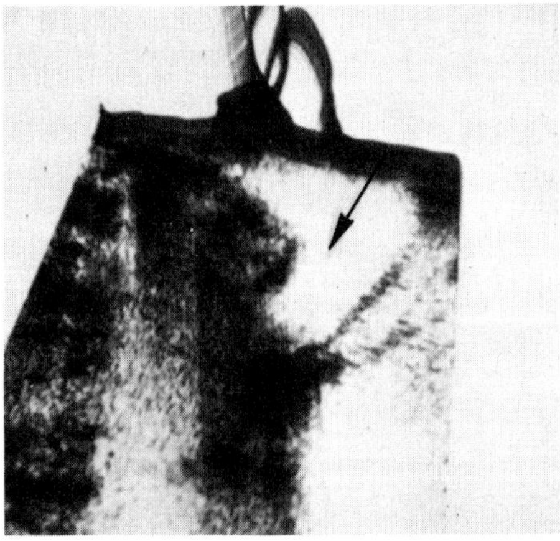

Figure 92–B. Longitudinal section 8R. Periappendicular abscess.

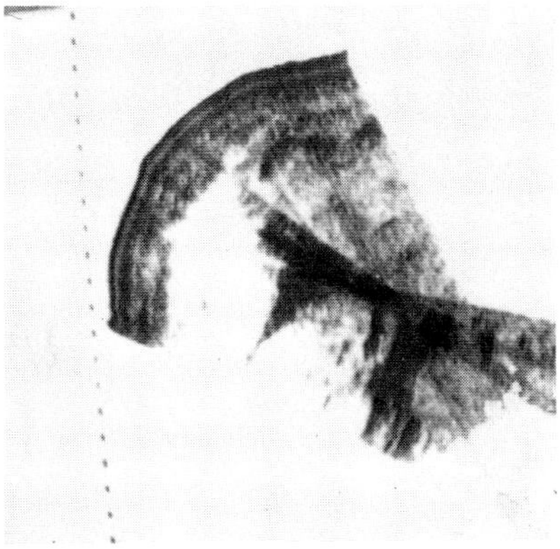

Figure 92–C. Transverse section +3. Right lower quadrant abscess from Crohn's disease.

REFERENCES

Doust BD, Quiroz F, Stewart JM: Ultrasonic distinction of abscesses from other intra-abdominal fluid collections. Radiology 125:213 (1977)

Jensen F, Pedersen JF: The value of ultrasonic scanning in the diagnosis of intra-abdominal abscesses and hematomas. Surg Gynec Obstet 139:326 (1974)

Case 93

CLINICAL INFORMATION

This 31-year-old woman collapsed following a severe attack of pelvic pain. She was brought to the hospital where she was awake but quite weak and pale, with a persistent dull ache in her pelvis. Six years before she had had a similar episode of pain (but without the collapse). This had proved to be secondary to an ovarian cyst, which was removed. She has two children and is uncertain whether she might now be pregnant — she has been amenorrheic for six weeks but has always had very irregular periods. Physical examination discloses a large mass in the left adnexal region which is "doughy" and extends into the cul-de-sac. Her temperature is 99.8, blood pressure 106/70, pulse 92, hematocrit 39, white cell count 9,400, and her pregnancy test is positive. Figure 93–A is a longitudinal scan through the center of the uterus (small arrows) and Figure 93–B a transverse scan through the body of the uterus (arrows).

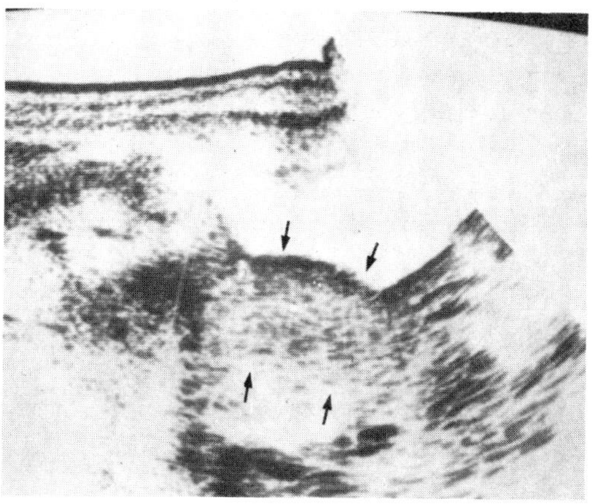

Figure 93–A. Longitudinal section 0L.

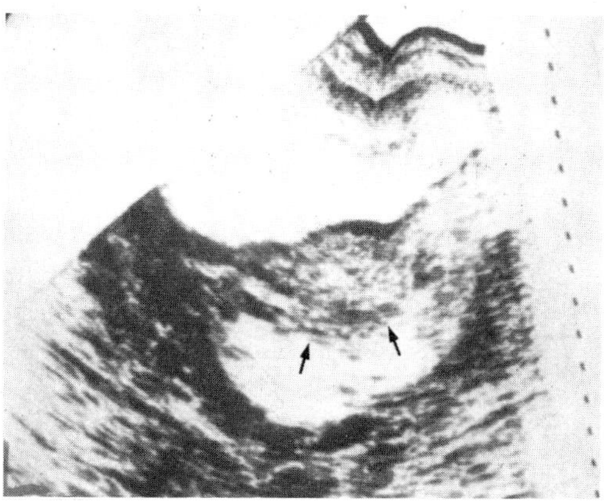

Figure 93–B. Transverse section −11.

DIAGNOSIS

Ectopic pregnancy (ruptured).

ULTRASOUND FINDINGS

In evaluating a possible ectopic pregnancy we first look for an intrauterine pregnancy; in this case the uterus is normal. Next we look at the adnexal regions for evidence of an ectopic pregnancy. In this case there is a complex mass inferior to the uterus in the cul-de-sac. The combination of a normal uterus, parauterine mass, and positive pregnancy test is almost pathognomonic for ectopic pregnancy. (We will discuss the exceptions in a moment.) If the mass is well-circumscribed and has the appearance of an early gestation, so much the better; but most ectopics do not come to medical attention until there has been disruption of the gestational sac, either as internal hemorrhage or frank rupture. When there has been rupture there will be free fluid (blood) in the most dependent portion of the pelvis, the cul-de-sac, or pouch of Douglas. The blood is easily seen in this case and confirms the clinical suspicion of ruptured ectopic pregnancy. Figure 93–C shows an unruptured ectopic pregnancy (straight arrow) near the neck of the uterus (curved arrow). Figure 93–D is another ruptured ectopic pregnancy, in which the bladder (*B*), uterus (*U*), ectopic fetus (*E*), and intraperitoneal hemorrhage (*H*) are all clearly seen.

The most important single piece of information in making this diagnosis is the result of the pregnancy test. There are very few false positive results, and a patient with a positive pregnancy test must be considered pregnant! If the pregnancy is not intrauterine, it must be nearby. As with any rule, there are exceptions: gestational trophoblastic disease (molar pregnancy) will give a positive pregnancy test, and a patient who has recently aborted may have a positive test for a few days. Figure 93–E shows such a case. This patient has a positive pregnancy test and pelvic pain. There is no pregnancy in the uterus — only a few loud echoes from the endometrium and a small cystic lesion in the right ovary. This is not an ectopic pregnancy but only a corpus luteum cyst. The patient had aborted an intrauterine pregnancy 72 hours earlier.

Figure 93–F is a scan from a patient with an interesting and frightening variation on this theme. She had missed several periods but at the time of the study was having bloody vaginal discharge. A recent pregnancy test was positive. The scan shows a slightly enlarged uterus with louder than normal echoes within it; no gestational sac is noted. A small ovarian cyst is also present. The pregnancy test remained positive, and she was reexamined three weeks later (Figure 93–G). Now the uterus is very much enlarged; but again no recognizable gestation is found, and the uterus contains a "snow storm" of loud echoes. The ovarian cyst is still present. When a D & C was done the next day, the diagnosis of hydatidiform mole was made. The unusual echoes in the uterus seen on the first scan were the changes of trophoblastic disease, the earliest signs of this process we have ever encountered. The important clue here again was the positive pregnancy test.

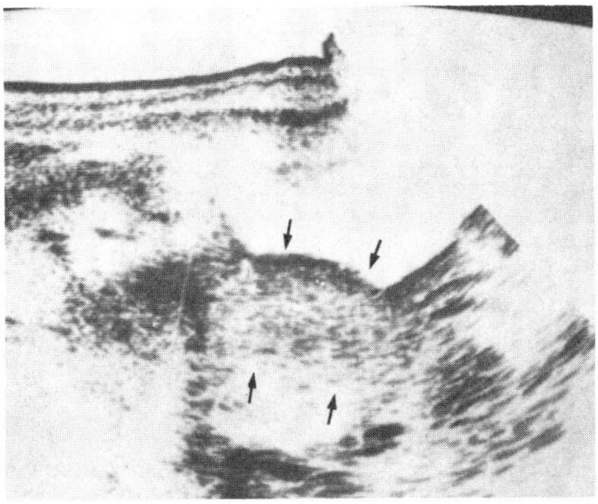

Figure 93–A. Longitudinal section 0L. Ruptured ectopic pregnancy.

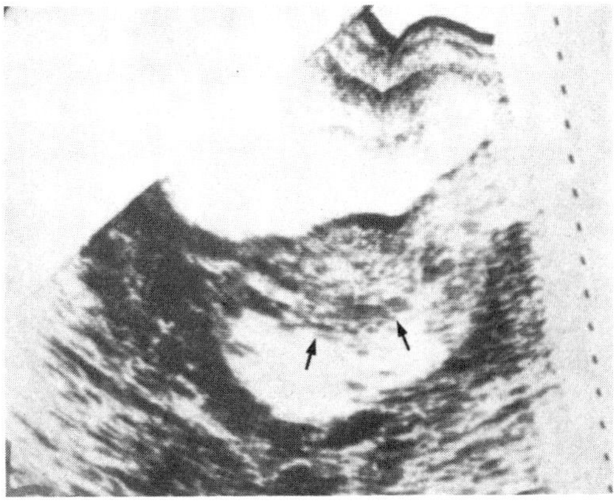

Figure 93–B. Transverse section –11. Ruptured ectopic pregnancy.

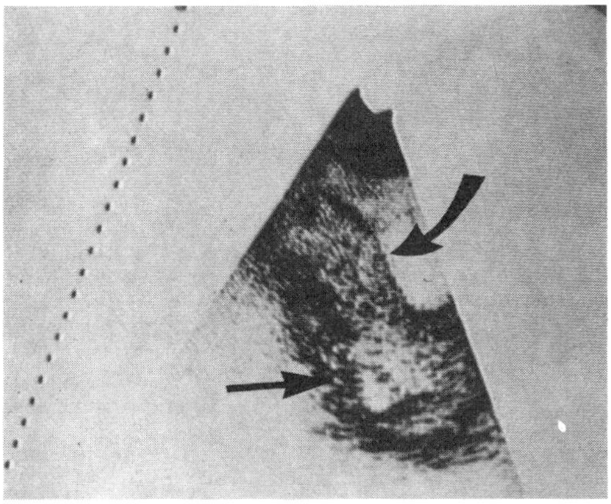

Figure 93–C. Longitudinal section 2L. Ectopic pregnancy.

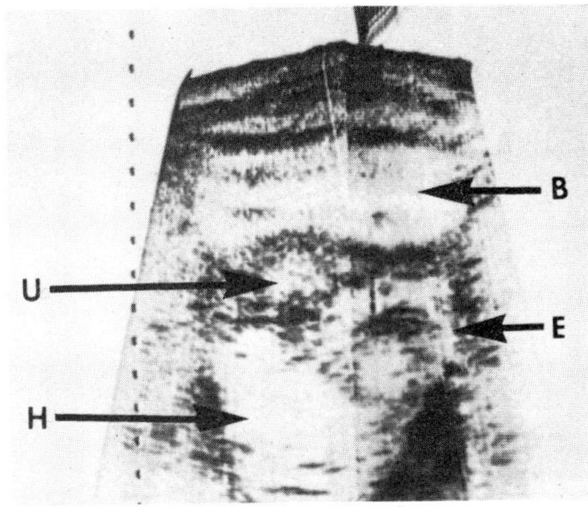

Figure 93–D. Transverse section –14. Ruptured ectopic pregnancy.

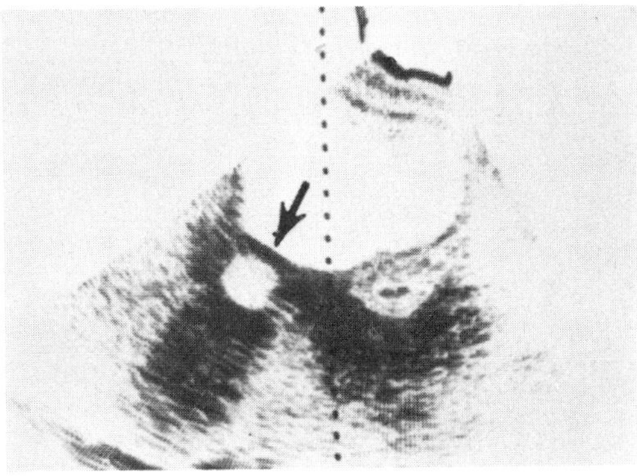

Figure 93–E. Transverse section – 13. Recent abortion and corpus luteum cyst.

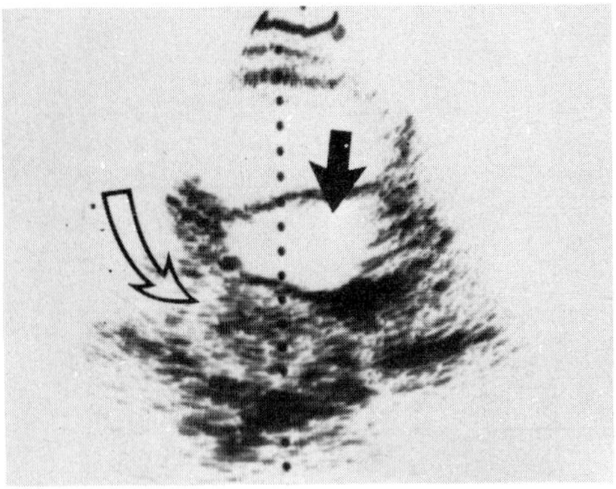

Figure 93–F. **Transverse section −10.** Molar pregnancy and theca lutein cyst.

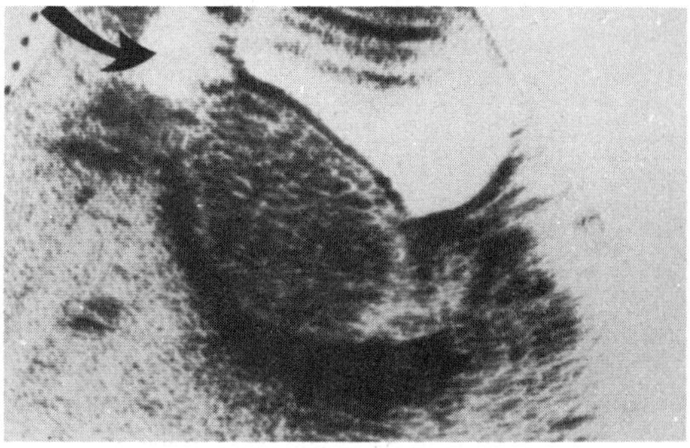

Figure 93–G. **Longitudinal section 1L.** Molar pregnancy and theca lutein cyst.

REFERENCE

Maklad NF, Wright CH: Grey scale ultrasonography in the diagnosis of ectopic pregnancy. Radiology 126:221 (1978)

Case 94

CLINICAL INFORMATION

This 85-year-old woman has episodic epigastric pain that is not characteristic of any particular disease. The pain is neither severe nor persistent, and she considers it more an annoyance than any serious problem. Her physical examination is normal except for a mild heart murmur; extensive laboratory tests have failed to turn up anything unusual. Her past health has been good and she has never had a serious illness or operation. She has never traveled outside the United States. An oral cholecystogram and barium enema are normal and an upper GI series reveals only a small hiatal hernia without reflux.

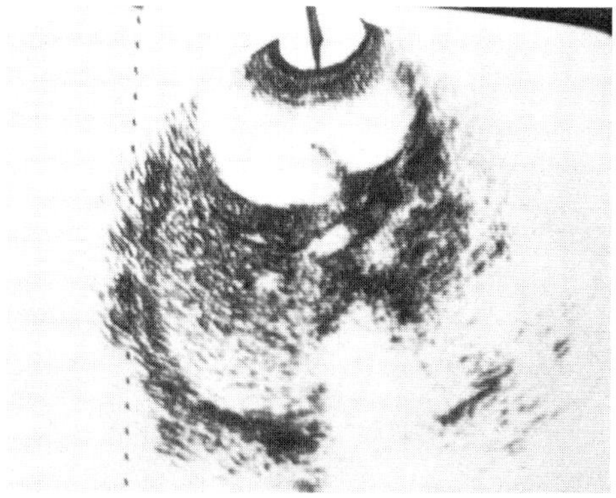

Figure 94–A. Transverse section +8.5.

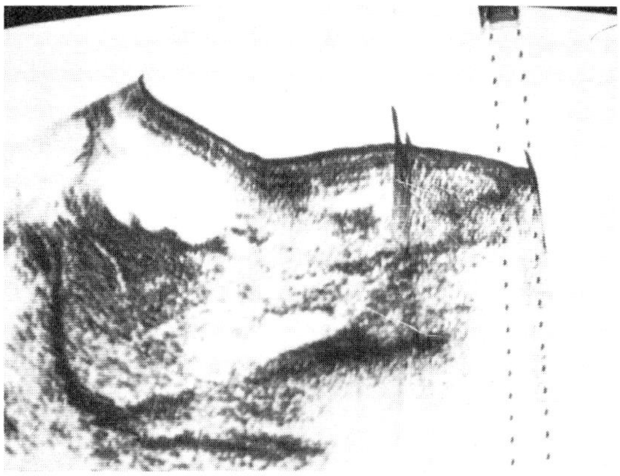

Figure 94–B. Longitudinal section 4R.

DIAGNOSIS

Benign hepatic cyst.

ULTRASOUND FINDINGS

There is an obvious echo-free mass occupying the superior portion of the right lobe of the liver and most of the left lobe as well. The lack of internal echoes and the smooth back wall tell us that we are dealing with a fluid-filled mass. First we must decide where the mass is located: although it appears to be within the liver we check the patient's recent barium enema and upper GI series to assure ourselves that we are not looking at a displaced portion of bowel. Once we have satisfied ourselves that this is a liver lesion, what can we do to make a more specific diagnosis? We feel that a percutaneous aspiration biopsy is the next logical procedure and is the quickest way to separate the possibilities of benign cyst, malignant cyst, pancreatic pseudocyst (pseudocysts can appear anywhere), or abscess. Percutaneous biopsy is an innocuous procedure and can be performed without fear of morbidity. (Theoretically, it may be unwise to puncture an echinococcus cyst; however, this lesion is not only exceedingly rare in this country but also has a different appearance, because it contains multiple internal echoes.) Biopsy yielded benign cyst fluid.

Liver cysts come in all shapes and sizes. Figure 94–C shows a patient with a single, large benign cyst near the porta hepatis. The scan in Figure 94–D is from a patient with an unbelievably large cyst, which not only has replaced most of the liver but is also causing dramatic abdominal distension. Figure 94–E shows multiple small cysts scattered throughout the liver (and kidneys) in a patient with polycystic kidneys. The complex cyst shown in Figures 94–F and 94–G is malignant — a cystic metastasis from a rare glucagonoma of the pancreas.

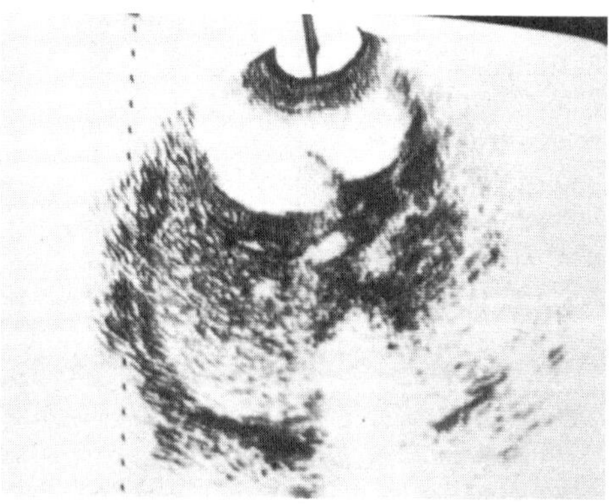

Figure 94–A. Hepatic cyst. **Transverse section +8.5.**

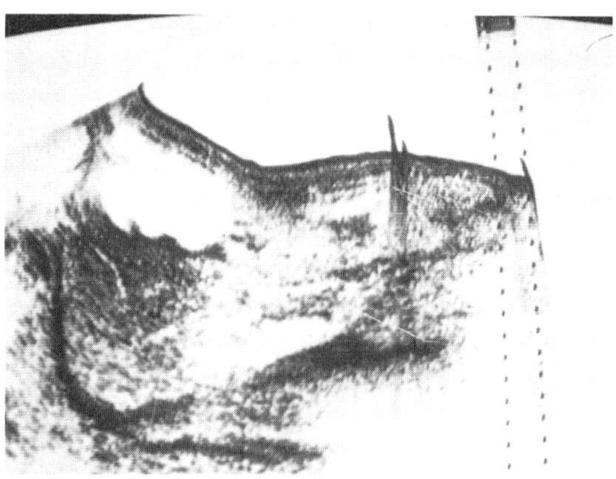

Figure 94–B. Hepatic cyst. **Longitudinal section 4R.**

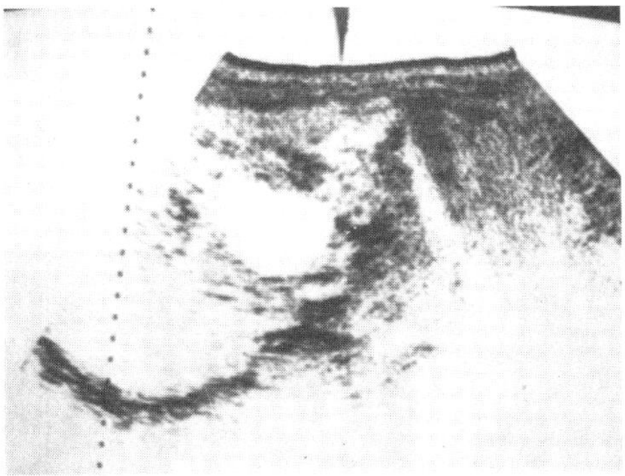

Figure 94–C. Transverse section +10.
Hepatic cyst.

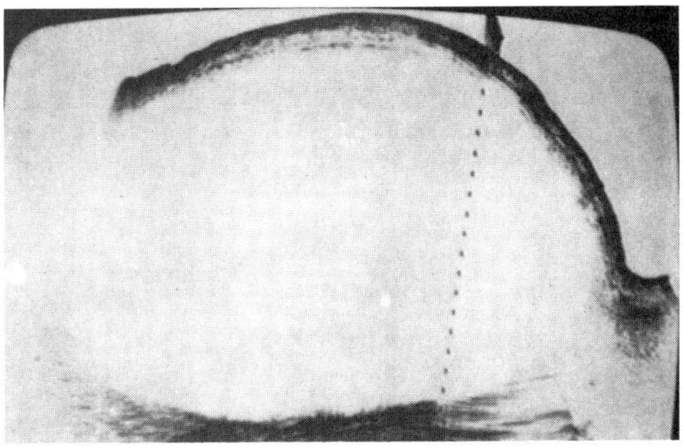

Figure 94–D. Longitudinal section 0L.
Massive hepatic cyst.

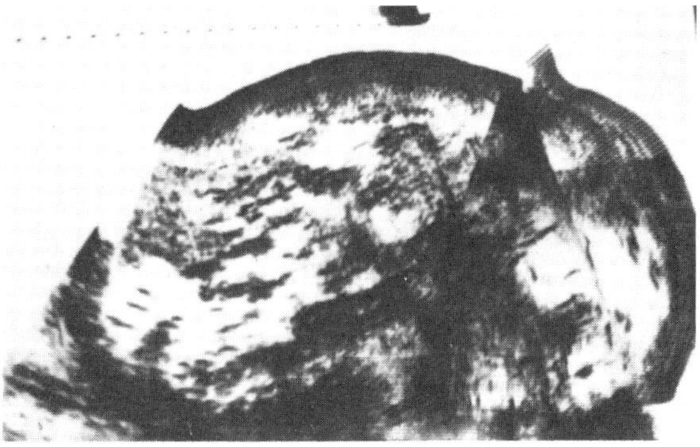

Figure 94–E. Transverse section +8.
Polycystic disease of liver and kidneys.

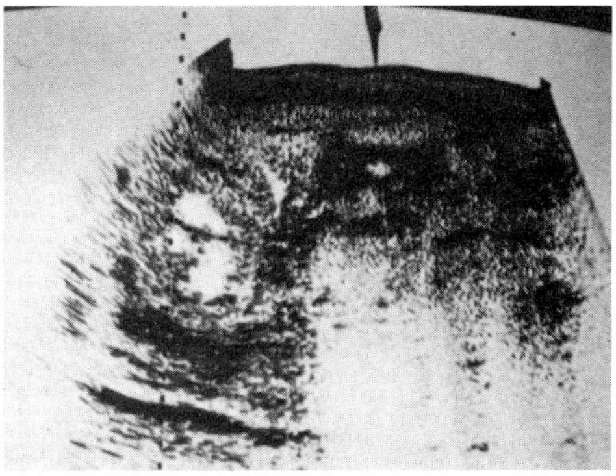

Figure 94–F. Transverse section +7.
Cystic metastasis.

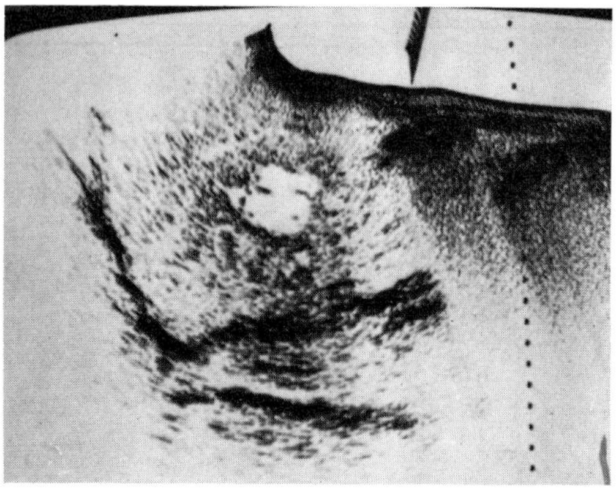

Figure 94–G. Longitudinal section 8R. Cystic metastasis.

REFERENCES

Albertson KW, Leopold GR: Liver, Chapter 5, in *Abdominal Gray Scale Ultrasonography*, BB Goldberg, editor. John Wiley & Sons, New York (1977)

King DL: Ultrasonography of echinococcal cyst. J Clin Ultra 1:64 (1973)

Case 95

CLINICAL INFORMATION

This 63-year-old man is suffering from dyspepsia. He has epigastric pain, which is not related to food intake in any particular fashion. The pain is episodic and not severe and is relieved by taking 4 "Rolaids" and a shot of whiskey. The pain does not radiate and does not go to his back. He is otherwise as "healthy as a horse." His physical examination is normal, and he is scheduled for an abdominal ultrasound examination to be followed by an upper GI series.

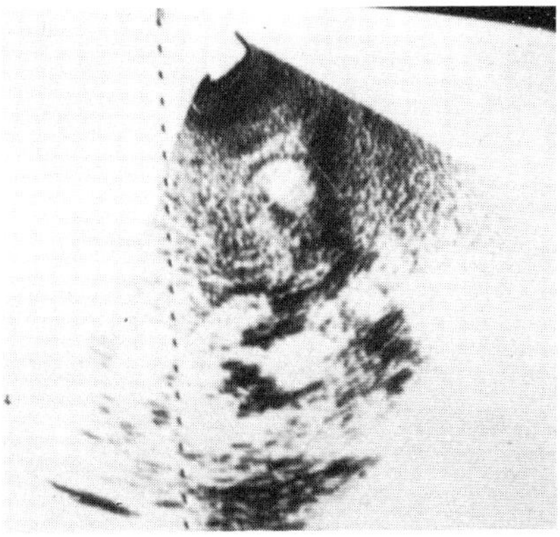

Figure 95–A. Transverse section +4.

DIAGNOSIS

Mild hydronephrosis (uretero-pelvic junction obstruction).

ULTRASOUND FINDINGS

This is a sector scan of the right upper quadrant of the abdomen. (It was necessary to sector the transducer to stay over an acoustic window.) We see a portion of the liver and the gallbladder in cross-section, both of which are unremarkable. In the renal sinus there is a fluid-filled (i.e., echo-free) mass, which is displacing the loud echoes from the calyces and blood vessels; a small "tongue" of the mass extends toward the renal cortex. This is the typical appearance of hydronephrosis. The only differential diagnostic possibility is a parapelvic renal cyst. The latter lesion is rare and has no extension into the renal parenchyma (Figure 95–B). Further investigation showed that this patient's hydronephrosis was secondary to uretero-pelvic junction obstruction rather than to a lesion in the ureter. We usually cannot make this distinction with ultrasound. Although there are instances when a dilated ureter can be seen, we have not been able to do this reliably and do not think that failure to demonstrate a dilated ureter excludes the presence of hydroureter. The site of an obstructing lesion must be identified by other means, such as IVP, retrograde pyelogram, or antegrade pyelogram. (Oh yes, we nearly forgot: The upper GI series on this patient showed a duodenal ulcer, the cause of his symptoms.)

Uretero-pelvic junction obstruction is a lesion of confusing etiology and significance. It can produce mild, asymptomatic dilatation of the renal pelvis but can also lead to massive enlargement of the renal collecting system and infection. Figure 95–C is another case of uretero-pelvic junction obstruction showing a huge, dilated renal pelvis, which proved to be infected (pyonephrosis).

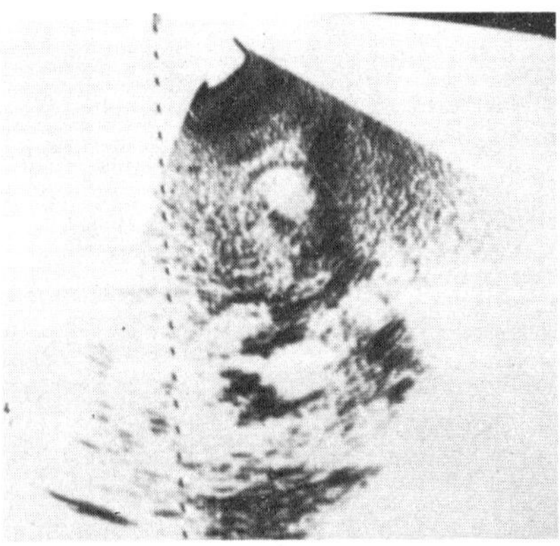

Figure 95–A. Transverse section +4. Hydronephrosis (uretero-pelvic junction obstruction).

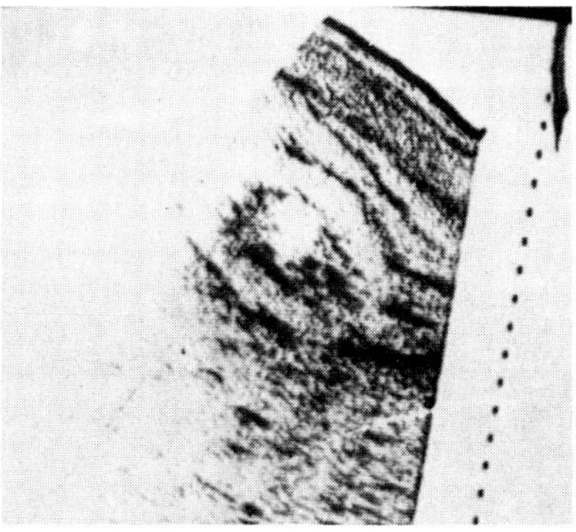

Figure 95–B. Longitudinal section 0L. Prone scan. Parapelvic cyst.

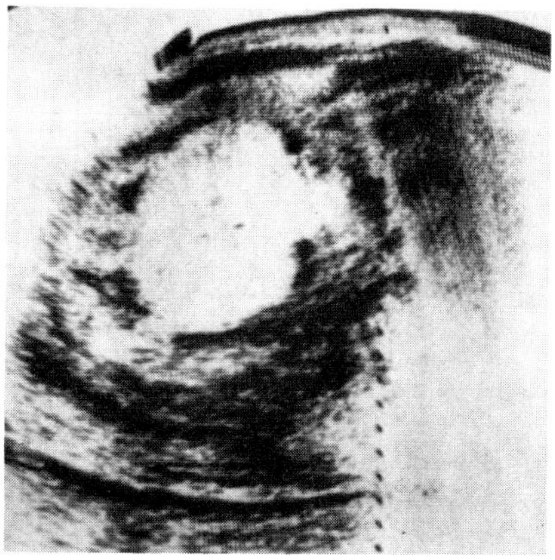

Figure 95–C. Transverse section +3. Pyonephrosis secondary to uretero-pelvic junction obstruction.

REFERENCES

Ellenbogen PH, Scheible FW, Talner LB, et al: Sensitivity of gray scale ultrasound in detecting urinary tract obstruction. Am J Radiol 130:731 (1978)

Sanders RC, Conrad MR: The ultrasonic characteristics of the renal pelvicalyceal echo complex. J Clin Ultra 5:372 (1977)

Case 96

CLINICAL INFORMATION

This 64-year-old woman has several medical problems. She has varicose veins and in the past has had two episodes of thrombophlebitis. During the most recent bout, nine months ago, there was some concern that she had had a minor pulmonary embolism, and therefore she was treated with anticoagulants for two weeks. She has had several episodes of obscure abdominal pain over the years, the cause for which has never been found. Ten days ago she suffered some type of cerebrovascular accident, which left her with some left-sided paresis; and two days after this she may have had a seizure. A CT scan performed then showed a large low-density area in the right parietal area, which had focal areas of enhancement — probably due to a recent cerebral infarction, but also considered to be compatible with a tumor. Her doctors have ordered this ultrasound scan to search for evidence of a primary neoplasm.

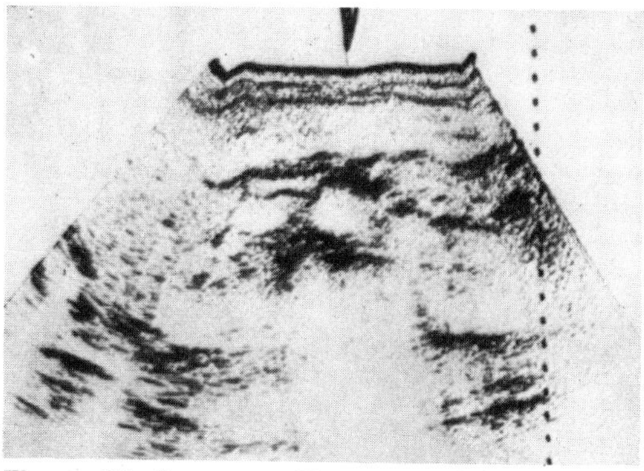

Figure 96–A. Transverse section +10.

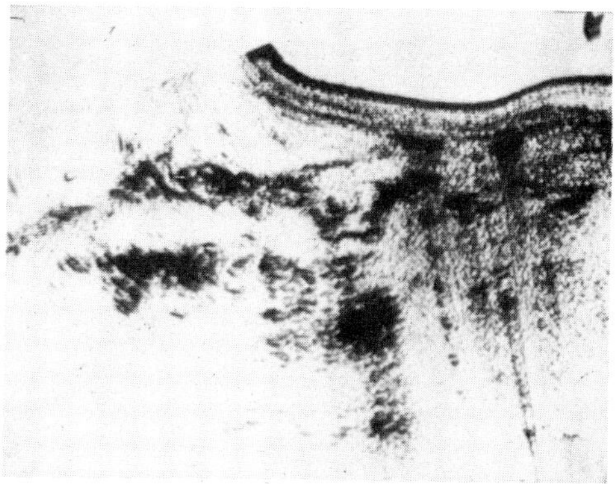

Figure 96–B. Longitudinal section 1L.

DIAGNOSIS

Prominent, but normal pancreas.

ULTRASOUND FINDINGS

We hoped that you missed the diagnosis and called this a tumor. We did, and we will again. The body and tail of the pancreas are very prominent and seem too large for their surroundings, while the head of the pancreas (seen on other sections) is quite small and is poorly seen. This combination should suggest a focal pancreatic mass, and the working diagnosis must be considered neoplasm until proven otherwise. Although most likely a pancreatic mass, this could also represent enlarged lymph nodes, either from metastatic disease or primary lymphoma.

The simplest way to sort out these diagnostic possibilities is to perform an aspiration biopsy of the suspicious area. This is safer, easier, and more specific than any of the other possible diagnostic imaging procedures such as CT scan, arteriogram, or lymphangiogram. After biopsy of this mass, the cytology report came back "Negative, normal pancreatic cells." We were not convinced, since there are false negative aspiration biopsies, and we repeated the procedure. The cytology report was the same. Everyone was somewhat reassured but still uneasy.

A barium enema, done as part of the tumor work-up, showed a questionable mass in the cecum. A repeat barium study two weeks after the first was unchanged. Since there were now two suspicious areas in the abdomen, an exploratory laparotomy was performed. Both the cecum and the pancreas were normal. The patient had a mild lumbar scoliosis, and the pancreas was somewhat twisted in position, with nearly all of the gland lying to the left of the spine. This probably accounts for the unusual ultrasonic appearance.

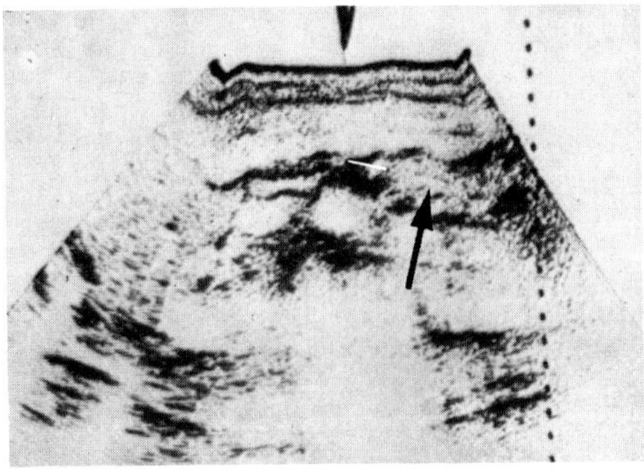

Figure 96–A. **Transverse section +10.** Normal pancreas.

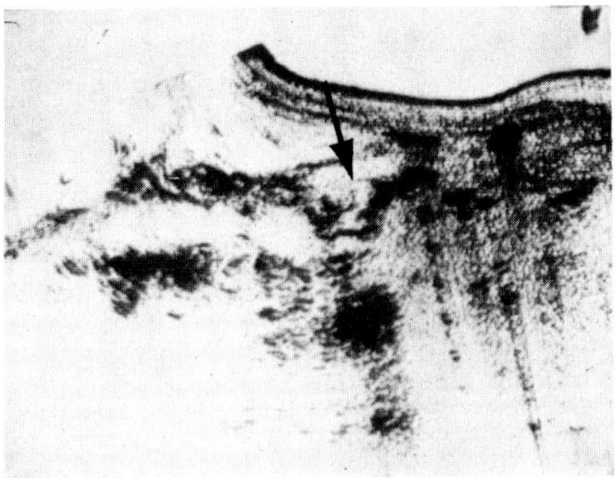

Figure 96–B. **Longitudinal section 1L.** Normal pancreas.

REFERENCES

Ghorashi B, Rector WR: Gray scale sonographic anatomy of the pancreas. J Clin Ultra 5:25 (1977)

Sample WF, Po JB, Gray RK, et al: Gray scale ultrasonography: Techniques in pancreatic scanning. App Radiol 4:63 (1975)

Bartrum RJ Jr, Crow HC: *Gray-Scale Ultrasound: A Manual for Physicians and Technical Personnel.* Chapter 13. WB Saunders Co, Philadelphia (1977)

Case 97

CLINICAL INFORMATION

This 29-year old woman is 36 weeks pregnant according to her dates and the uterine size. She has been leaking amniotic fluid for 12 hours and her obstetrician wishes to assess fetal lung maturity with an amniotic fluid L/S ratio. Unfortunately the placenta seems to be located directly anteriorly. Where do you recommend making the puncture?

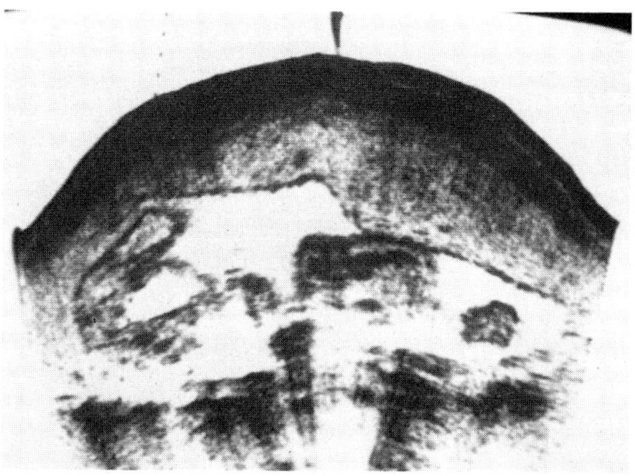

Figure 97–A. Transverse section 0.

DIAGNOSIS

Anterior placenta — the puncture can be made through a thin portion of the placenta.

ULTRASOUND FINDINGS

Most obstetricians prefer to use ultrasonic guidance when performing amniocentesis, because it has been shown that the incidence of successful taps is improved with this technique. We believe that it is important that the amniocentesis be performed in the ultrasound room with the patient in the same position as when she was scanned. If the patient moves or gets off the table (even to empty her bladder) the uterine contents often shift dramatically and the optimal puncture site will also change. *Ultrasonic guidance of amniocentesis is of little value if the puncture is made after the patient has moved.* Unfortunately, this precaution is not observed in many institutions, and this probably accounts for the poor results some investigators have reported.

The goal is to avoid puncturing the placenta or baby. Usually this is easy, but when the placenta is located anteriorly, as in this case, there is no choice but to go through it. This is not dangerous as long as a thin needle (22 gauge) is used and the thick central portion, where the major vessels are located, is avoided. We do not think that the placenta should be punctured in an Rh-negative mother because of the danger of placental transfusion and Rh sensitization. (You noticed that this is a baby girl, of course!)

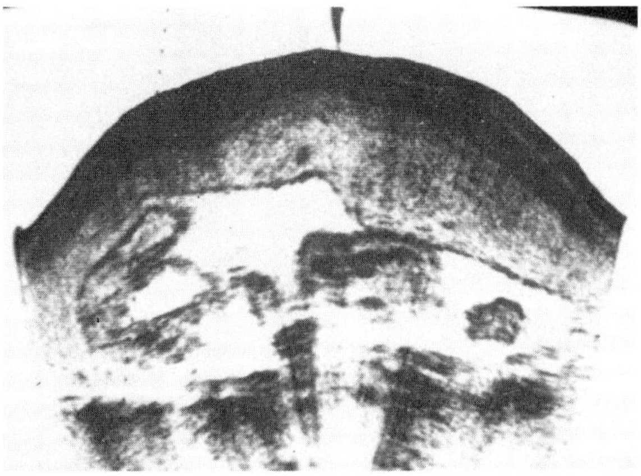

Figure 97–A. **Transverse section 0.**
Anterior placenta.

REFERENCE

Hobbins JC, Winsberg F: *Ultrasonography in Obstetrics and Gynecology.* Chapter 7. Williams & Wilkins Co., Baltimore (1977)

Case 98

CLINICAL INFORMATION

This 87-year-old woman slipped while getting up from a chair and fell about three feet to the floor. Although she thought she was only "bruised" at the time, within three days she began having severe back pain, which has persisted without change. Extensive radiographs reveal osteoporosis and several compression fractures of thoracic and lumbar vertebrae; however, all these fractures were present on films taken two years earlier, and there are no definite new fractures. In the films we can see that her aorta is heavily calcified and very tortuous, with a suggestion of an abdominal aneurysm. On physical examination the aorta is easily palpated and seems to be enlarged. What do you think has happened to this patient?

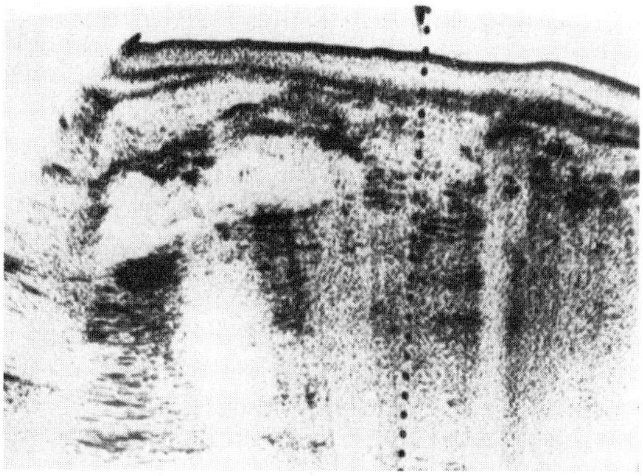

Figure 98–A. Longitudinal section 0L.

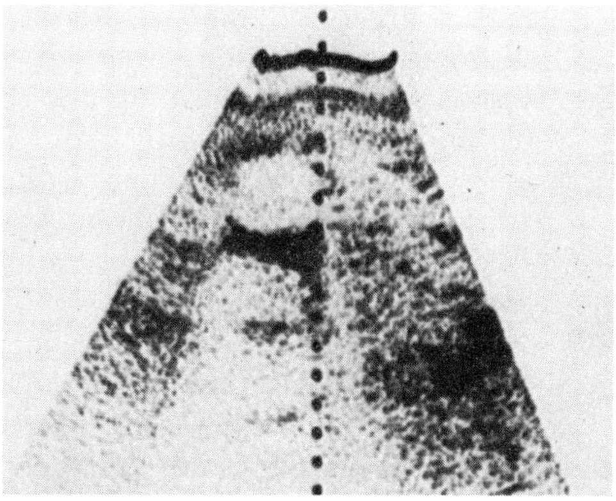

Figure 98–B. Transverse section +1.

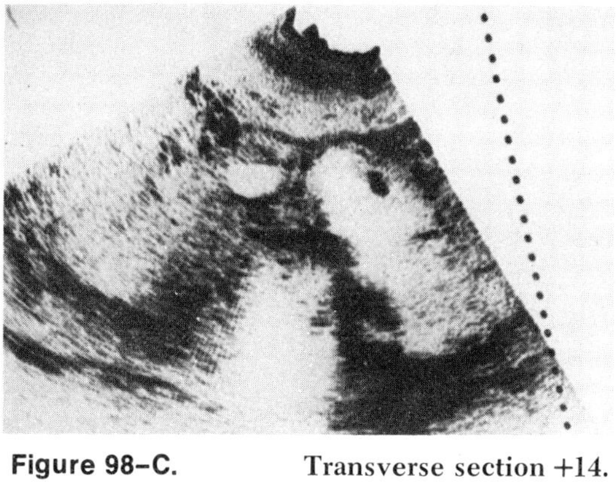

Figure 98–C. Transverse section +14.

DIAGNOSIS

Extremely tortuous, ectatic aorta.

ULTRASOUND FINDINGS

We missed this case. The aorta is very easily seen on the ultrasound scan and is certainly dilated (Figure 98–A). Most of the enlargement seems confined to the upper abdominal aorta and a cross-section made just above the bifurcation (Figure 98–B) shows that the diameter is normal at that level. Of particular concern is the area just beneath the left hemidiaphragm (Figure 98–C). Here the enlarged aorta merges imperceptibly with a large, irregular retroperitoneal "mass." Despite careful scanning we were never able to define clearly the aortic lumen and could not tell whether there was any clot present. Because of the bizarre appearance and the large retroperitoneal "mass" that was intimately associated with the aorta, we concluded that the patient must have an actively dissecting and probably leaking aneurysm.

Because of the potentially serious consequences of this diagnosis, the patient was taken immediately to surgery without having an aortogram. Her aorta was found to be extremely tortuous and very ectatic, but there was no discrete aneurysm nor any evidence of para-aortic bleeding. The "mass" underneath the left hemidiaphragm was a loop of the tortuous aorta which had knuckled into the splenic fossa.

What lessons can be learned from this case? First, all patients in whom aortic surgery is planned should have an aortogram (unless they are in extreme danger of exsanguinating). Second, it is important to identify the aortic lumen on the ultrasound examination, so that the path and course of the vessel can be charted accurately. Third, aneurysms limited to the upper abdominal aorta are unusual and such a diagnosis should be made with caution. The presence of a normal infrarenal segment should have raised our suspicion in this case, but would not have allowed us to exclude the presence of dissection.

Aortas can and do leak; and para-aortic hematomas can be easily identified with ultrasound. Figure 98–D shows a different patient with a large para-aortic hematoma (false aneurysm). This patient had previously had an aneurysm resected and a graft inserted; the suture line subsequently broke down, and a large false aneurysm developed. Figures 98–E and 98–F show another patient with an iatrogenic hematoma following translumbar aortography.

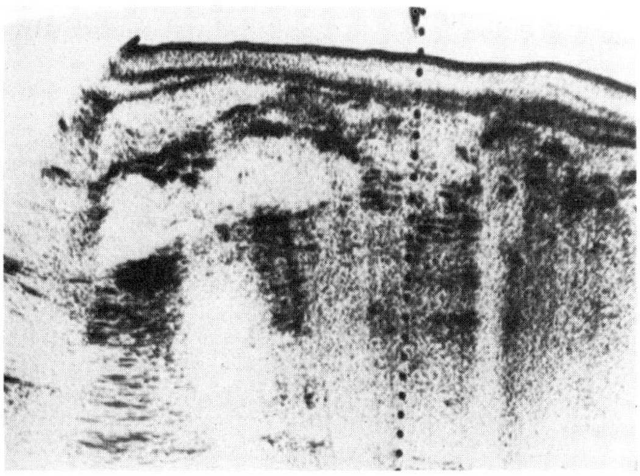

Figure 98–A. Longitudinal section 0L. Ectatic, atherosclerotic aorta.

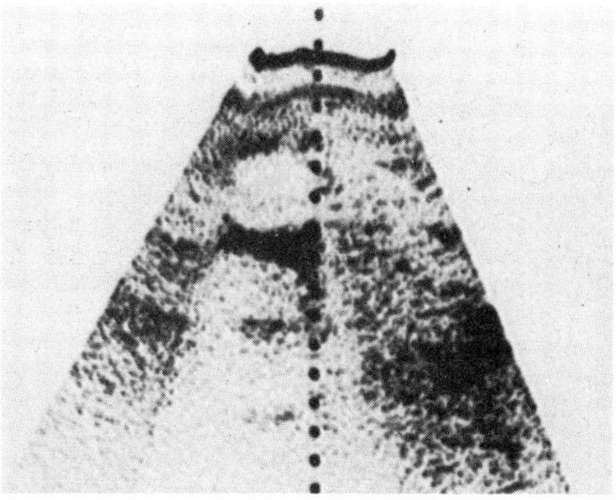

Figure 98–B. Transverse section +1. Ectatic aorta.

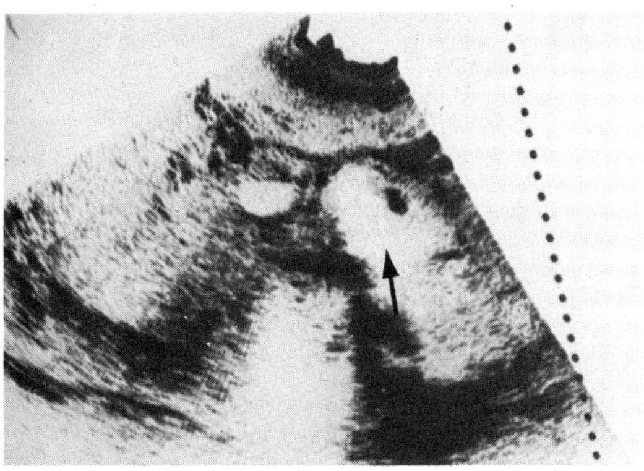

Figure 98–C. Transverse section +14. Tortuous, ectatic aorta.

Figure 98–D. Longitudinal section 3L. False aneurysm. There was a leak at the anastomosis between the aorta (solid arrows) and the graft (open arrows).

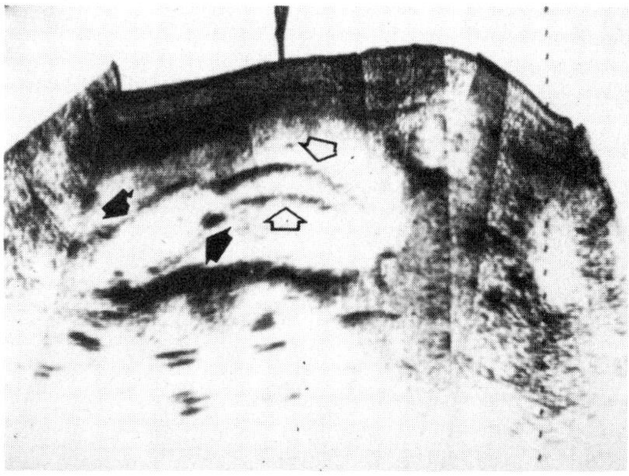

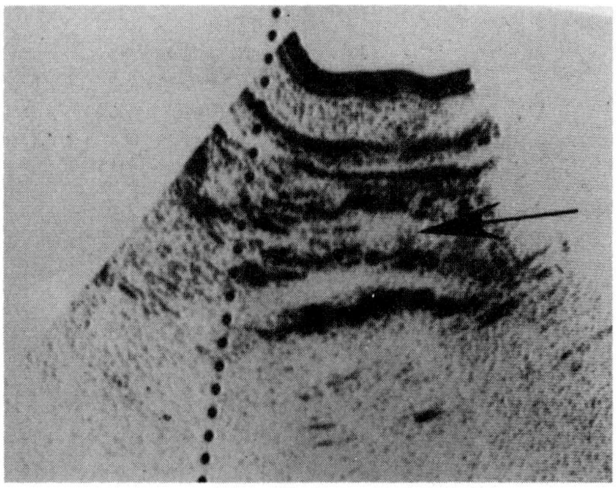

Figure 98–E. **Longitudinal section 0L.** Para-aortic hematoma (arrow).

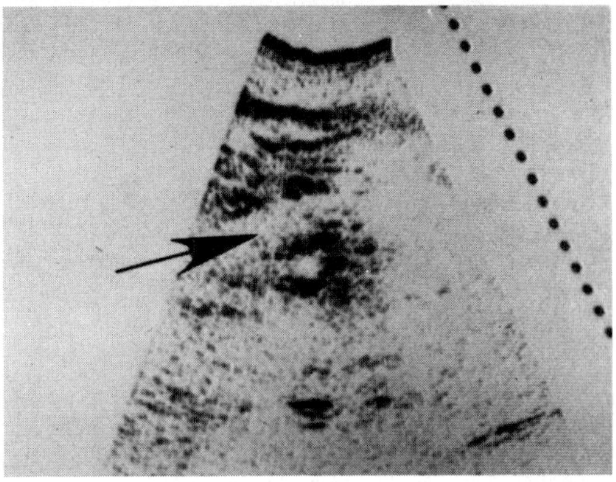

Figure 98–F. **Transverse section +3.** Para-aortic hematoma (arrow).

REFERENCE

Filly RA, Goldberg BB: Abnormal vessels, Chapter 4 in *Abdominal Gray Scale Ultrasonography*, BB Goldberg, editor. John Wiley & Sons, New York (1977)

Case 99

CLINICAL INFORMATION

This 23-year-old woman has had several bouts of pelvic pain in the past, and for the past two days she has had extremely severe abdominal and pelvic pain, has been vomiting, and has had sweats and chills. She has never been pregnant, and does not use oral contraceptives or an IUD. It is possible she could be pregnant now, although she had a normal period three and a half weeks ago. She has never had any surgery but three years ago she was hospitalized for gonorrhea. Her periods have always been regular and not uncomfortable. The patient cannot tolerate a complete pelvic examination because of extreme tenderness. There is a cervical discharge but no bacteria are identified by Gram stain. Her pulse and blood pressure are normal, her temperature is 102, hematocrit 43, white cell count 14,700, pregnancy test negative, and urinalysis normal. Here are longitudinal and cross-sectional scans through the pelvis; B = urinary bladder.

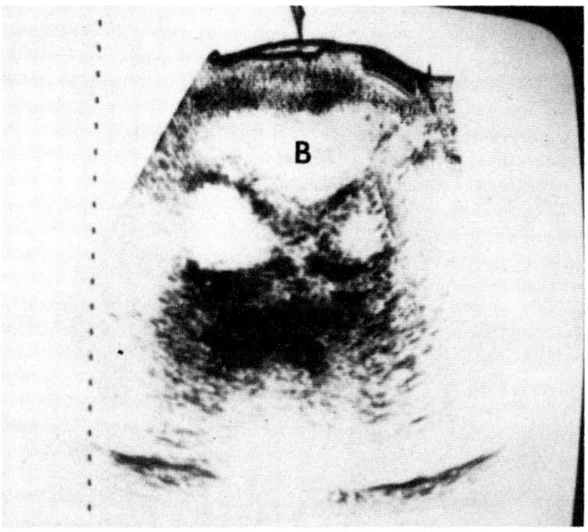

Figure 99–A. Transverse section –12.

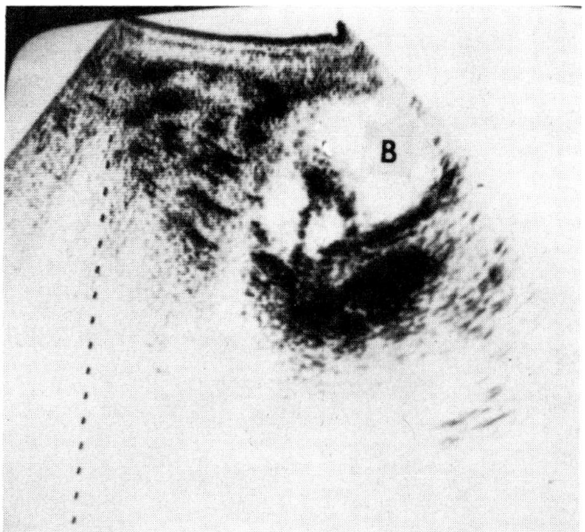

Figure 99–B. Longitudinal section 0L.

DIAGNOSIS

Bilateral tubo-ovarian abscesses (pelvic inflammatory disease).

ULTRASOUND FINDINGS

There are small echo-free masses in both adnexal regions (Figure 99–A); the lesion on the left is multilocular (Figure 99–B). The differential diagnosis includes benign ovarian cysts, tubo-ovarian abscesses, ectopic pregnancy (possibly with an ovarian cyst), and ovarian carcinoma. Carcinoma is unlikely in a patient this age with this clinical history, and we can eliminate it from further consideration. An ectopic pregnancy is unlikely bacause of the normal pregnancy test, vital signs, and hematocrit. These could be bilateral ovarian cysts and, in particular, endometrioma is a good possibility. The overall clinical setting, however, is more suggestive of tubo-ovarian abscess.

The general term "pelvic inflammatory disease" (PID) covers a wide spectrum of pelvic infection, ranging from acute inflammatory gonorrhea to chronic indolent tubal abscess. In acute infections the ultrasound findings are minimal or normal. As the disease becomes chronic with abscess formation and reactive inflammatory tissue proliferation, masses with variable degrees of homogeneity will appear. The proportion of fluid to solid elements varies from case to case and there is no characteristic ultrasound appearance. Careful consideration of the ultrasound findings in conjunction with the clinical setting will usually permit an accurate diagnosis.

Figures 99–C and 99–D show another tubo-ovarian abscess which has only a small cystic component. The abscess in Figure 99–E has a thick rind of inflammatory tissue.

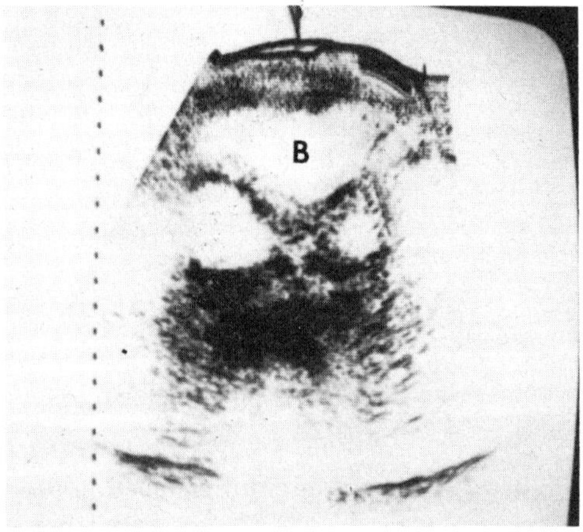

Figure 99–A. Transverse section −12. Tubo-ovarian abscess.

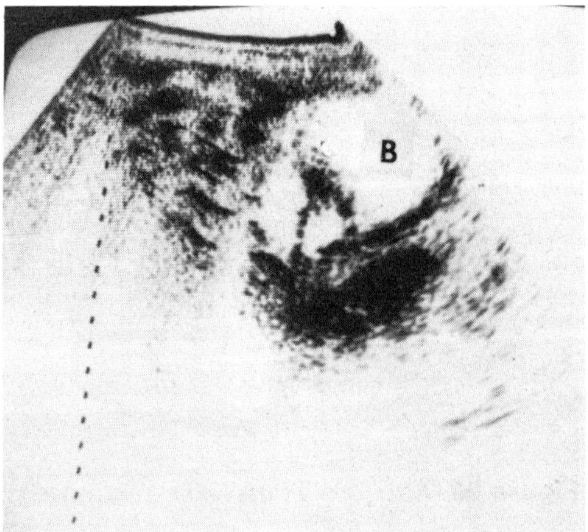

Figure 99–B. Longitudinal section 0L. Tubo-ovarian abscess.

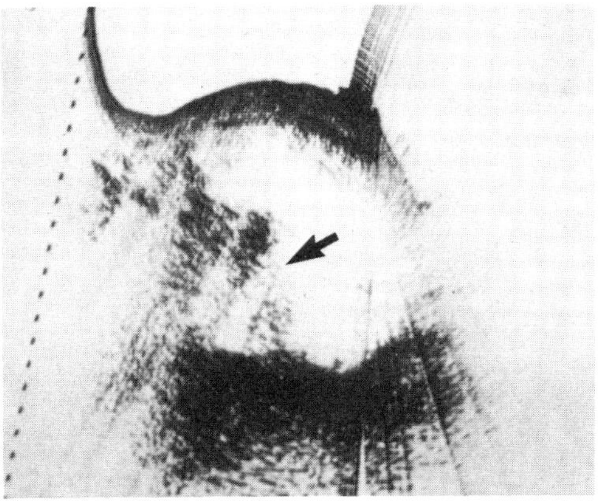

Figure 99–C. Transverse section –2. Tubo-ovarian abscess.

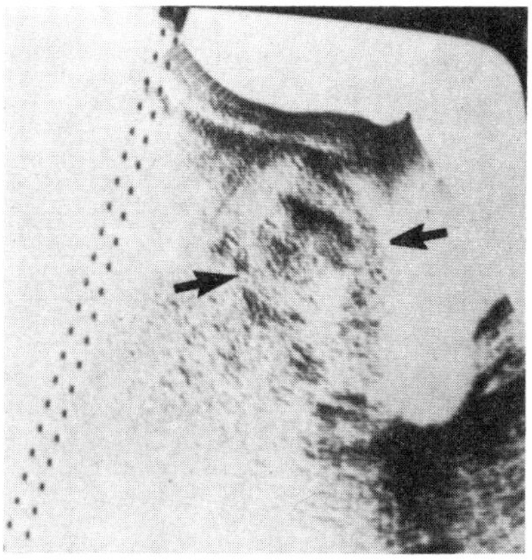

Figure 99–D. Longitudinal section 4L. Tubo-ovarian abscess.

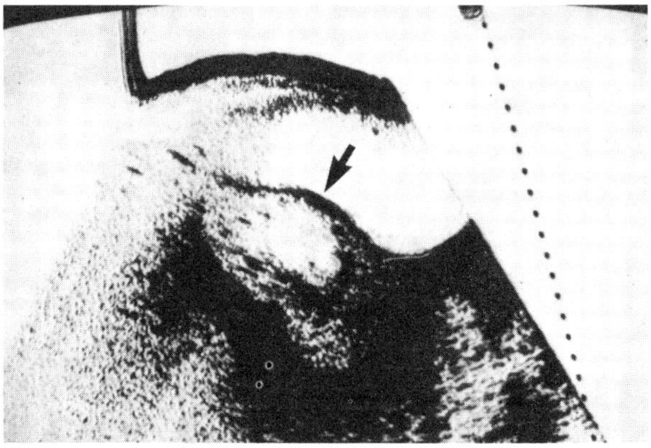

Figure 99–E. Longitudinal section 2L. Tubo-ovarian abscess.

REFERENCES

Uhrich PC, Sanders RC: Ultrasound characteristics of pelvic inflammatory disease. J Clin Ultra 4:199 (1976)

Sample WF: Pelvic inflammatory disease, Chapter 28 in *Ultrasonography in Obstetrics and Gynecology*, Sanders and James, editors. Appleton-Century-Crofts, New York (1977)

Case 100

CLINICAL INFORMATION

This patient is a 43-year-old man with abdominal pain. The only other thing we will tell you is that six years earlier he had a cholecystectomy for gallstones. What are the arrows pointing at?

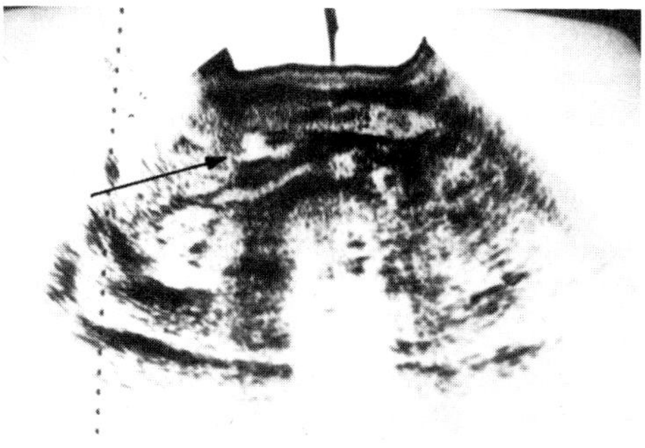

Figure 100–A. Transverse section +8.

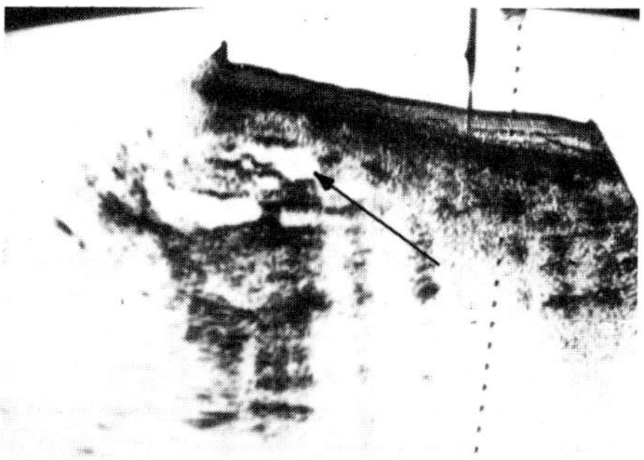

Figure 100–B. Longitudinal section 4R.

DIAGNOSIS

Dilated common bile duct.

ULTRASOUND FINDINGS

The arrows in Figures 100–A and 100–B are pointing to the common bile duct, which is considerably dilated. How do we know this is the common duct? First, it is located where the common duct should be found, and courses slightly posteriorly as it passes from the liver toward the papilla. Second, the portal vein is the only differential diagnostic consideration and we have located this on other scans. Finally (for those of little faith), we performed a CT scan (Figure 100–C) after infusing the patient with intravenous *biliary* contrast material (Cholografin), which opacified the common duct (arrow). A dilated common bile duct should be identifiable with ultrasound. The trick is knowing where to look for it. Take a few minutes to study an anatomy text and you will be ready to orient the scan plane so that it passes through the long axis of the duct. This involves a longitudinal scan with a slight obliquity so that the scan plane is roughly parallel to the line connecting the patient's right shoulder with his left foot. We like to start scanning laterally and work our way toward the midline, keeping a sharp eye for a tubular structure with somewhat irregular walls (Figure 100–D) — the common duct. The duct is always located anterior to the vena cava and is usually anterior and to the right of the portal vein. The first reference discusses a scanning procedure which allows visualization of both the duct and portal vein on a single scan. When the duct is normal in size it may be very difficult to see (we do not lose any sleep over this). When enlarged, it stands out like a sore thumb. One final point: in the longitudinal view the more cranial part of the duct usually lies more anterior than the caudal portion; sometimes, in fact, the angulation is much more obvious than in this case.

Although ultrasound is an accurate method of identifying an enlarged common duct, we are not asked to evaluate this problem often. Maybe this is because differentiation of obstructive from nonobstructive jaundice is not a frequent problem in clinical medicine or maybe it reflects the popularity in our hospital of other diagnostic procedures, such as thin-needle percutaneous cholangiography. We won't argue about such preferences. We do feel, however, that ultrasound is almost always a more useful diagnostic procedure than intravenous cholangiography (and hope that the latter procedure will fall into well-deserved obsolescence).

There have been reports in the ultrasound literature of the successful diagnosis of common duct stones. So far we have not been able to do this accurately — prospectively, at least.

The patient in this case had a benign stricture at the ampulla of Vater, presumably from the previous passing of gallstones.

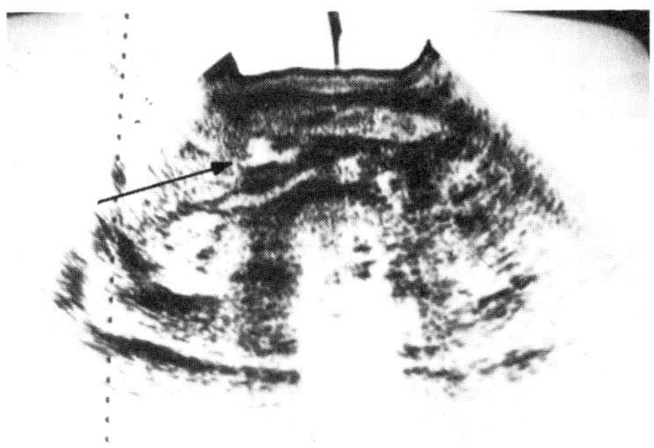

Figure 100–A. Transverse section +8. Dilated common bile duct.

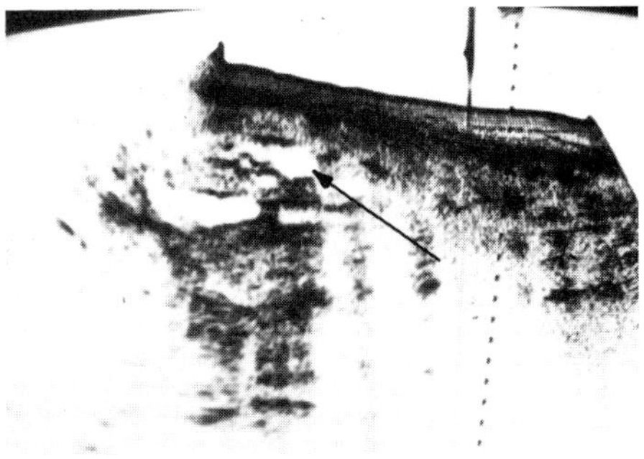

Figure 100–B. Longitudinal section 4R. Dilated common bile duct.

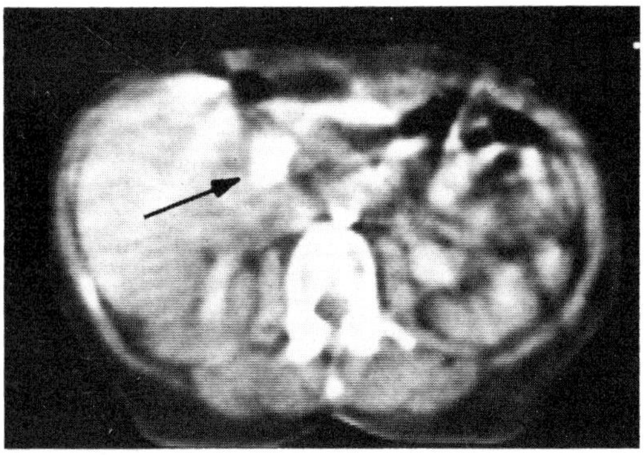

Figure 100–C. Dilated common bile duct. CT scan of patient in Figure 100–A performed with intravenous biliary contrast agent.

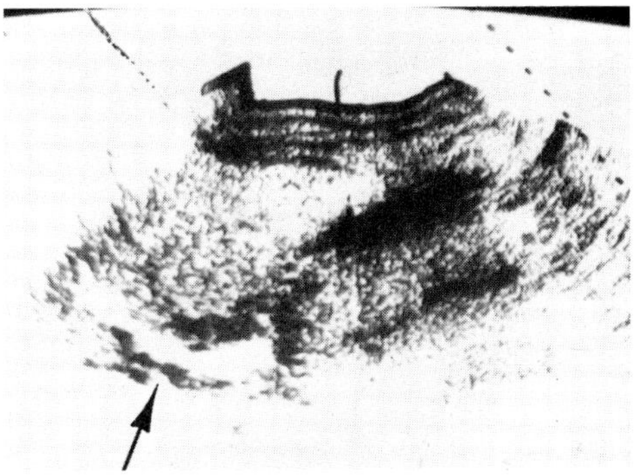

Figure 100–D. Longitudinal section 1R. Dilated common bile duct.

REFERENCES

Behan, M., Kazam E: Sonography of the common bile duct: value of the right anterior oblique view. Am J Radiol 10:701 (1978)

Malini S, Sabel J: Ultrasonography in obstructive jaundice. Radiology 123:429 (1977)

Lee TG, Henderson SC, Ehrlich R: Ultrasound diagnosis of common bile duct dilatation. Radiology 124:793 (1977)

Goldberg BB: Ultrasonic cholangiography: gray scale B-scan evaluation of the common bile duct. Radiology 118:401 (1976)

Case 101

CLINICAL INFORMATION

This 66-year-old woman is constipated. Twenty-six months ago she had a total abdominal hysterectomy and bilateral salpingo-oophorectomy for endometrial cancer. The tumor was felt to be confined to these organs, and all the resection margins as well as removed lymph nodes were free of tumor. She received a course of external radiation and has been well for the past year. Her physical examination, including rectal and pelvic examinations, is normal. Laboratory studies, including liver function tests, are unrevealing. There is no evidence of metastases on isotope liver or brain scan, and the chest film is clear.

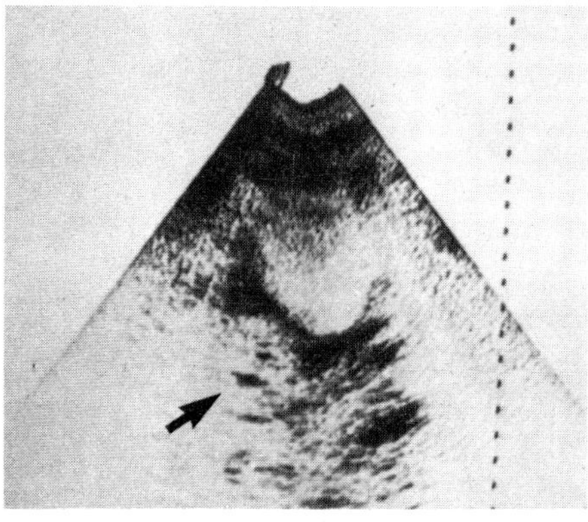

Figure 101–A. Longitudinal section 0L.

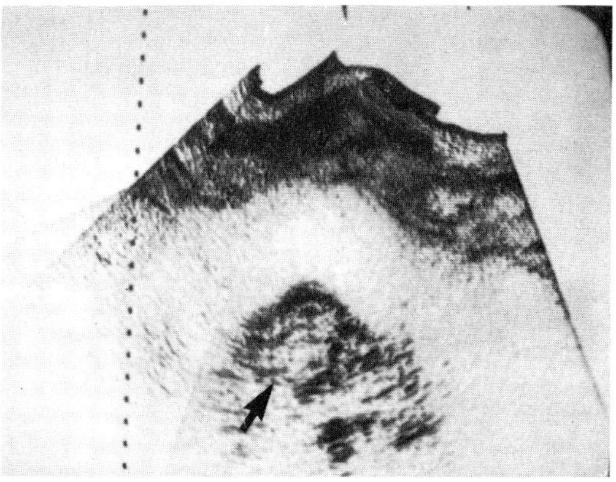

Figure 101–B. Transverse section −13.

DIAGNOSIS

Carcinoma of the colon.

ULTRASOUND FINDINGS

Both the longitudinal and cross-sectional scans show a solid mass behind the bladder (arrows). What is it? The ovaries and uterus have been removed. With this history our first choice is recurrent uterine cancer. We must also consider whether this could be bowel — a sigmoid colon filled with feces or tumor. There is no sure way to tell ultrasonically. We could give the patient a gentle water enema and repeat the scan. The enema will fill the colon and define its margin; but if the mass does turn out to be intracolonic we still would not be able to separate tumor from feces. We think it is, therefore, best to proceed directly to a barium enema (which is inevitable in this clinical setting). The enema results are surprising — the mass is an adenocarcinoma of the colon, not recurrent ovarian carcinoma (Figure 101–C).

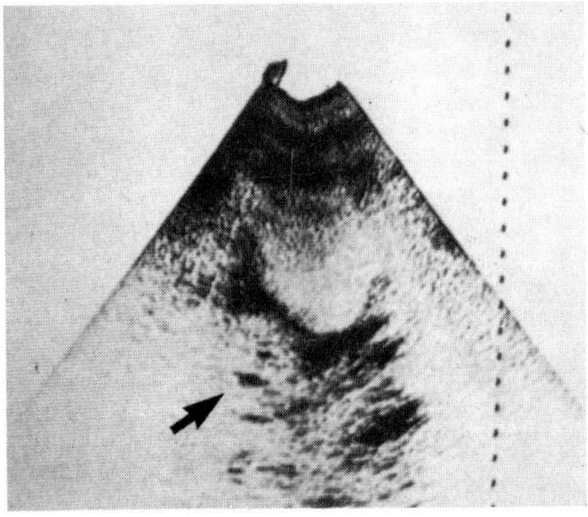

Figure 101A. Longitudinal section 0L. Carcinoma of sigmoid colon.

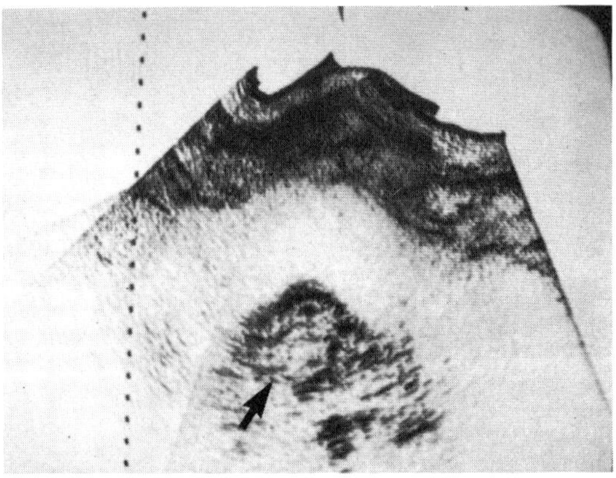

Figure 101–B. Transverse section −13. Carcinoma of sigmoid colon.

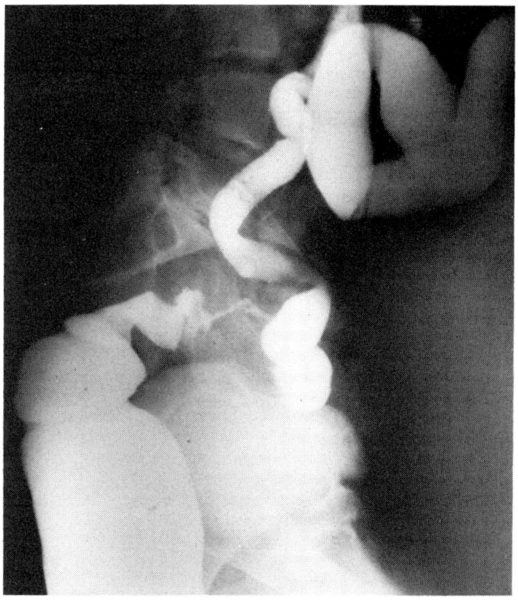

Figure 101–C.
Carcinoma of sigmoid colon. Radiograph from barium enema of patient in Figures 101–A and 101–B.

REFERENCES

Sample WF, Lippe BM, Gyepes MT: Gray scale ultrasonography of the normal female pelvis. Radiology 125:477 (1977)

Rubin C, Kurtz AB, Goldberg BB: Water enema: a new ultrasound technique in defining pelvic anatomy. J Clin Ultra 6:28 (1978)

Case 102

CLINICAL INFORMATION

This 17-year-old woman is 37 weeks pregnant by dates and biparietal measurement. There is some question of cephalopelvic disproportion and the obstetrician wants an estimation of the baby's weight.

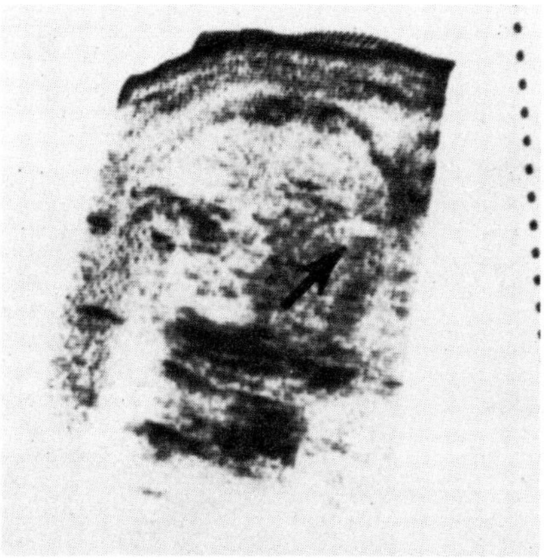

Figure 102–A.
Scan of fetal abdomen.

DIAGNOSIS

The estimated weight is 2,650 grams.

ULTRASOUND FINDINGS

Since the advent of obstetrical ultrasound, researchers have been striving to find measurements that will reflect the fetal condition accurately. By far the most successful has been the biparietal diameter, which correlates well with fetal maturation through most of pregnancy. Attempts to estimate fetal weight have not been as successful and are not a routine part of the obstetrical ultrasound practice. On occasion, you may, however, be asked to estimate fetal weight, so here is the current preferred method.

Fetal weight has been shown to be most closely related to the circumference of the fetal abdomen measured at the level of the umbilical vein (arrow in Figure 102-A). Sometimes the vein is obscured by the spine, and then we try to obtain a scan at the level of the kidneys (Figure 102-B). The circumference can be measured with either a small map-mileage measuring device or a piece of string. In Figure 102-A the measurement is 30.8 centimeters. Consulting published tables, we learn that this baby weighs 2,650 grams. Before getting too excited, we must remember that this is only a crude estimation; the 95 per cent confidence limit is about 15 per cent of the estimated weight, so that in this case we could be off by 370 grams. The references below go into this business in considerably more detail for those of you who want to know more.

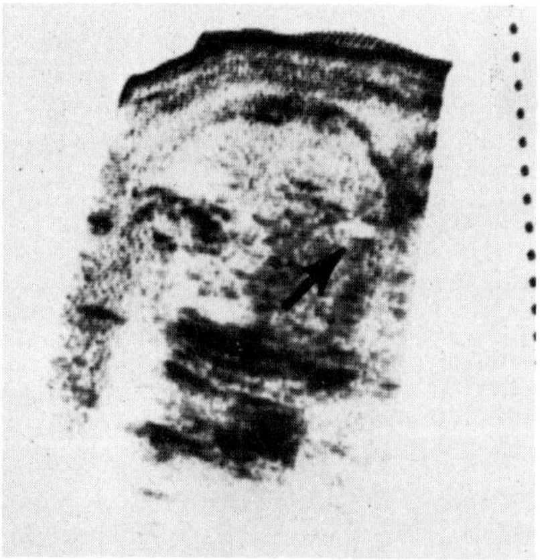

Figure 102–A.
Fetal abdomen showing umbilical vein.

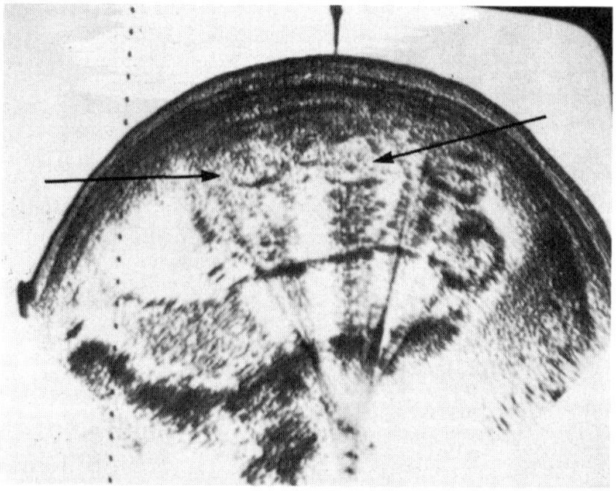

Figure 102–B.
Fetal abdomen showing spine and kidneys.

REFERENCES

Campbell S, Wilkin D: Ultrasonic measurement of fetal abdominal circumference in the estimation of fetal weight. Br J Obstet Gyn 82:689 (1975)

Morrison J, McLennan MJ: The theory, feasibility, and accuracy of an ultrasonic method of estimating fetal weight. Br J Obstet Gynecol 83:833 (1976)

Case 103

CLINICAL INFORMATION

This 30-year-old man feels "lousy." He is chronically tired, has frequent chills and headaches, and seems to "get every cold around." Recently he has become nauseated and feverish. On physical examination there is a questionable mass in the abdomen. Routine laboratory studies, including liver function tests and serum amylase, are normal. Chest film, upper GI series, and barium enema are also unrevealing. These scans were performed at high gain (accounting for the echo density in the liver) so that the para-aortic area could be adequately evaluated.

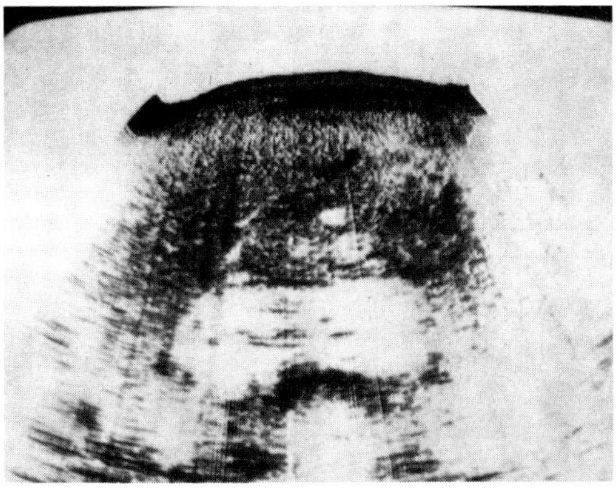

Figure 103–A. Transverse section +3.

Figure 103–B. Longitudinal section 1L.

DIAGNOSIS

Lymphoma (Hodgkin's disease).

ULTRASOUND FINDINGS

There is a prominent, nearly echo-free mass in the para-aortic and retroperitoneal areas. In initial scans it seemed that this mass was echo-free, but careful scanning at higher gain demonstrated the presence of low-level internal echoes. The diagnosis is limited to two conditions: acute pancreatitis and lymphadenopathy — most commonly due to lymphoma. (Horseshoe kidney might also be considered in a complete differential diagnosis.) The mass is located in the region of the pancreas, but pancreatitis can be excluded by the normal amylase and lack of pain. Thus we are left with lymphadenopathy.

Retroperitoneal lymphoma is a homogeneous tumor with few internal density differences to create acoustic interfaces. It can easily be confused wtih acute pancreatitis or even a cyst. Whenever we encounter an echo-free retroperitoneal mass we must make careful scans at high gain to avoid being misled. Lymphadenopathy due to lymphoma tends to develop on either side of the spine; frequently there is no significant retroaortic component. The appearance is more like a "dumbbell" or "saddle-bag" rather than the "doughnut" which is characteristic of lymph node metastases. This is not a hard and fast rule, however, because lymphoma can assume any configuration. Figure 103–C shows a focal mass of lymphoma, while Figure 103–D shows a patient with Hodgkin's disease and a classic "doughnut" encircling the aorta.

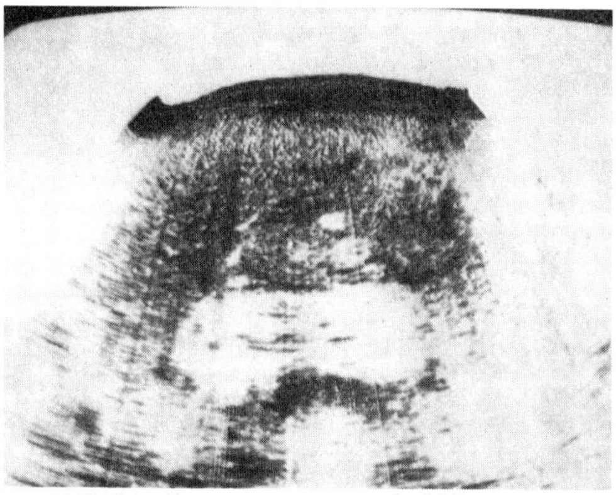

Figure 103–A. Transverse section +3. Hodgkin's disease.

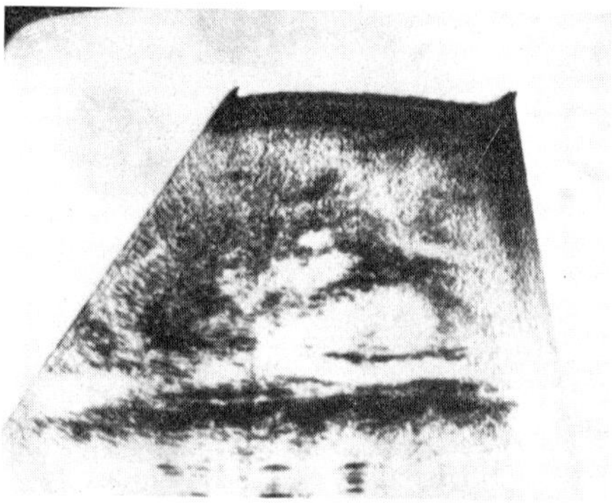

Figure 103–B. Longitudinal section 1L. Hodgkin's disease.

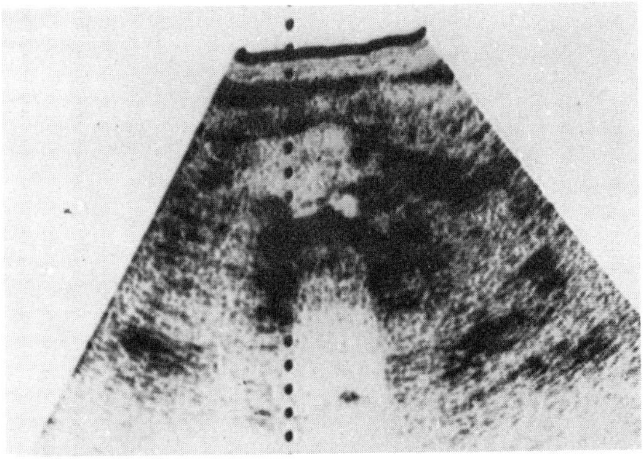

Figure 103–C. Transverse section +1. Lymphoma.

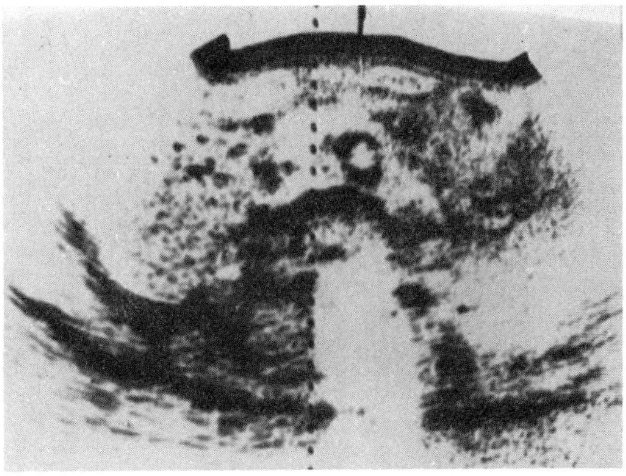

Figure 103–D. Transverse section +7. Hodgkin's disease.

REFERENCES

Brascho DJ, Durant JF, Green LE: The accuracy of retroperitoneal ultrasonography in Hodgkin's disease and non-Hodgkin's lymphoma. Radiology 125:485 (1977)

Rochester D, Bowie JD, Kunzmann A, et al: Ultrasound in the staging of lymphoma. Radiology 124:483 (1977)

Filly RA, Morglin S, Castellino RA: The ultrasonographic spectrum of abdominal and pelvic lymphoma. Cancer 38:2143 (1976)

Case 104

CLINICAL INFORMATION

This 33-year-old woman is only 25 weeks pregnant by dates but her uterus is very large. The general practitioner who is caring for her has sent her to you for help. Why is her uterus so big?

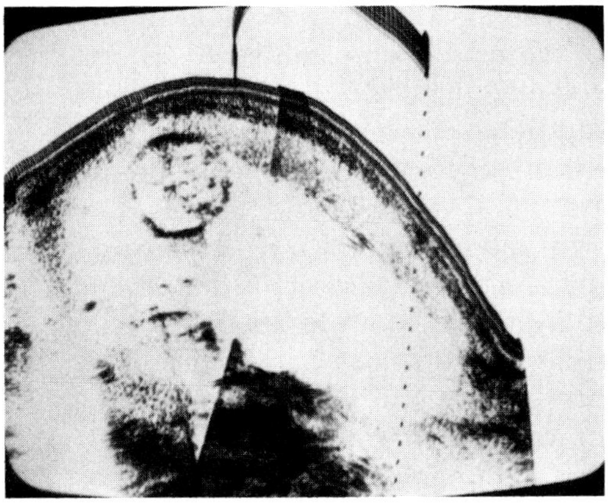

Figure 104–A. Longitudinal section 10R.

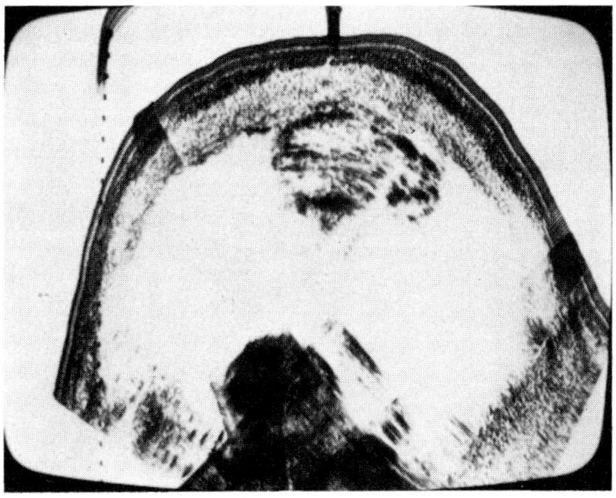

Figure 104–B. Transverse section −4.

DIAGNOSIS

Polyhydramnios.

ULTRASOUND FINDINGS

This enormous uterus barely fits into the picture, and we can see that it contains primarily amniotic fluid. This is an incontestable case of polyhydramnios. Polyhydramnios is an amorphous, ill-defined, wastebasket term for a variety of pathologic fetal conditions. Technically it means "too much amniotic fluid" but what is "too much"? Obviously this patient has "too much," but how about the patient shown in Figures 104–C and 104–D? The clinician who examined this patient thought that she had polyhydramnios, and who are we to disagree. Polyhydramnios has been associated with virtually every type of fetal maldevelopment and anomaly, as well as with most conditions that lead to spontaneous abortion. The whole business has always struck us as being silly. First, no one can agree on what constitutes "polyhydramnios." (It would seem that the most consistent criterion is the presence of a fetal anomaly or spontaneous abortion.) Second, assuming we could diagnose it, so what? What is the therapy? Until there are well-established criteria for defining this disease and a demonstrated effective therapy for it, we don't think you should bother with it. If the obstetrician wants to diagnose polyhydramnios we let him, but unless the case is as obvious as this one we do not make any statement about it on our reports.

A similar situation is intrauterine growth retardation (IUGR). This was a rather nebulous concept in the recent past, fraught with circuitous definitions and lack of agreement as to what, if anything, could be done about it. With the advent of ultrasound, considerable scientific attention has been focused on this problem, and some fairly objective criteria for making the diagnosis have been formulated. If either you or your obstetricians are interested in this sort of thing, we suggest you consult the reference below for a more complete discussion of the topic. It is a tough problem, but fortunately one that does not occur often, even in a busy general ultrasound laboratory.

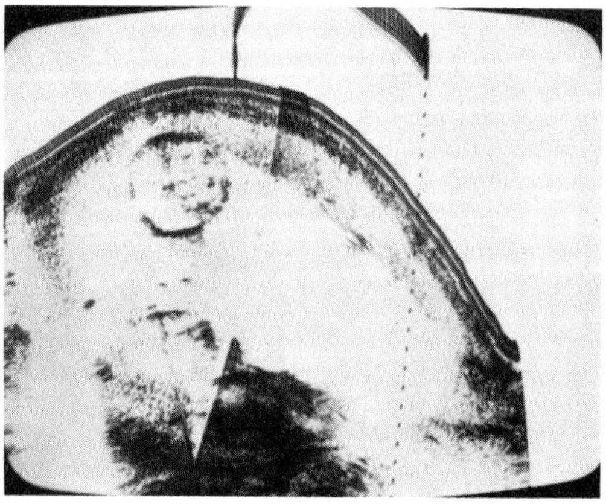

Figure 104–A. **Longitudinal section 10R.** Polyhydramnios.

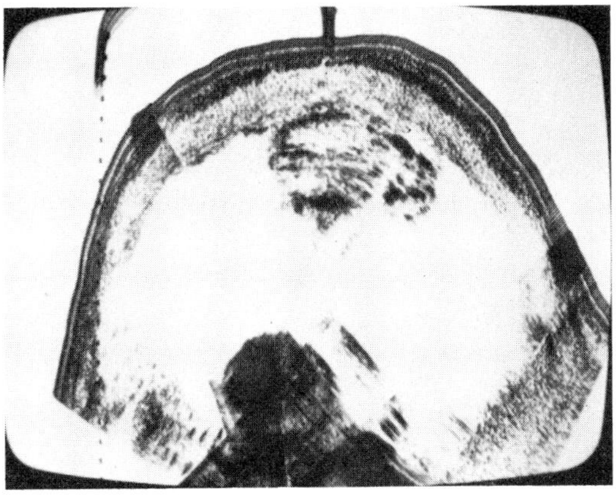

Figure 104–B. Transverse section −4. Polyhydramnios.

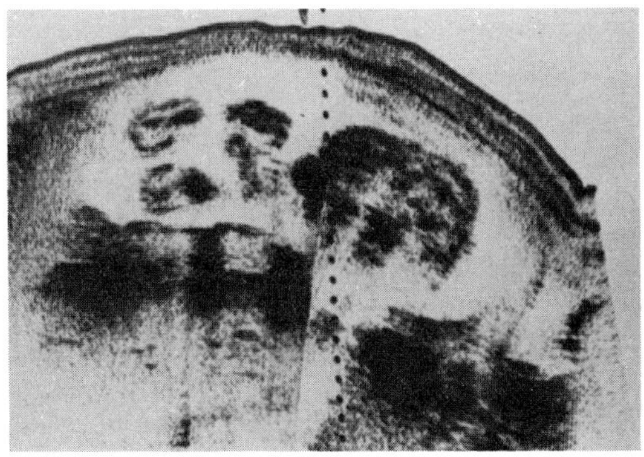

Figure 104–C. Longitudinal section 0L. Clinical polyhydramnios.

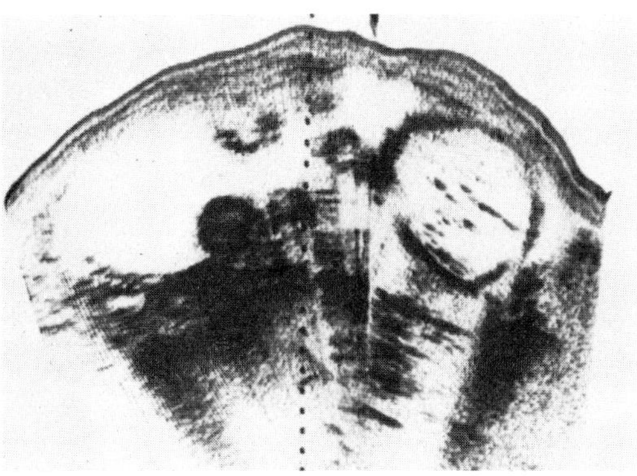

Figure 104–D. Transverse section 0. Clinical polyhydramnios.

REFERENCES

Phillips JF, Goodwin DW, Thomason SB, et al: The volume of the uterus in normal and abnormal pregnancy. J Clin Ultra 5:107 (1977)

Hobbins JC, Winsberg F: *Ultrasonography in Obstetrics and Gynecology.* Chapter 6. Williams & Wilkins Co., Baltimore (1977)

Case 105

CLINICAL INFORMATION

This 47-year-old man is a known alcoholic who has managed to keep his drinking under reasonable control; he has never had to be hospitalized and holds a steady job. Over the past few weeks he has had increasing difficulty with epigastric pain and burning. The symptoms are sometimes relieved by eating but frequently remain unabated. His appetite has remained good and he has not lost any weight. The physical examination is unremarkable and laboratory tests reveal mild anemia and borderline abnormal liver function tests.

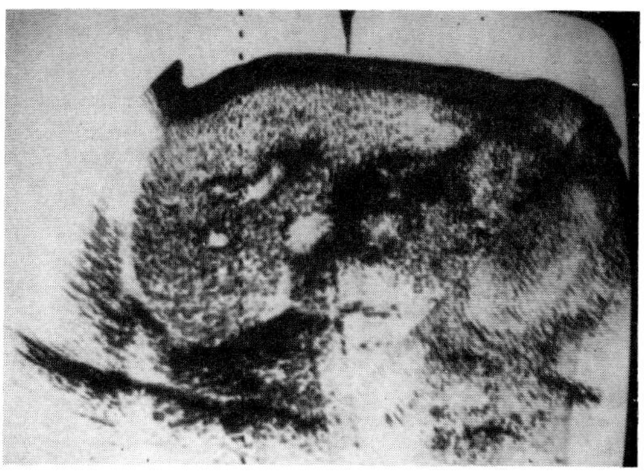

Figure 105–A. Transverse section +7.

DIAGNOSIS

Cirrhosis with early ascites.

ULTRASOUND FINDINGS

The liver generates loud echoes in a rather homogeneous distribution; this is the typical appearance of early generalized liver disease. Although any generalized hepatocellular disease can give this appearance, for all practical purposes this pattern represents cirrhosis of the liver. Of greater interest than the patient's cirrhosis is the presence of a thin, echo-free rim outlining the right lobe of the liver and separating it from the abdominal wall. This represents fluid in the peritoneal space — minimal ascites.

Gross ascites is easily recognized by ultrasound, but when there is only a small amount of fluid present, we must search carefully to find it. A small amount of ascites usually obeys the laws of gravity and collects in dependent portions of the peritoneal cavity. When the patient is supine we look around the tip of the right lobe of the liver (Figures 105–A and 105–B), or beneath the left lobe (Figure 105–C). The most dependent portion of the peritoneum is the pelvis, and if the bladder is full so that there is an acoustic window, small amounts of fluid can be identified behind the pelvic organs. Figure 105–D is an example of minimal ascites in a woman. The arrow points to the ascitic fluid collection lying behind the uterus and broad ligaments. (The bladder is the large echo-free space anterior to the uterus.)

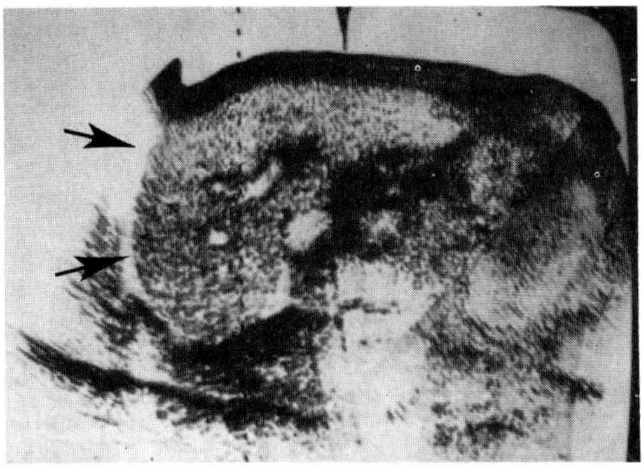

Figure 105–A. Transverse section +7. Cirrhosis and minimal ascites.

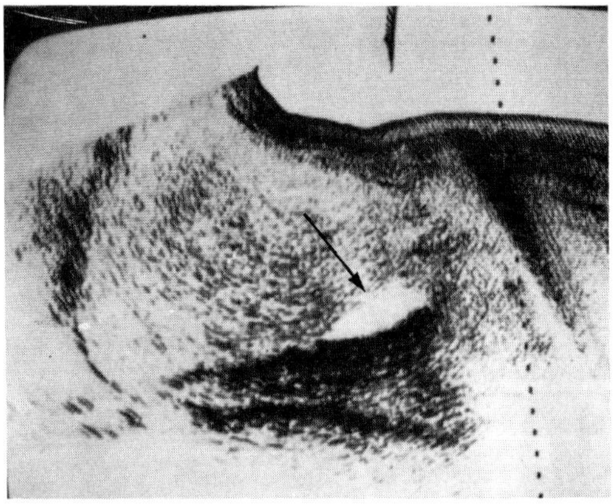

Figure 105–B. Longitudinal section 8R. Minimal ascites.

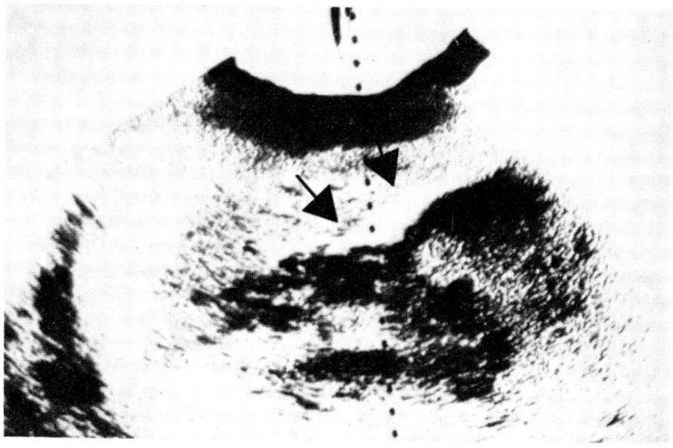

Figure 105–C. Transverse section +13. Minimal ascites.

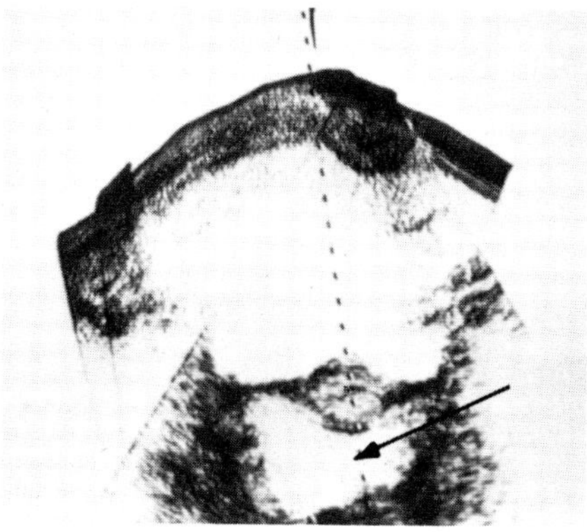

Figure 105–D. Transverse section −10. Minimal ascites.

REFERENCES

Yeh HC, Wolf BS: Ultrasonography in ascites. Radiology 124:783 (1977)

Goldberg BB, Goodman GA, Clearfield HR: Evaluation of ascites by ultrasound. Radiology 96:15 (1970)

Case 106

CLINICAL INFORMATION

This 47-year-old woman seems to be plagued with abdominal problems. For several years she had episodes of pain in her right upper abdomen and back. This was finally diagnosed as gallstones, and she had a cholecystectomy. The surgery and postoperative course were smooth, and for three years she has felt well and been pain-free (her gallbladder did contain gallstones). Three weeks ago she had another episode of pain, which was very severe but fortunately lasted only two hours. The pain was again centered in her right upper abdomen and back and spread so that it involved her entire right side and even her leg. Her doctor, examining her in his office, found nothing suspicious. Aside from a nicely healed cholecystectomy scar, her physical examination was normal. He decided that an ultrasound scan was indicated.

The technician brings you cross-sectional (Figure 106–A) and longitudinal (Figure 106–B) scans and asks what is going on in the right upper quadrant. She can easily identify the liver (large arrowhead), but wonders what she is seeing beneath it (small arrows). She is asked to get a supine abdominal x-ray film (Figure 106–C).

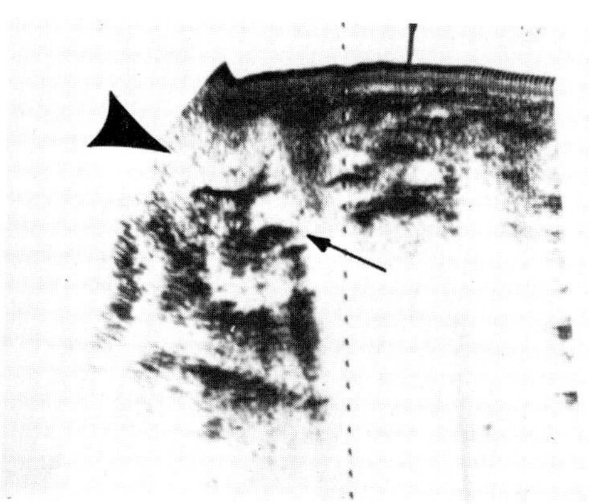

Figure 106–A. Transverse section +5.

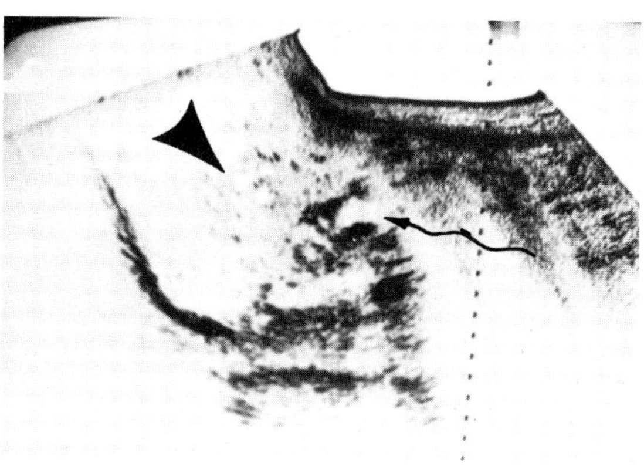

Figure 106–B. Longitudinal section 6R.

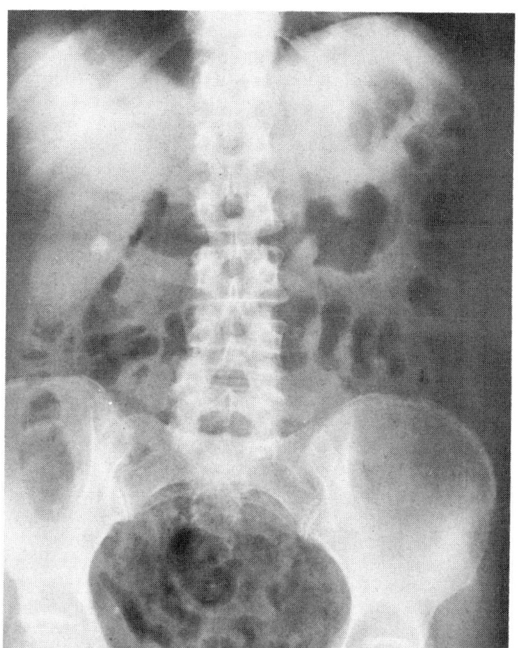

Figure 106–C. Supine abdominal radiograph.

DIAGNOSIS

Stone in renal pelvis.

ULTRASOUND FINDINGS

The small arrows are pointing to a fluid-filled (echo-less) area, which contains a dominant, high-level echo-producing structure. Looking carefully, we notice that the echo-producing structure is casting a faint acoustic shadow. This would be a typical appearance of gallstones, but the patient has had a cholecystectomy. To confirm our suspicion of a stone, we look at the abdominal film; and, sure enough, there is a fairly large stone present. But where is it? The gallbladder is gone. Could this be a stone in the common bile duct or a cystic duct remnant? Although possible, this diagnosis seems unlikely because the location is too far to the right for the biliary duct system. We look again at our ultrasound pictures and notice that the stone seems to be sitting within the right kidney! Checking the abdominal film further, we see that, indeed, the stone is projected over the kidney and confidently make the diagnosis of "right renal stone, probably in the collecting system." A subsequent retrograde pyelogram confirms our findings.

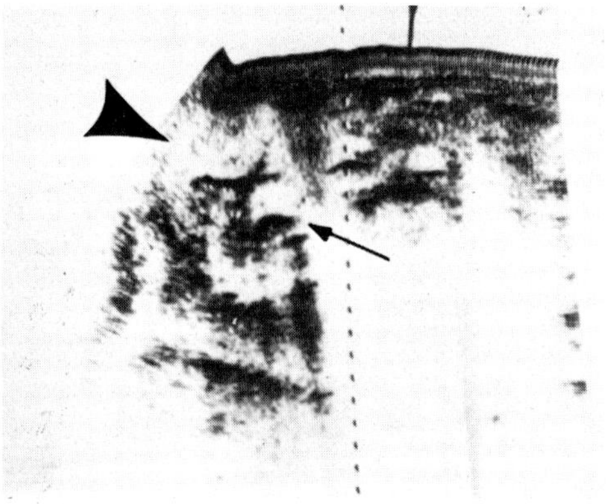

Figure 106–A. Transverse section +3. Renal stone.

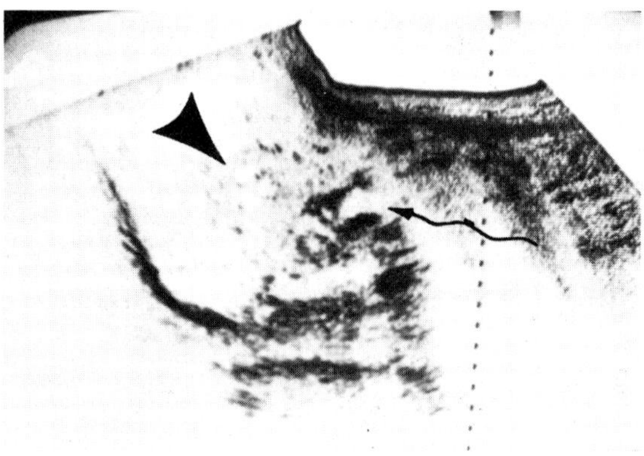

Figure 106–B. Longitudinal section 6R. Renal stone.

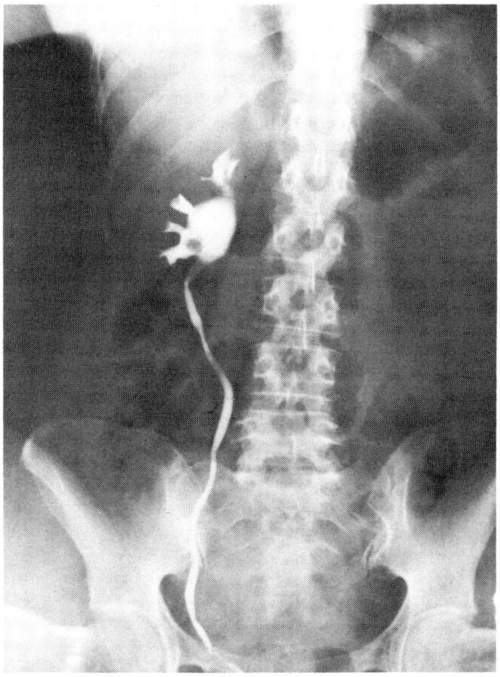

Figure 106–D.
Retrograde pyelogram from patient in Figures 106–A through 106–C.

REFERENCES

Pollack HM, Arger PH, Goldberg BB, et al. Ultrasonic detection of nonopaque renal calculi. Radiology 127:233 (1978)

Edell S, Zegel H: Ultrasonic evaluation of renal calculi. Am J Radiol 130:261 (1978)

Case 107

CLINICAL INFORMATION

This 32-year-old woman has systemic lupus erythematosus, which has been under good control with low-dose steroids. She has been asymptomatic for five years, but in the last six months has had three episodes of midepigastric pain lasting about 12 hours. The pain has not been particularly severe, is not colicky, and is not associated with nausea or vomiting. In each episode the pain has resolved spontaneously with no specific treatment. Physical examination, routine laboratory studies, including liver function tests and serum amylase, and oral cholecystogram and an upper GI series are all normal. In error, the patient was not instructed to skip breakfast and arrived for her ultrasound examination two hours after she had eaten a bowel of cereal, two pieces of toast, and two cups of coffee. Despite the full stomach we went ahead with the scan, and somewhat to our surprise the image turned out to be quite nice. The technician reported the examination as normal except for the presence of an unusual tubular structure (arrow).

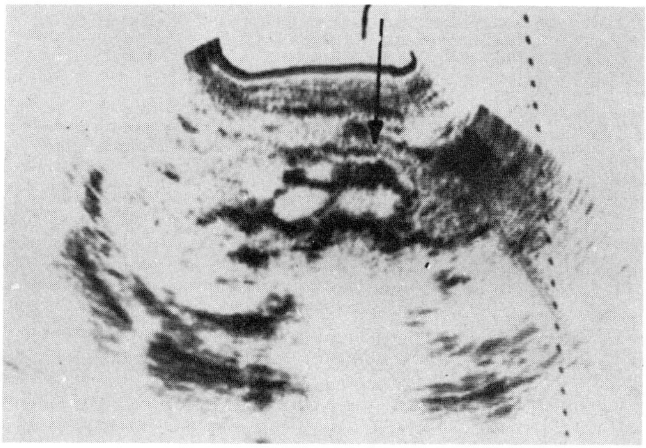

Figure 107–A. Transverse section +14.

DIAGNOSIS

Pancreatic duct (probably normal).

ULTRASOUND FINDINGS

The first thing we notice is the patient's breakfast — the partially echo-less mass in the left upper quadrant. In this case there was not much air in the stomach, and the sound beam passed through readily, to give us a good look at the underlying tail of the pancreas, just at the upper limits of normal in size. The body of the pancreas is very clearly seen and lies just underneath the thin left lobe of the liver. Then there is the unusual tubular structure running along the posterior margin of the pancreas. What can it be? It is not the splenic vein, because we can identify that structure lying just above the inferior vena cava and just to the left of the portal vein. (We verify this by following the splenic vein out toward the spleen in other scans.) Could it be the splenic artery? We follow the mystery vessel on serial scans to see if it connects with the aorta, celiac, or superior mesenteric arteries; it does not. Therefore, we are left with the conclusion that this must be the pancreatic duct.

It is very unusual to identify the pancreatic duct on an abdominal scan. In the few cases which have been reported the duct has been obstructed, either from tumor or gallstones. Nevertheless, this patient appears to have neither tumor nor gallstones, and the remainder of the pancreas seems to be normal, both anatomically and clinically. The episodes of abdominal pain have not recurred, and she remains asymptomatic and healthy (except for her underlying lupus). Talking with other ultrasonographers, we have learned that most large hospitals have a few cases like this: apparently healthy people in whom the pancreatic duct is visualized. While we do not feel that visualization of the pancreatic duct necessarily indicates pathology, it should alert the ultrasonographer to search carefully for evidence of either gallstones or an ampullary tumor.

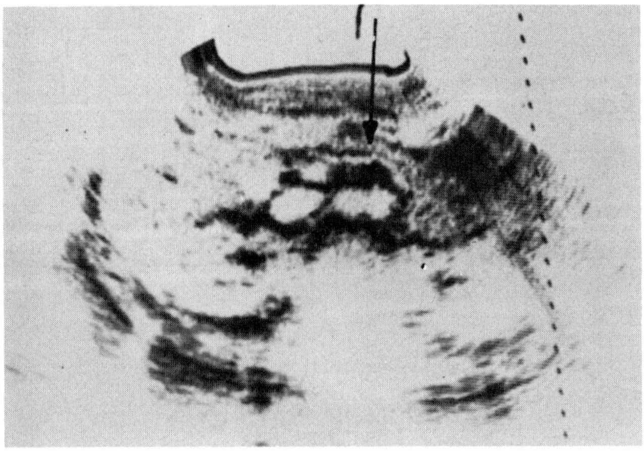

Figure 107–A. **Transverse section +14.** Pancreatic duct.

REFERENCE

Gosink BB, Leopold GR: The dilated pancreatic duct: Ultrasonic evaluation. Radiology 126:475 (1978)

Case 108

CLINICAL INFORMATION

This patient is an acutely ill 21-year-old college senior who is normally very robust. When he appeared at the college infirmary several days ago, however, he was nearly prostrate and complained of fever, chills, extreme malaise, left-sided chest pain on inspiration, and a productive cough. The resident who examined him discovered both rales and dullness to percussion in the left chest; a chest x-ray film showed pneumonia in the left lower lobe with effusion; and his WBC was 18,500 with a marked shift to the left. Sputum culture has subsequently grown out *Staphylococcus aureus*, and attempts to remove the fluid (which is clearly visible on the lateral chest film) have met only with frustration: one attempt produced 25 ml. of pink-tan pus, and a second was completely unproductive. The man's response to antibiotic therapy has been slow, and the clinicians would like to drain as much of the pleural fluid as possible without resorting to a chest tube. Also, they are a bit suspicious of the radiologist's repeated statements that there is still pleural fluid present. They wonder if the increased density on the x-ray film (Figure 108–A) could be consolidation in the lung. Figure 108–B is a longitudinal scan with the patient upright; the large arrow points to the previous thoracentesis site.

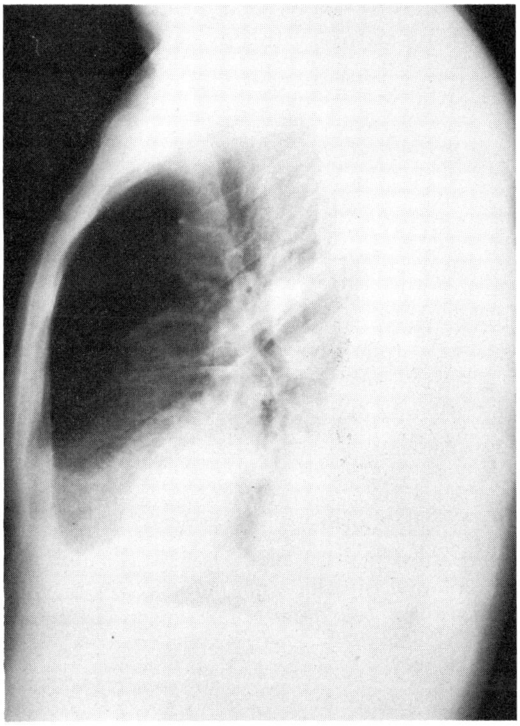

Figure 108–A.
Lateral chest radiograph.

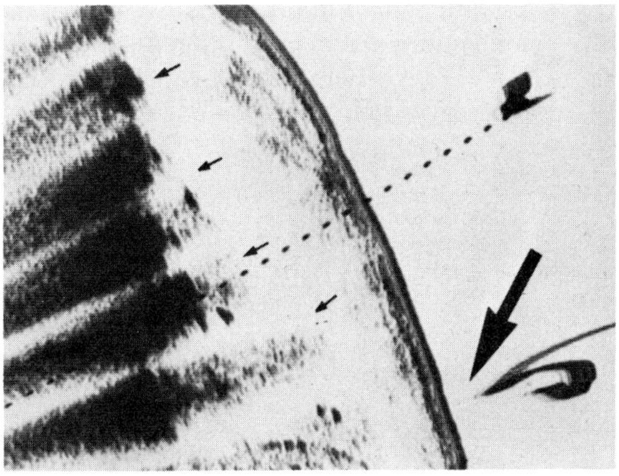

Figure 108–B.
Upright scan of left hemithorax performed from the back.

DIAGNOSIS

Pleural effusion: previous thoracentesis site was too low.

ULTRASOUND FINDINGS

The marker on Figure 108–B (large arrow) shows the level of the previous thoracentesis. The smaller arrows delineate the margin of the lung. A pocket of pleural fluid is seen as an echo-free space above this, just inside the chest wall. The earlier taps failed because they were made at a site where the fluid collection is very shallow. It is a simple matter to mark the patient's skin where a thoracentesis seems likely to be productive, and try again. If the procedure is not going to be done in the ultrasound laboratory, be certain to indicate how the patient was positioned when you scanned and marked him.

We do *not* advocate ultrasound scanning before thoracentesis except in problem cases. We would be run ragged if we had to do this for every patient. We request that the physician first attempt the tap guided by the chest film and physical findings. If his attempts fail, we will then help out. Also, ultrasound is not an effective way to look for small effusions; in general, if fluid cannot be demonstrated by conventional and decubitus x-ray films, ultrasound scanning will not find it.

In Figure 108–B the very strong bands of heavy black echoes are reverberations within aerated lung. Figure 108–C is a similar scan of the left chest of a senior radiologic technologist who was discovered to have a totally opaque hemothorax on chest x-ray. The scan demonstrates a large effusion associated with complete collapse of the left lung (arrows); in this case there are no reverberations because there is no aerated lung.

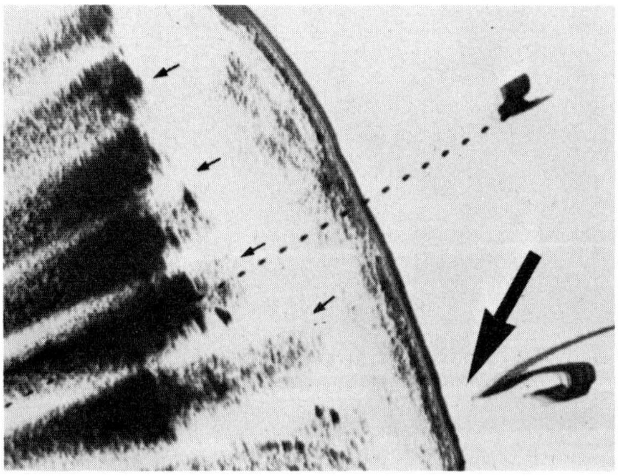

Figure 108–B.
Pleural effusion. Upright scan of left hemithorax performed from the back.

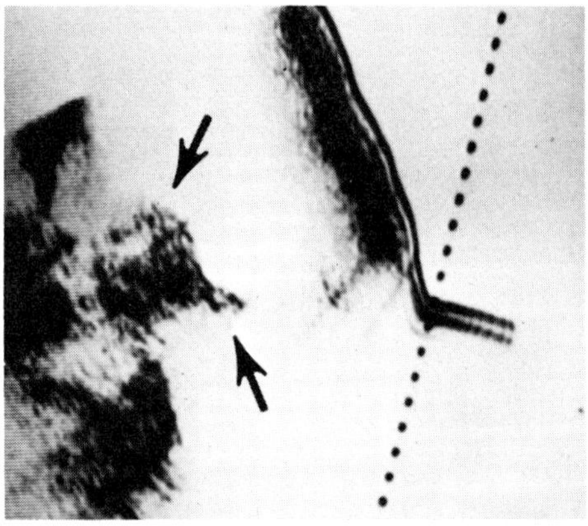

Figure 108–C.
Pleural effusion and atelectasis. Upright scan of left hemithorax performed from the back.

REFERENCE

Doust BD, Baum JK, Maklad NF, et al: Ultrasonic evaluation of pleural opacities. Radiology 114:135 (1975)

Case 109

CLINICAL INFORMATION

This 60-year-old woman has had a lump in her abdomen for several years, but over the past two months it has throbbed on several occasions. She has no idea where the lump came from; she has always been healthy, has never had a major illness and has not suffered any trauma. Except for the recent pain, the lump has always been asymptomatic and, to her belief, has not changed in size. Her physician feels a 10-centimeter mass in the left upper quadrant that he thinks is an enlarged spleen. The remainder of the physical examination is normal, as are all routine laboratory tests. An upper GI series shows in the left upper quadrant a large mass which displaces the stomach to the right and anteriorly; it is thought to be an enlarged spleen.

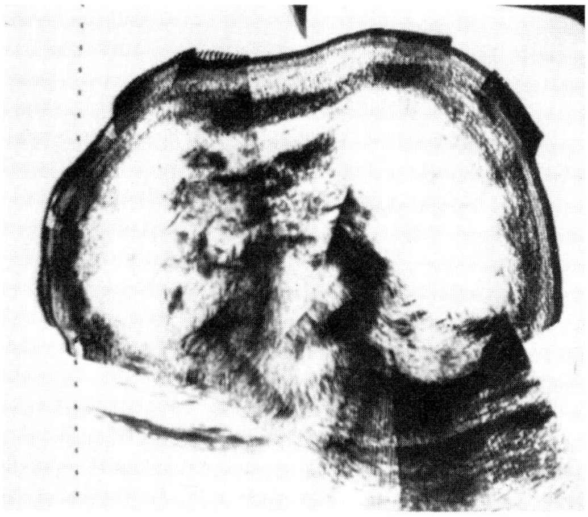

Figure 109–A. Transverse section +9.

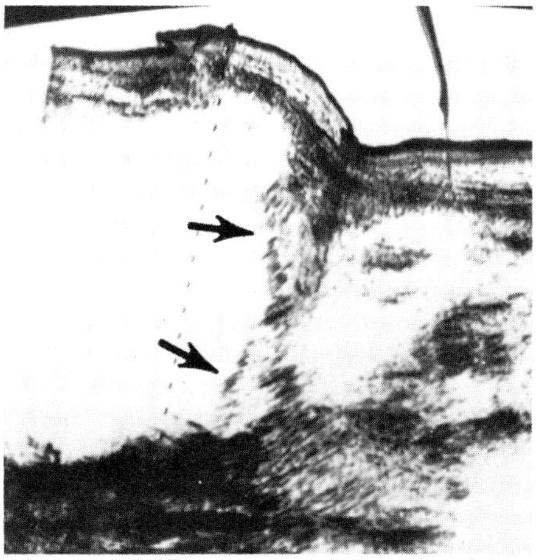

Figure 109–B. Longitudinal section 8L.

DIAGNOSIS

Splenic cyst.

ULTRASOUND FINDINGS

The ultrasound scans show that the mass is echo-free and, hence, fluid-filled. On the longitudinal section (Figure 109–B) a small amount of compressed splenic tissue is apparent just above the kidney (arrows). These findings are very suggestive of a benign splenic cyst. A benign cyst of the left lobe of the liver must also be considered, but we can exclude a renal cyst on the longitudinal scan (Figure 109–B), and a pancreatic pseudocyst is unlikely in this clinical setting. A CT scan (Figure 109–C) confirms the ultrasound diagnosis — the "density" of the mass is 8 Hounsfield units. Because of the pain, this cyst was surgically removed.

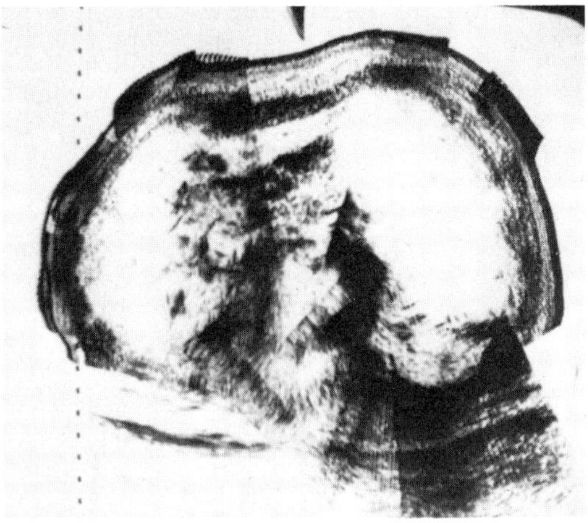

Figure 109–A. Transverse section +9. Splenic cyst.

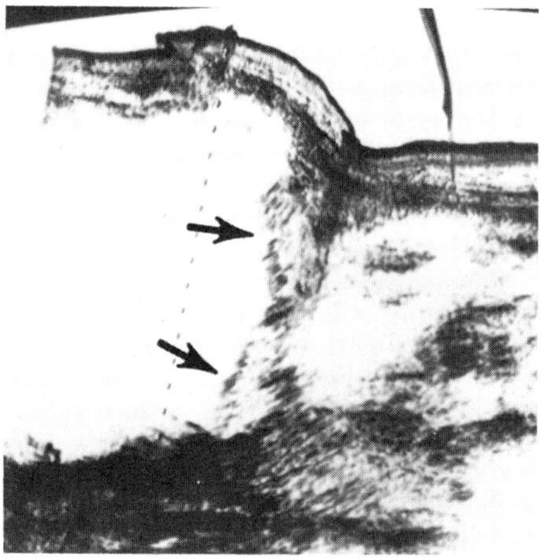

Figure 109–B. Longitudinal section 8L. Splenic cyst.

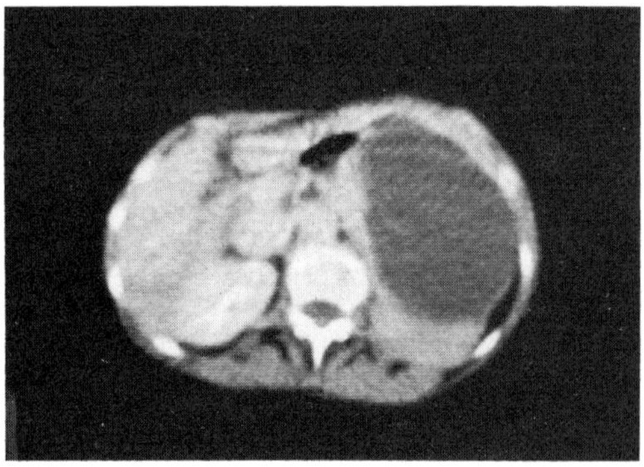

Figure 109–C. Transverse section +9. Splenic cyst. CT scan.

REFERENCE

Bhimji SD, Cooperberg PL, Naimon S, et al: Ultrasound diagnosis of splenic cysts. Radiology 122:787 (1977)

Case 110

CLINICAL INFORMATION

This elderly farmer has aching in the lower leg after exercise. The vascular surgeon to whom he was referred discovered a painless lump behind the man's left knee and suspects an aneurysm of the popliteal artery. Pulses in both feet are diminished in strength. An arteriogram shows decreased "run-off" in the calf, but the popliteal artery is relatively normal in caliber (Figure 110–A). Figure 110–B is a longitudinal ultrasound scan of the region.

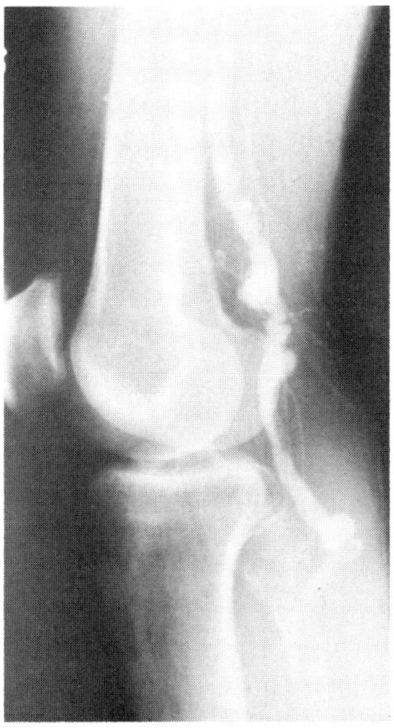

Figure 110–A.
Lateral arteriogram of left popliteal artery.

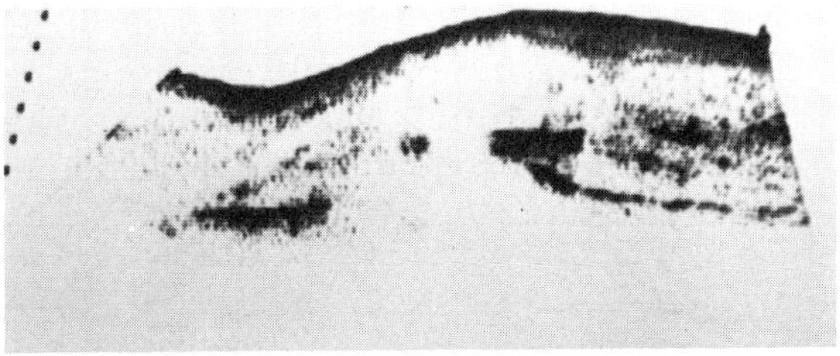

Figure 110–B.
Longitudinal scan of popliteal artery.

DIAGNOSIS

Popliteal artery aneurysm.

ULTRASOUND FINDINGS

Here is another example of the advantage which ultrasound has over arteriography in evaluating blood vessels: the aneurysm in the popliteal artery is clearly seen because ultrasound shows the exterior of the vessels, whereas the arteriogram shows only the lumen. The course of the artery is irregular on the radiograph so that we suspect that something is amiss in the popliteal region, but we cannot be certain that there is an aneurysm. The ultrasound scan confirms our suspicions. For comparison, Figure 110–C shows the opposite leg, where the popliteal artery is normal. The popliteal fossa is easy to examine, and, with care, the artery can usually be seen. We use a 3.5 MHz, 13 mm diameter transducer with a medium focus for better detail and scan longitudinally. Rarely do we gain anything by scanning transversely. Figure 110–D is a scan from another patient who has an aneurysm of the distal femoral artery extending to, but not really involving, the popliteal artery.

It is not possible to evaluate flow in the vessel by B-mode scan; an arteriogram or Doppler ultrasound is necessary to get at this problem.

Figure 110–E is a scan from yet another patient with a mass behind the knee. The scan shows a relatively echo-free zone behind the femoral condyle, a part of which extends into the upper calf. This picture is typical of a large popliteal cyst (Baker's cyst) with extension downward. The portion in the calf may be ruptured. Ultrasound scanning provides an easy, noninvasive means of making this differentiation and avoids the necessity for an invasive study, such as venography or arteriography.

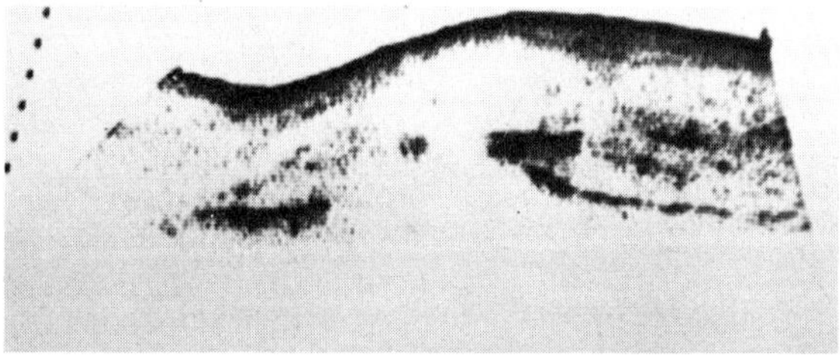

Figure 110–B.
Popliteal artery aneurysm. Longitudinal scan of popliteal artery.

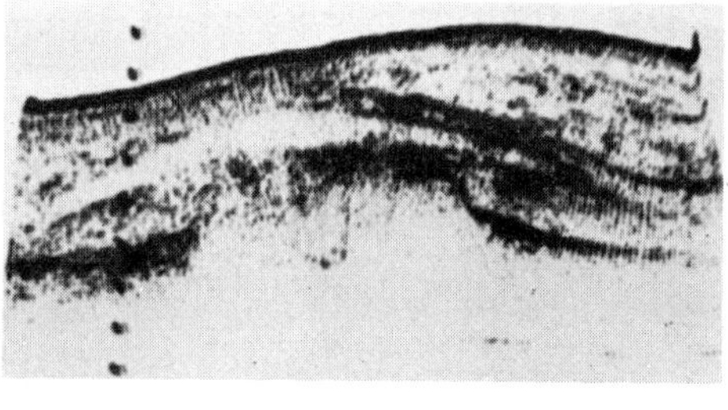

Figure 110–C.
Normal popliteal artery. Longitudinal section.

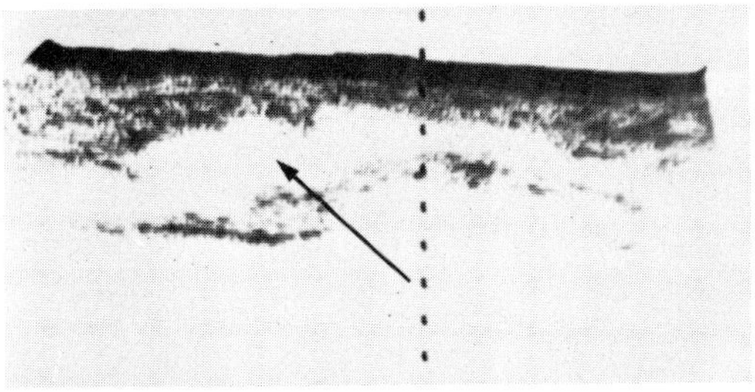

Figure 110–D.
Femoral artery aneurysm. Longitudinal section.

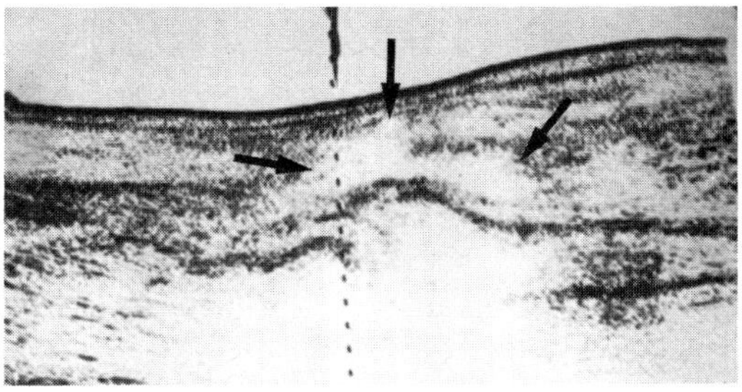

Figure 110–E.
Popliteal cyst (Baker's cyst). Longitudinal section through popliteal fossa.

REFERENCES

Silver TM, Washburn RL, Stanley JC, et al: Gray scale ultrasound evaluation of popliteal artery aneurysms. Am J Radiol 129:1003 (1977)

Davis RP, Neiman HL, Yao JST, et al: Ultrasound scan in diagnosis of peripheral aneurysms. Arch Surg 112:55 (1977)

Goldberg BB: Ultrasonic evaluation of superficial masses. J Clin Ultra 3:91 (1975)

Case 111

CLINICAL INFORMATION

Five days ago this 17-year-old boy was involved in an automobile accident. His injuries included a fracture of the left femur, several fractured ribs, a contused lung, and a ruptured spleen. When brought to the Emergency Ward, he was in frank shock and was immediately taken to the Operating Room where a badly lacerated spleen was removed. Once the bleeding was controlled, his condition stabilized rapidly, and for the next three days he continued to improve, was afebrile, and took oral feedings. On the fourth postoperative day he had some abdominal pain with a low-grade fever. In the past 24 hours the pain has increased in severity and his temperature has risen to 103 degrees. The abdomen is diffusely tender with guarding and there are no bowel sounds. The surgical incision is healing normally. The white blood count is 17,000 with a marked left shift; the liver function tests and electrolytes are normal, but the serum amylase is 735. The ultrasound technician reports that the abdomen is extremely tender, and it is difficult to make a smooth pass with the transducer because of the patient's discomfort.

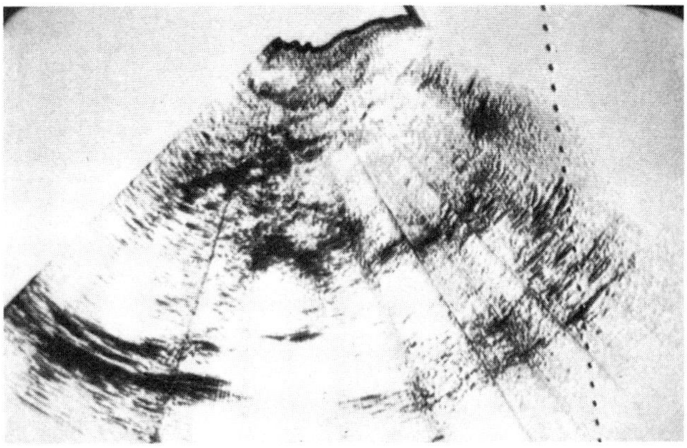

Figure 111–A. Transverse section +15.

DIAGNOSIS

Traumatic pancreatic pseudocyst.

ULTRASOUND FINDINGS

There is a large, mostly echo-free mass in the left upper abdomen. (The very low-level "echoes" in this mass are artifacts caused by electrical interference.) This mass has an irregular and poorly defined wall, indicating it is not a simple cyst (of the liver or spleen, for example). The clinical setting suggests two diagnoses: intraperitoneal abscess and pancreatic pseudocyst. Although the patient is definitely a candidate for an abscess, having had emergency surgery and considerable hemorrhage and trauma to this area, the rapid and fulminant onset of symptoms is somewhat unusual for this process. On the other hand, the clinical setting is quite suggestive of an acute pancreatic inflammation or traumatic pseudocyst.

There are two ways the pancreas can be traumatized in a patient such as this. The most obvious is direct blunt trauma from the auto accident, but another, and not unheard of, cause is iatrogenic trauma from emergency splenectomy. In many patients the tail of the pancreas is not securely attached to the retroperitoneum and can be lifted up along with the splenic artery and vein when the surgeon mobilizes the spleen. This is particularly likely in emergency operations where rapid bleeding obscures the operative field, and the surgeon's prime concern is to control the hemorrhage. A large clamp is quickly placed across the splenic pedicle; and if the tail of the pancreas is in the area, it can easily be caught in the clamp and traumatized.

In this case an aspiration biopsy verified the diagnosis, and with conservative therapy the pseudocyst resolved over the course of the next month.

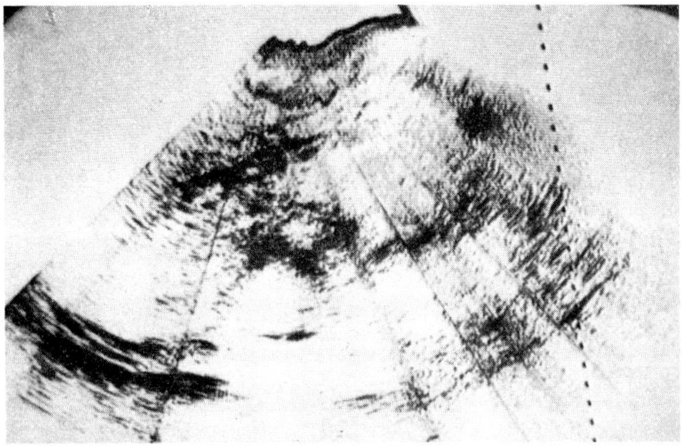

Figure 111–A. Transverse section +15. Pancreatic pseudocyst.

REFERENCES

Conrad MR, Landay MJ, Khoury M: Pancreatic pseudocysts: unusual ultrasound features. Am J Radiol 130:265 (1978)

Van Hurden JA, ReMine WH: Pseudocysts of the pancreas: review of 71 cases. Arch Surg 110:500 (1975)

Hancke S, Pedersen JF: Percutaneous puncture of pancreatic cysts guided by ultrasound. Surg Gynec Obstet 142:551 (1976)

Case 112

CLINICAL INFORMATION

This 24-year-old woman is 19 weeks pregnant by dates and is having a difficult time. She had severe morning sickness and frequent headaches in the first trimester. The emesis has worsened in the past few weeks, and three days ago there was a brief episode of vaginal bleeding. Her uterus is 24 weeks in size, and there is a tender mass in the left adnexal region.

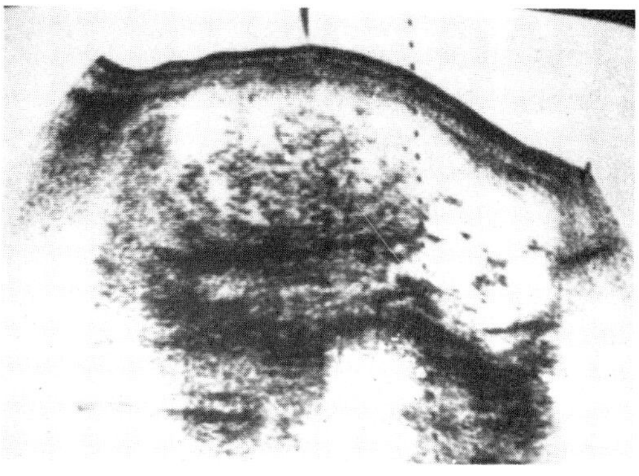

Figure 112–A. Longitudinal section 4L.

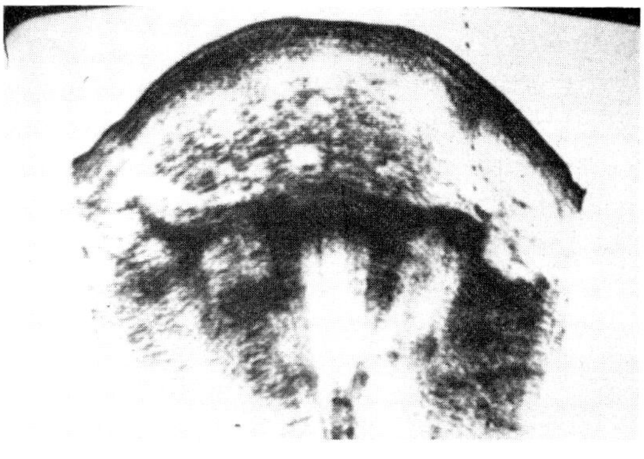

Figure 112–B. Transverse section +1.

DIAGNOSIS

Gestational trophoblastic disease (molar pregnancy).

ULTRASOUND FINDINGS

There is no fetus in this uterus; instead we see a heterogeneous collection of echoes interspersed with cystic areas. In the left adnexal region there is a large, multilocular cyst (arrows). These findings are characteristic of trophoblastic disease or molar pregnancy.

Trophoblastic disease is a term applied to a group of conditions resulting from abnormal placental development. The most common is molar pregnancy or "hydatidiform mole," a benign degeneration of the placenta. "Invasive mole" is a locally invasive form of trophoblastic disease, which some authorities consider to be a low-grade malignancy. Choriocarcinoma is the malignant member of the group and accounts for about 2.5 per cent of the cases. All forms of this disease are characterized by an extremely high output of chorionic gonadotropin, which stimulates large theca lutein cysts in the ovaries. The classic ultrasound appearance is the combination of (1) a bizarre, heterogeneous mass in the uterus, (2) no identifiable fetus, and (3) a large, often multilocular theca lutein cyst in the ovary. The majority of cases are not difficult to diagnose, but occasionally problems are encountered. Sometimes the intrauterine molar tissue is uniform and can be mistaken for a normal placenta; on other occasions there is very little tissue, and the presence of considerable fluid and blood causes the uterus to resemble that in a normal pregnancy. Careful serial scans through the entire uterus will usually clarify the situation. Figure 112–C is another example of molar pregnancy.

There is no way to distinguish a simple from an invasive mole on the basis of the ultrasound scan appearance. Similarly, it is not possible to tell choriocarcinoma from a mole by looking only at the tissue mass. In general, choriocarcinomas produce extremely high levels of hormone, and the resulting ovarian cysts are huge. Figure 112–D is a choriocarcinoma (arrow) with a massive theca lutein cyst. When choriocarcinoma is suspected, the liver should be scanned to search for metastases. If a hepatic lesion is found, we are dealing with a malignant form of trophoblastic disease (Figure 112–E).

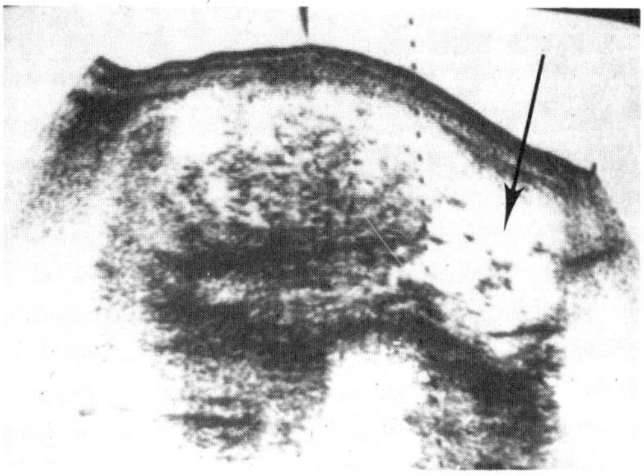

Figure 112–A. Longitudinal section 4L. Molar pregnancy with theca lutein cyst.

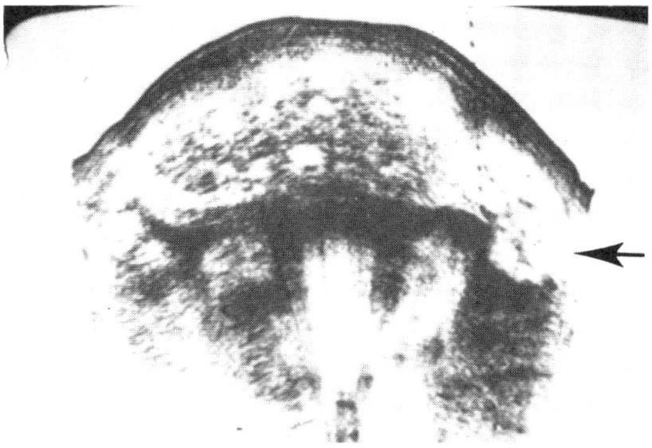

Figure 112–B. **Transverse section +1.** Molar pregnancy with theca lutein cyst.

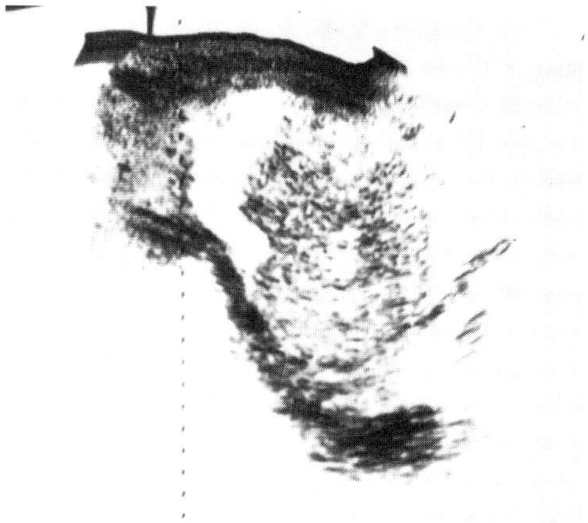

Figure 112–C. **Longitudinal section 0L.** Molar pregnancy.

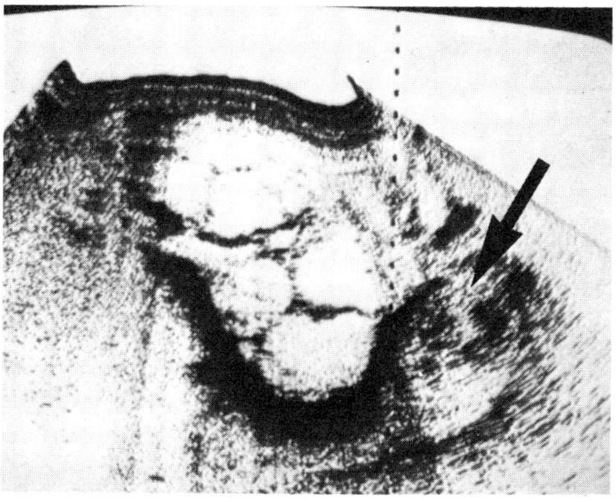

Figure 112–D. **Longitudinal section 2R.** Choriocarcinoma with theca lutein cyst.

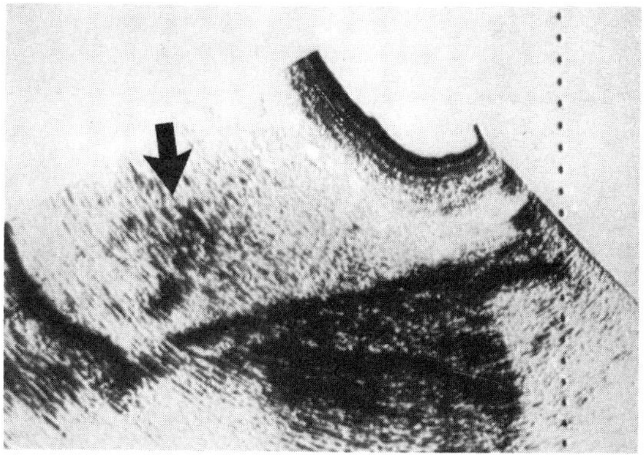

Figure 112–E. **Longitudinal section 7R.** Liver metastases in patient shown in Figure 112–D.

REFERENCES

Fleischer AC, James AE Jr, Krause DA, et al: Sonographic patterns in trophoblastic diseases. Radiology 126:215 (1978)

Sauvage JP, Crane JP, Kopta NM: Difficulties in the ultrasonic diagnosis of hydatidiform mole. Obstet Gynecol 44:546 (1974)

Case 113

CLINICAL INFORMATION

This 47-year-old man complains of right flank and back pain, which he characterizes as an ache or burning sensation. It started six weeks ago and at first occurred episodically. Recently, however, there have been very few pain-free intervals. He has no other symptoms. On physical examination there is a question of either an enlarged liver or a right upper quadrant mass. All routine laboratory tests are negative except for urinalysis, which shows microscopic hematuria. An IVP is performed, but the right kidney is not seen well at ten minutes, even with tomography (Figure 113–A). A film taken two hours after the injection of contrast shows a dense nephrogram on the right, but the collecting system is never visualized (Figure 113–B). The radiologist feels that this is most likely an obstruction of the right ureter. These ultrasound scans were made through the longitudinal axis of the kidney with the patient in the supine position.

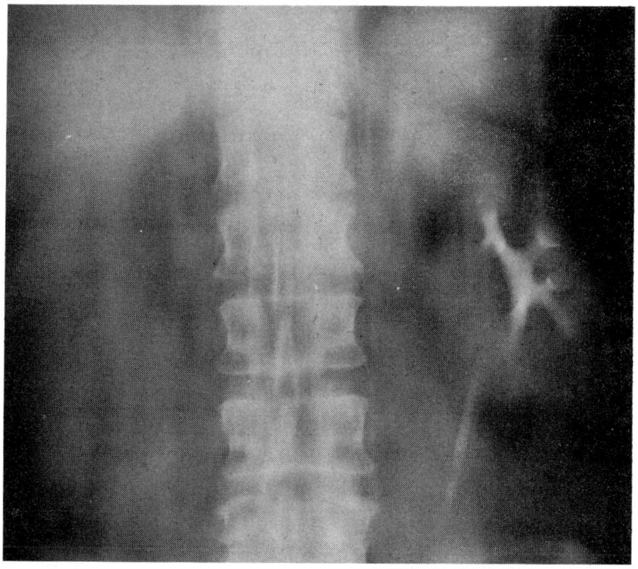

Figure 113–A.
IVP, 10-minute compression tomogram of kidneys.

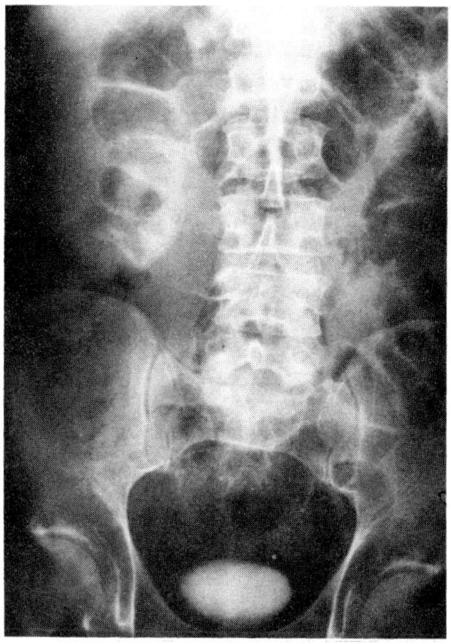

Figure 113–B.
IVP, 2-hour film.

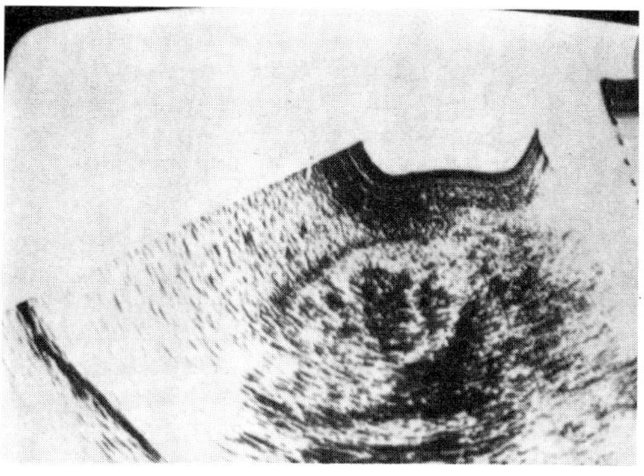

Figure 113–C. **Longitudinal section 6R.**
Supine scan.

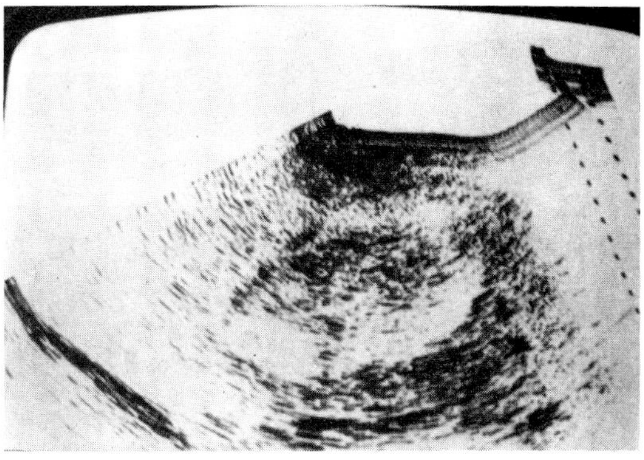

Figure 113–D. **Longitudinal section 9R.**
Supine scan.

DIAGNOSIS

Hypernephroma (renal cell carcinoma) involving the entire kidney.

ULTRASOUND FINDINGS

The mass in the right upper quadrant that was found on physical examination is the enlarged right kidney. Because the kidney was readily accessible we performed the scans with the patient in the supine position. The liver serves as an excellent acoustic window; the scanning was not hindered by overlying ribs and iliac crests, so often encountered when the patient is examined prone. The kidney is bizarre in appearance. Its overall size is increased; the central high-intensity echoes from the renal sinus are much more diffuse than normal, and the upper pole contains a mass of complex echo composition. The presence of the upper pole mass, in conjunction with hematuria, is very suggestive of renal tumor. The appearance of the lower pole is much more difficult to explain. The prominent central sulcus echoes are suggestive of renal fibrolipomatosis, but this certainly would not explain the increased renal size or lack of visualization of the collecting system on the IVP. There is no ultrasonic evidence of hydronephrosis. The only explanation for all these findings is a diffuse infiltrative process involving the entire kidney and compromising renal function. The mass seen clearly in the upper pole makes a tumor the prime candidate. Were it not for this finding, an alternate diagnosis might be xanthogranulomatous pyelonephritis or angiomyolipomatosis. A selective renal arteriogram confirms the impression of malignancy (Figure 113–E). The renal vein was completely obstructed and we would have predicted this from the poor renal function.

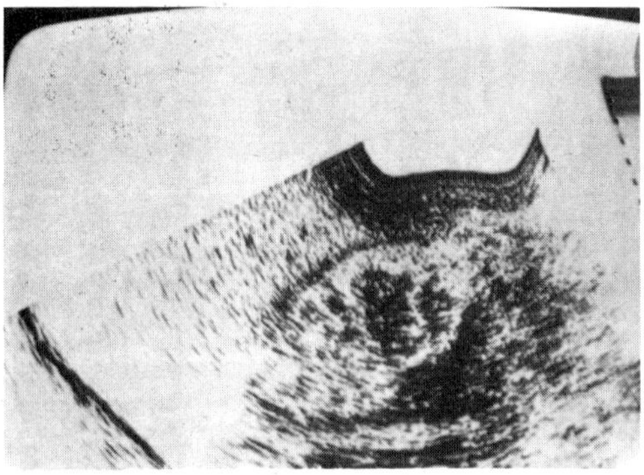

Figure 113–C. Longitudinal section 6R. Supine scan. Hypernephroma.

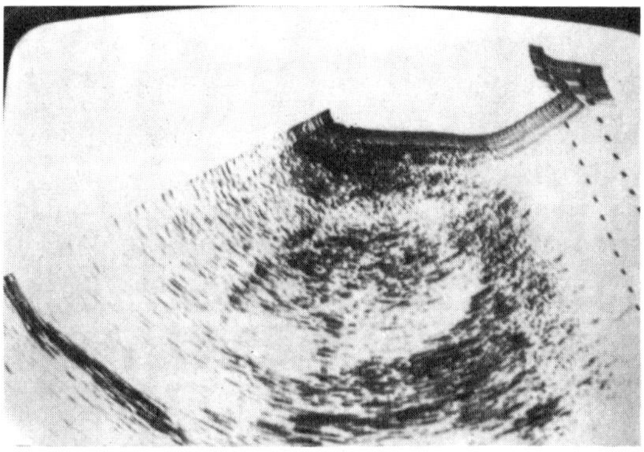

Figure 113–D. Longitudinal section 9R. Supine scan. Hypernephroma.

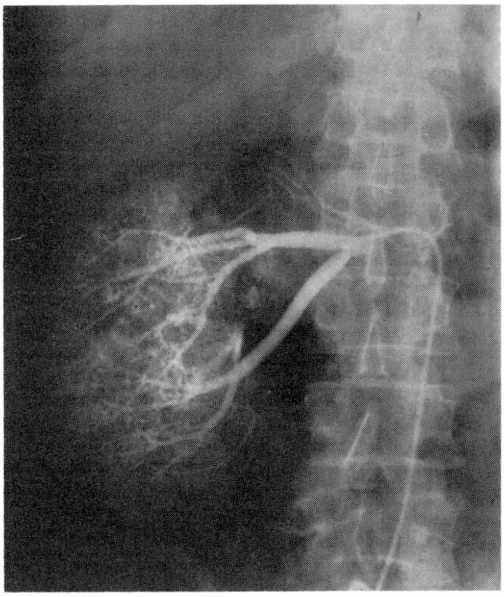

Figure 113–E.
Hypernephroma. Selection arteriogram of kidney in Figures 113–A through 113–D.

REFERENCES

Morangola JB, Bryan PJ, Azimi F: Ultrasonic evaluation of unilateral non-visualized kidney. Am J Radiol 126:853 (1976)

Makland NF, Chuang VP, Doust BD, et al: Ultrasonic characterization of solid renal lesions. Radiology 123:733 (1977)

Case 114

CLINICAL INFORMATION

This 37-year-old man has atypical angina, but a thorough cardiac evaluation has failed to turn up any concrete evidence of heart disease. A coronary arteriogram is under consideration but before that invasive procedure is performed his doctor wants to be certain that there is no abdominal explanation for the pain. In reviewing the scans that the technician has obtained we notice a loud echo with a small, but real, acoustic shadow (arrows) just to the left of the gallbladder on the +8 section. The gallbladder itself is well seen on other sections and is normal.

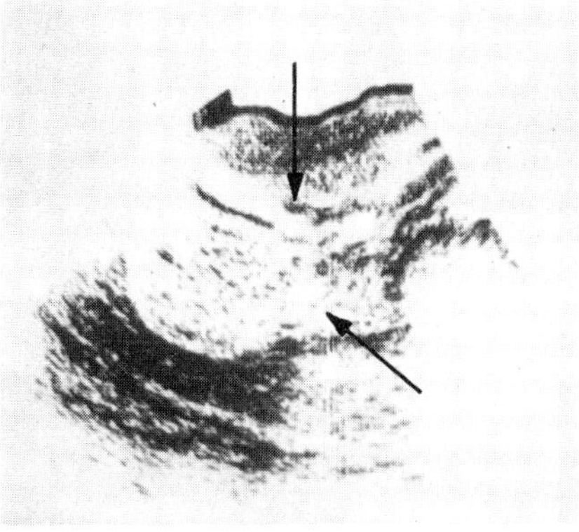

Figure 114–A. Transverse section +8.

DIAGNOSIS

Normal abdomen — air in the duodenal bulb.

ULTRASOUND FINDINGS

The presence of a loud echo that produces an acoustic shadow should immediately make us think of a stone or other calcified lesion. In this case the echo is not arising from the gallbladder but certainly could be due to a stone impacted in the infundibulum of the gallbladder or lying in the cystic or common duct. Before we settle on this diagnosis we must be certain there are no other explanations for the finding. It is easy to forget about the GI tract when interpreting ultrasound scans. (After all, we do our best to avoid it when making the scan.) Before accepting a finding as evidence of disease, however, we must always ask ourselves if what we are seeing could be an artifact caused by bowel. The presentations are protean: if the gut contains fluid, it can look like a cyst; if it is filled with air, there will be loud echoes with acoustic shadows and other artifacts.

The loud echo and shadow that we see in Figure 14–A are caused by air in the duodenal bulb. This patient had had an oral cholecystogram just before his ultrasound examination (Figure 114–B); the gallbladder is normal and the large, air-filled bulb can be seen just to the left of the gallbladder neck. Whenever we see this echo and shadow in this position we are almost always looking at air in the bulb (Figure 114–C shows the same appearance in another patient). The way to verify this is to go back and repeat the scan several times: the normal peristalsis of the duodenum will soon move the air along and the echo will disappear. Some ultrasonographers have recommended scanning the patient in the right lateral decubitus position, which fills the bulb with fluid instead of air. We have not found this approach particularly useful since it is difficult to reproduce the scan plan exactly when the patient has shifted position, and we are never certain that we are looking at the same area. As high resolution real time scanners become commercially available, we may be able to observe the motion of air and fluid in the duodenum directly.

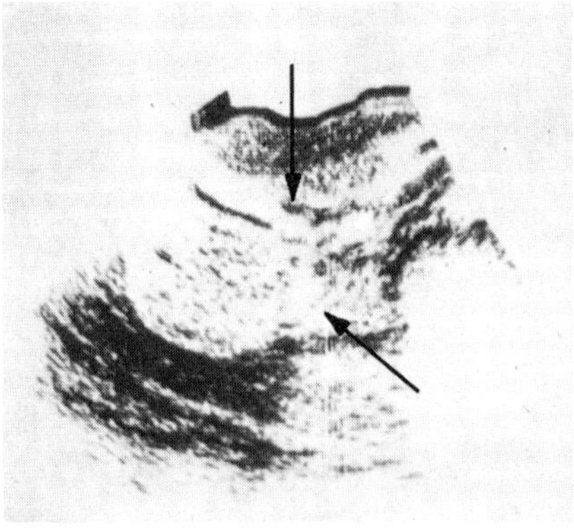

Figure 114–A. Transverse section +8. Air in duodenum.

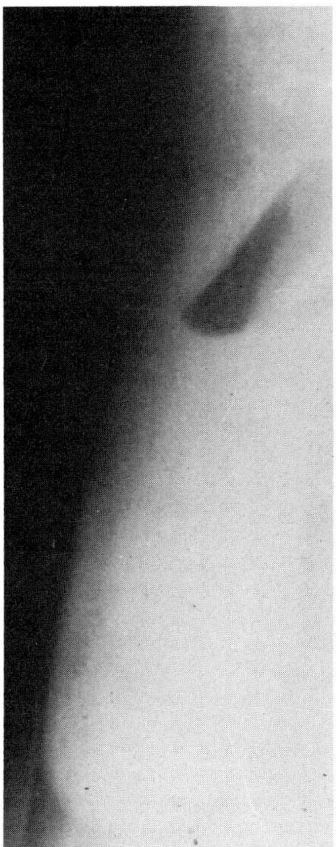

Figure 114–B.
Air in duodenum. Film from oral cholecystogram on patient in Figure 114–A.

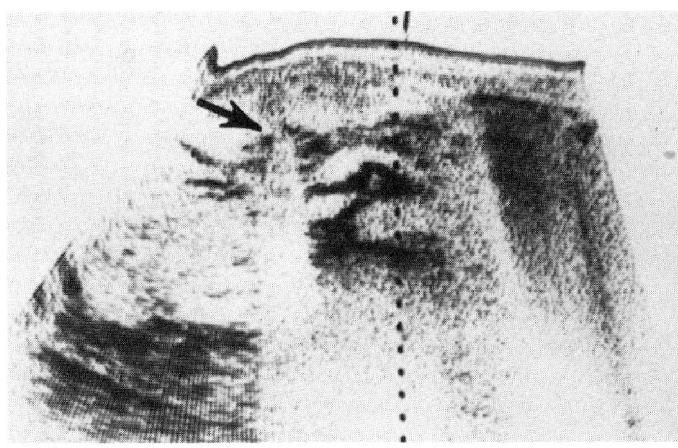

Figure 114–C.
Air in duodenum.

Case 115

CLINICAL INFORMATION

This 87-year-old woman, who is senile and confined to a nursing home, has been complaining of abdominal pains for the past several days. She was transferred to the hospital and admitted through the Emergency Ward, where a physical examination and simple laboratory tests were said to be normal for any abdominal problems (she does have congestive heart failure). She was sent to the ward by way of the ultrasound laboratory.

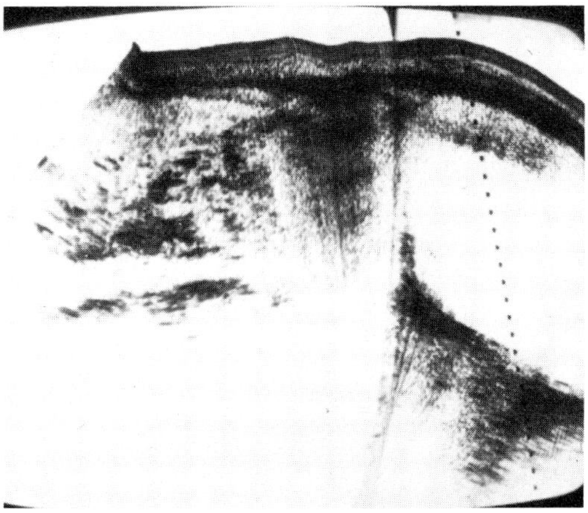

Figure 115–A. Longitudinal section 4R.

DIAGNOSIS

Urinary retention.

ULTRASOUND FINDINGS

Not hard was it? Although this was a scan of the abdomen our attention is drawn to the gigantic echoless mass in the pelvis and lower abdomen. A scan of the pelvis (Figure 115–B) does not demonstrate the urinary bladder and we begin to wonder — could this giant cyst actually be the bladder? We catheterized the patient, and removed 1500 milliliters of urine (leaving an even greater amount to be removed in small increments to avoid hemorrhagic cystitis). We trust that you also took note of the dilated inferior vena cava (Figure 115–A) secondary to her heart failure.

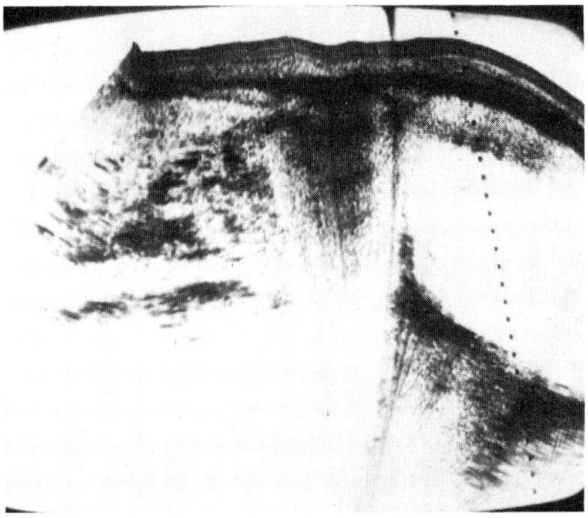

Figure 115–A. Longitudinal section 4R. Urinary retention. Dilated inferior vena cava.

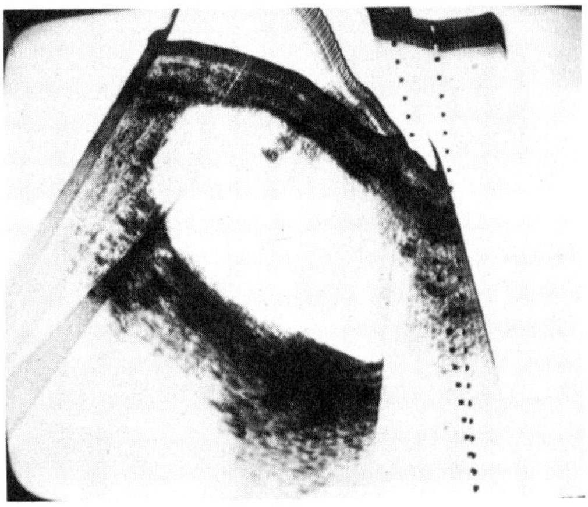

Figure 115–B. Longitudinal section 0L. Urinary retention.

REFERENCE

Lee TG, Reed TA: Ultrasonic diagnosis of the bladder as a symptomatic pelvic mass. J Urol 117:283 (1977)

Case 116

CLINICAL INFORMATION

This 22-year-old man came to the referring physician because one day while taking a shower he discovered a painless enlargement of his left testicle. On physical examination the physician confirmed the presence of the testicular mass, but found no other abnormality. We were asked to scan the scrotum to characterize the mass as cystic or solid. This is the picture of the left testicle we obtained — what is your diagnosis?

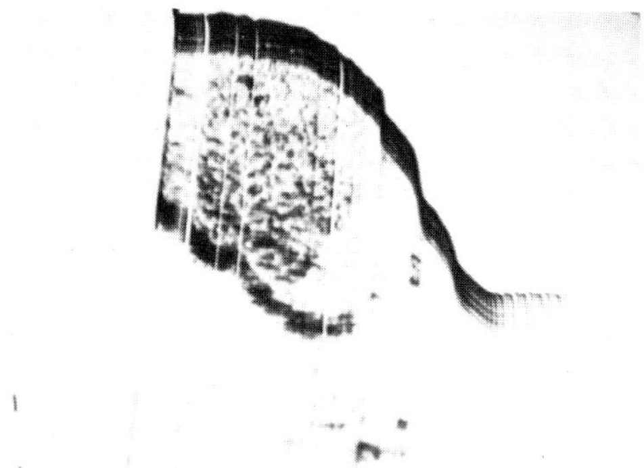

Figure 116–A.
Longitudinal scan of left testicle.

DIAGNOSIS

Teratocarcinoma of testicle.

ULTRASOUND FINDINGS

The mass, which was easily felt, is represented by the central area of relatively louder echoes, on either side of which are the low-level echoes of normal testicular tissue. There is no question here; this is not a cyst, and it is not external to the gland. It is a primary tumor of the testis and as such has to be considered to be malignant. While the echo pattern is clearly different from that of normal tissue, it is not specific for any one type of tumor. Seminoma, embryonal cell carcinoma, teratocarcinoma, and lymphoma can all cause alteration in the appearance of the testicle. Any change of the internal echo pattern is suspect; neoplasm can cause either an increase or a decrease in the intensity or the number of echoes produced.

What about benign disease producing a change in echo pattern? Edema of the testicle caused by orchitis can be localized, especially if associated with epididymitis. In such a case a region of diminished echo production can be seen and might be confused with neoplasm. The clinical presentation will, however, be quite different because the illness will be relatively rapid in onset, and the patient will have considerable pain.

When we had demonstrated the testicular tumor we scanned this man's abdomen. Figure 116–B shows that the aorta is surrounded by marked lymphadenopathy. Notice that the aorta is displaced anteriorly, away from the spine. Although this is not a beautiful scan, it demonstrates that this unfortunate patient already has widespread metastatic tumor.

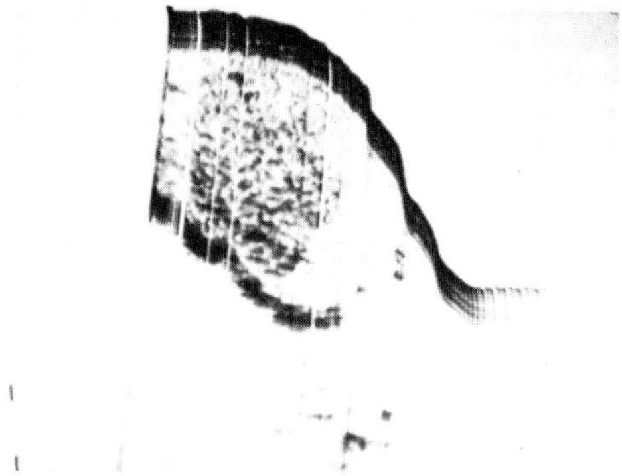

Figure 116–A.
Teratocarcinoma. Longitudinal scan of left testicle.

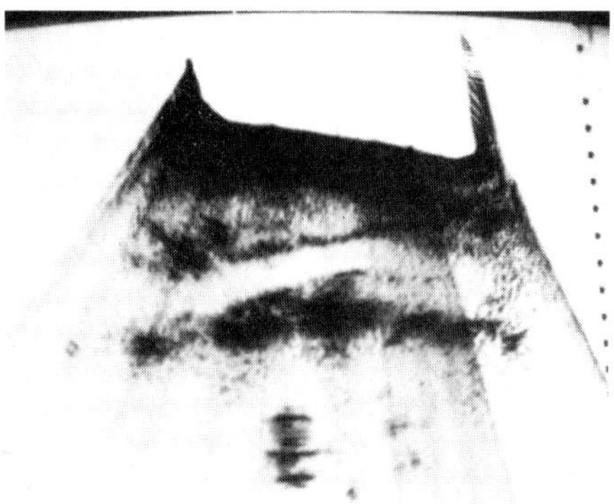

Figure 116–B.
Longitudinal section 1L. Retroperitoneal metastases (teratocarcinoma).

REFERENCES

Sample WF, Gottesman JE, Skinner DG, et al: Gray scale ultrasound of the scrotum. Radiology 127:225 (1978)

Miskin M, Busckspan M, Bain J: Ultrasonographic examination of scrotal masses. J Urol 117:185 (1977)

Case 117

CLINICAL INFORMATION

This 54-year-old man had chronic peptic ulcer disease and two weeks previously underwent surgery for a pyloroplasty and vagotomy. His convalescence had gone smoothly; he was eating, was up and about, and had only minor abdominal pain. The abdominal sutures had been removed for five days, and the surgery intern reported that the incision seemed well-healed and was not unduly tender. He was ready for discharge but had been held up because of a low-grade fever (100 to 101 degrees) and a slightly elevated white blood cell count (11,000 to 12,000). Chest x-rays, films, urinalysis, and blood cultures were all normal.

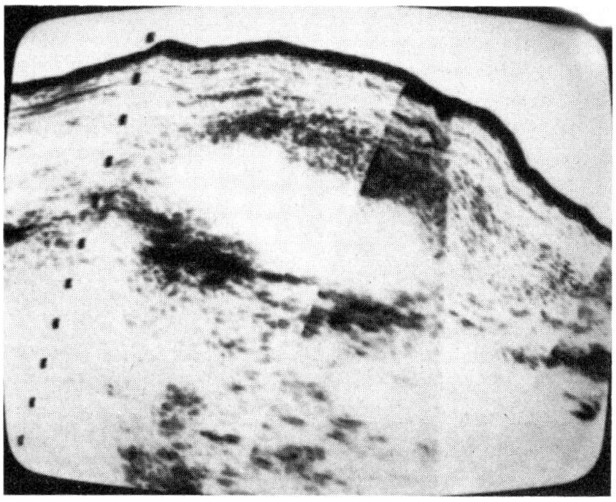

Figure 117–A. Transverse section +5.

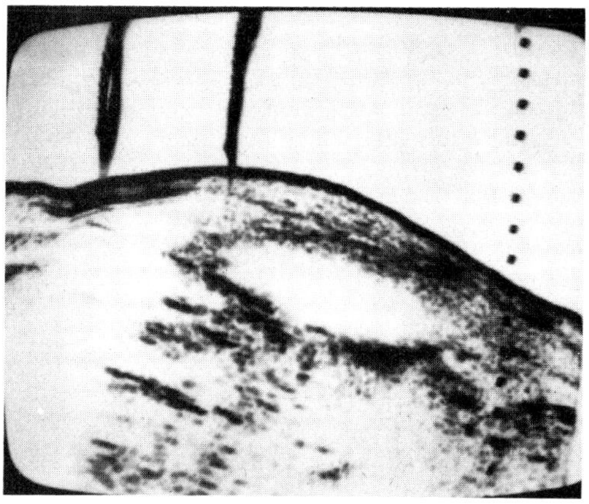

Figure 117–B. Longitudinal section 0L.

DIAGNOSIS

Superficial wound abscess.

ULTRASOUND FINDINGS

The cross-sectional and longitudinal scans show an elliptical, echo-free area in the anterior abdominal wall that is located just beneath his incision and clearly represents an abnormal fluid collection. Although this might be an aseptic seroma or resolving hematoma, it could just as easily represent a superficial abscess. Because we could not tell the difference from the scan pictures alone, we performed an aspiration biopsy while the patient was still in the ultrasound department, using a regular syringe and 20 gauge needle. In this case we got out pus and sent the specimen for culture and gram stain. We have made a complete diagnosis in short order with little threat or discomfort to the patient.

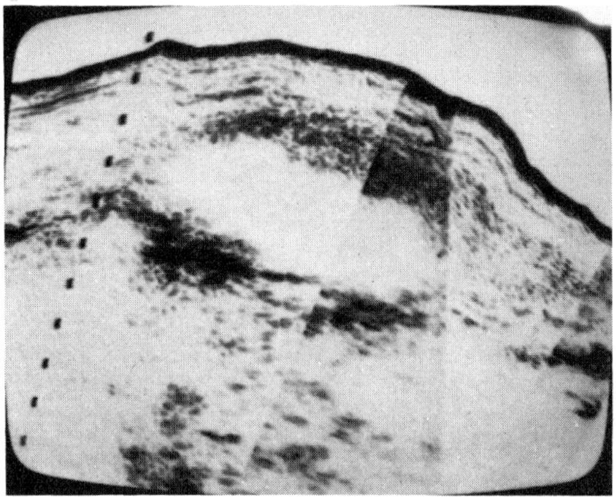

Figure 117–A. Transverse section +5. Superficial abscess.

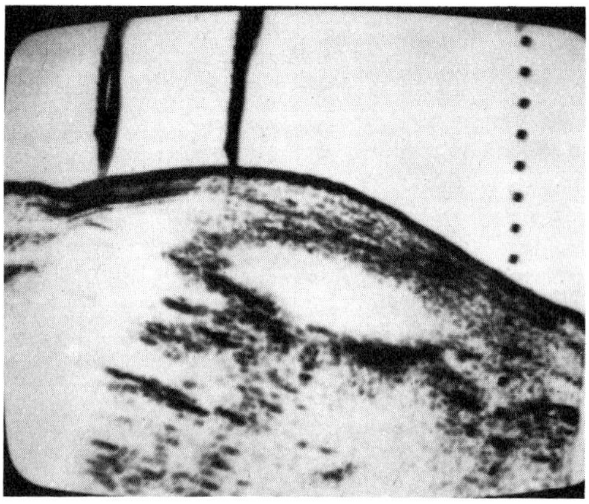

Figure 117–B. Longitudinal section 0L. Superficial abscess.

REFERENCES

Goldberg BB: Ultrasonic evaluation of superficial masses. J Clin Ultra 3:91 (1975)
Smith EH, Bartrum RJ Jr: Ultrasonically-guided percutaneous aspiration of abscesses. Am J Radiol 122:308 (1974)

Case 118

CLINICAL INFORMATION

This 26-year-old man had always been in excellent health until eight months ago. At that time he was involved in an automobile accident and suffered a severe contusion of his right thigh as well as a cerebral concussion. Although these injuries did not require hospitalization (except for overnight observation for the concussion) and he was back at work in 10 days, he has never felt completely "right" since. He has headaches and feels weak much of the time. He is always tired and just does not have his old "get up and go." He sometimes has night sweats. His physical examination and routine laboratory studies were normal, but he has a low-grade fever (99.6 degrees) which has been documented on three separate occasions. His doctor suspects infectious mononucleosis and carefully re-examines him: there is no pharyngitis, lymphadenopathy, or splenomegaly. A blood smear and heterophile titer are normal. An ultrasound scan is requested because of the fever of unknown origin. The upper portions of the abdomen, including the liver, spleen, and pancreas, are normal; however, the technician is concerned about the appearance of the midabdomen. (The arrows on Figures 118–A and 118–B point to the inferior vena cava and aorta; the boundaries of these vessels have been circled for clarity.)

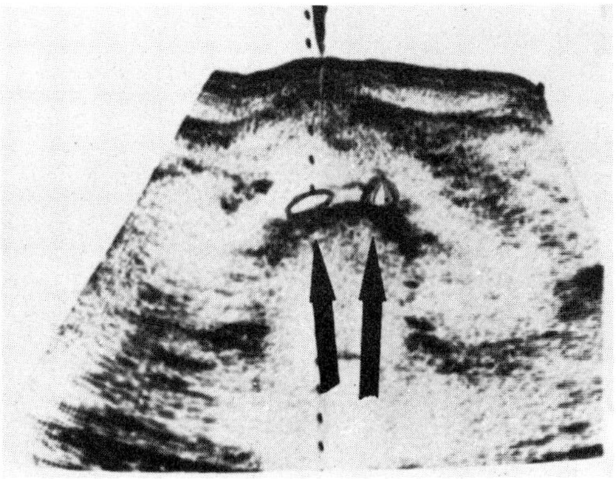

Figure 118–A. Transverse section +6.

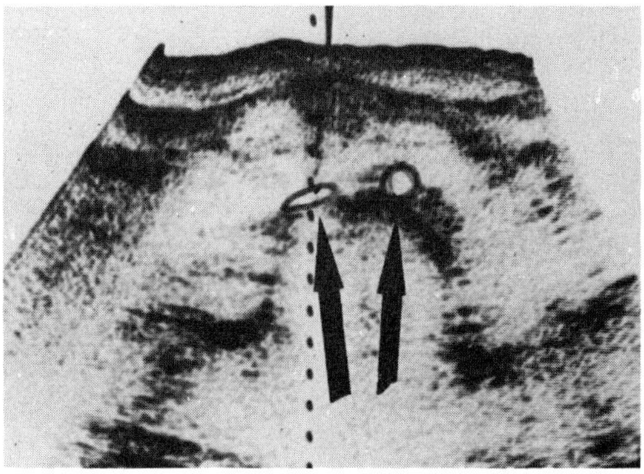

Figure 118–B. Transverse section +4.

DIAGNOSIS

Horseshoe kidney.

ULTRASOUND FINDINGS

There is a mass, demonstrated by low-level echoes, which straddles the spine and the great vessels in the midabdomen; on the +6 scan (Figure 118–A) it has two rather bulbous projections on either side of the spine. The mass lies over the aorta and IVC and does not surround them. Whenever we see a mass such as this draped over the spine and containing only weak internal echoes, we immediately think of enlarged lymph nodes and specifically lymphoma. This appearance is almost (but obviously not quite) pathognomonic for lymphadenopathy. Moreover, when the enlarged nodes are draped *over* the great vessels rather than *surrounding them,* the nodes are more likely involved with lymphoma than with metastases (this is a useful rule of thumb, but is not infallible). There are only two differential diagnostic possibilities. Acute, edematous pancreatitis can mimic this appearance in the upper abdomen; the clinical setting enables us to make this distinction in most cases. In the midabdomen (where these scans were obtained) we must always consider horseshoe kidney. Careful scanning shows that we are unable to identify the kidneys in their normal position — the paraspinous masses are the lower poles of the kidneys! An IVP (Figure 118–C) verifies this suspected diagnosis. (No cause was ever found for this patient's fever; six weeks after these scans were performed both his fever and symptoms had disappeared.)

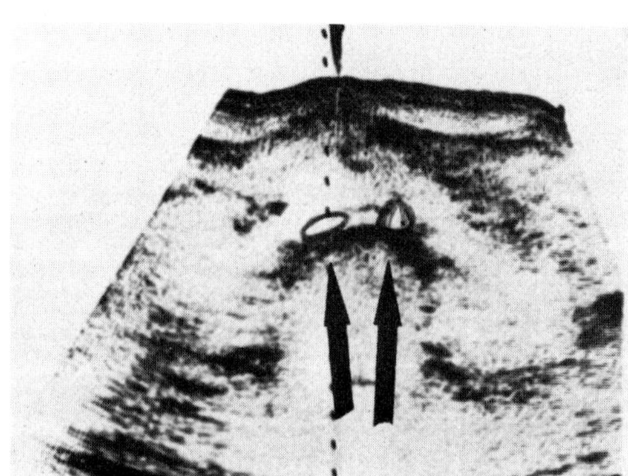

Figure 118–A. **Transverse section +6.** Horseshoe kidney.

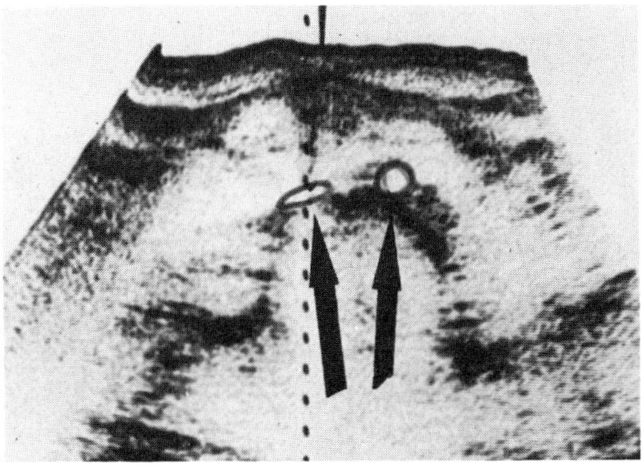

Figure 118–B. Transverse section +4. Horseshoe kidney.

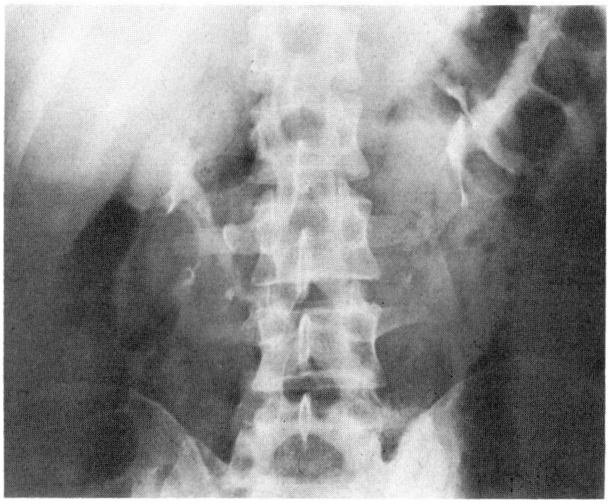

Figure 118–C. Horseshoe kidney. IVP film from patient in Figures 118–A and 118–B.

REFERENCES

Mindell HJ, Kupic EA: Horseshoe kidney: Ultrasonic demonstration. Am J Radiol 129:526 (1977)

Trackler RT, Resnick ML, Leopold GR: Pelvic horseshoe kidney: ultrasound findings and case report. J Clin Ultra 6:51 (1978)

Case 119

CLINICAL INFORMATION

This 63-year-old woman has a painful mass in her abdomen. She comes to the ultrasound laboratory with no clinical information as an outpatient from another town. She could not relate much about her history except that her abdomen has been painful for several years and "is getting real bad." She had not noticed a mass on the left extending from the costal margin down to the iliac crest; the mass seems to be separate from the liver. The patient is quite uncomfortable on the examining table and was continually shifting position and sitting up. She could not lie on her stomach nor hold her breath for more than a few seconds. Figures 119–A and 119–B are cross- and longitudinal sections through the mass. What is a reasonable differential diagnosis? What other ultrasound scans would you like to see? What other imaging studies would be helpful?

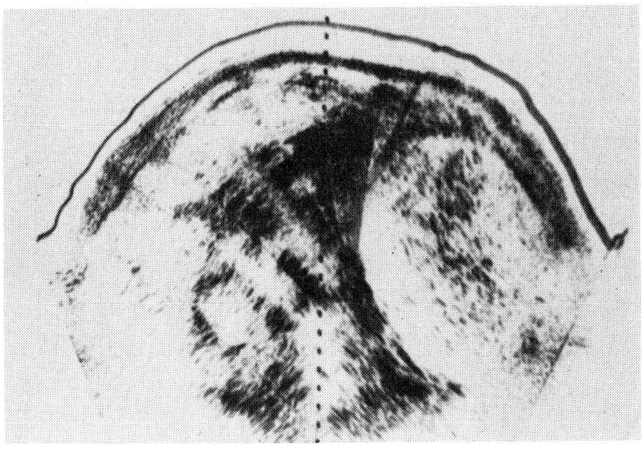

Figure 119–A. Transverse section +5.

Figure 119–B. Longitudinal section 7L.

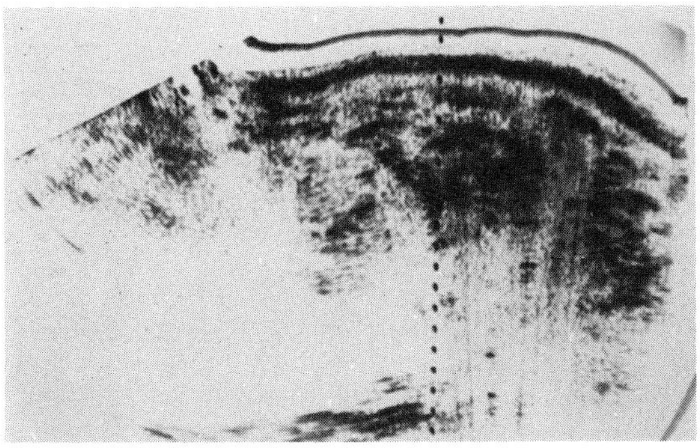

DIAGNOSIS

Adrenocortical carcinoma.

ULTRASOUND FINDINGS

This ultrasound examination is not of good quality but it is still possible to glean a considerable amount of useful information from it. First, we can define the size and extent of the palpable mass: it is approximately spherical with a diameter of 14 centimeters. Second, it contains multiple internal echoes and is therefore a solid tumor. We are not dealing with a splenic or hepatic cyst, a pseudocyst, a renal cyst or hydronephrosis, or an aneurysm. Third, it is not part of the liver, a hepatoma or metastatic tumor.

This is as far as we can go; since the patient could not assume a prone position, we cannot identify the spleen and hence cannot exclude the possibility of splenomegaly. In addition, we can never find the left kidney despite extensive searching. Either the kidney is displaced far into the pelvis where it is obscured by overlying bowel gas or it is totally invaded and destroyed by the mass. The mass could also be arising from any of the retroperitoneal structures in this area, such as the tail of the pancreas, lymph nodes, or adrenal gland. This case illustrates how dependent the ultrasonographer is upon ancillary data such as a good history, laboratory studies, and the results of other imaging procedures. Without them we can offer only a number of possibilities for differential diagnosis.

Our general philosophy is that all obscure masses such as this should have an aspiration biopsy, and this is our usual recommendation. Although aspiration biopsy is an innocuous procedure, we do observe one caveat: we do not like to puncture a mass which might be a pheochromocytoma. In most patients this possibility can be quickly eliminated by the clinical data, but in this case we knew too little about the patient to proceed directly to biopsy. We felt, therefore, that an IVP should be obtained to exclude the possibility of an adrenal tumor (and possibly to permit a specific diagnosis as well).

Figure 119–C is the subsequent IVP, showing the mass to be of suprarenal origin. Clinical laboratory data (now available) also suggested an endocrine dysfunction. At surgery an adrenocortical carcinoma was removed.

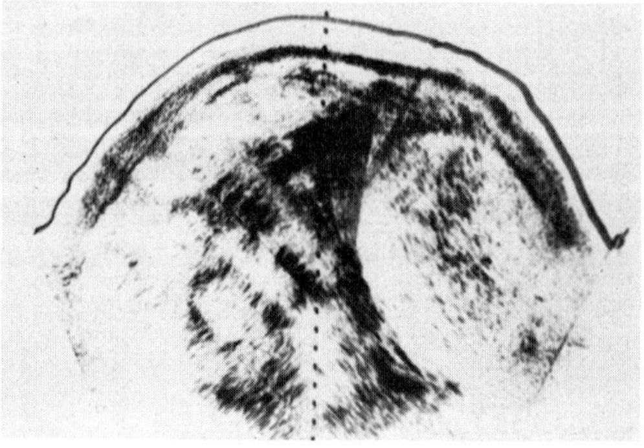

Figure 119–A. Transverse section +5. Adrenocortical carcinoma.

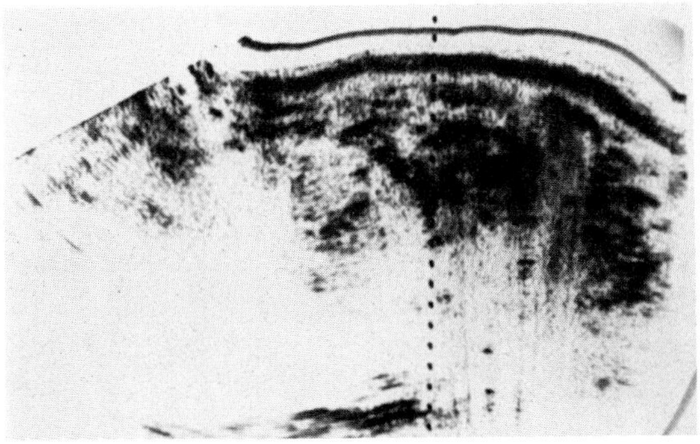

Figure 119–B. Longitudinal section 7L. Adrenocortical carcinoma.

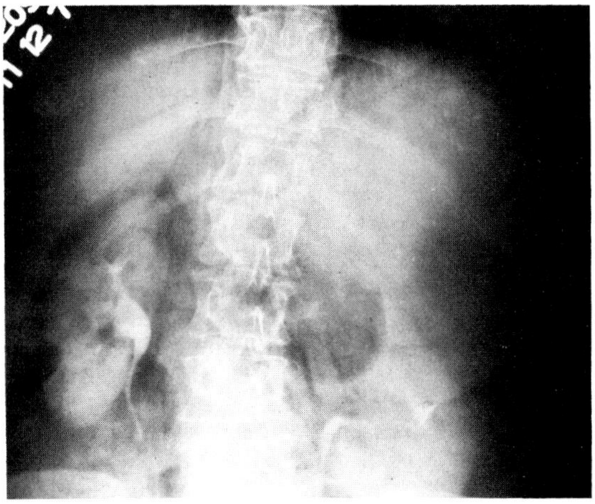

Figure 119–C. Adrenocortical carcinoma. Five-minute IVP film from patient in Figures 119–A and 119–B.

REFERENCES

Bernardino ME, Goldstein HM, Green B: Gray scale ultrasonography of adrenal neoplasms. Am J Radiol 130:741 (1978)

Sample WF: A new technique for the evaluation of the adrenal gland with gray scale ultrasonography. Radiology 124:463 (1977)

Behan M, Martin EC, Muecke EC, et al: Myelolipoma of the adrenal: two cases with ultrasound and CT findings. Am J Radiol 129:993 (1977)

Case 120

CLINICAL INFORMATION

This 27-year-old mother of two is 29 weeks pregnant and has a rising Rh antibody titer. She is Rh-negative and her second child was Rh-positive. The obstetrician is worried about the baby and has requested an ultrasound scan to look for fetal hydrops.

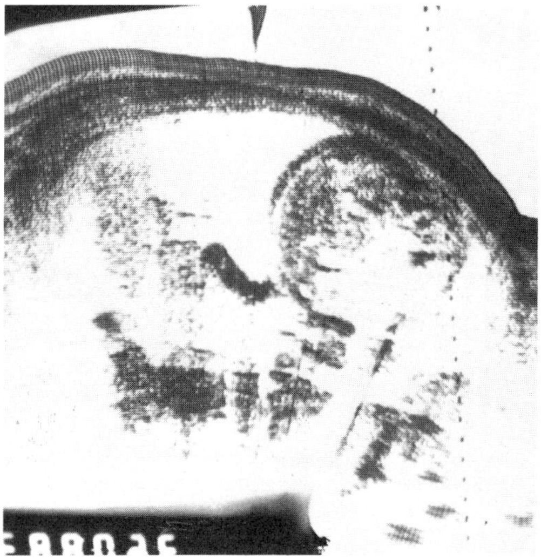

Figure 120–A. Longitudinal section 2L.

DIAGNOSIS

Erythroblastosis fetalis (Rh disease, fetal hydrops).

ULTRASOUND FINDINGS

This scan gives us a transverse view of the fetal abdomen; we see a thin, echoless rim between the abdominal wall and the liver — evidence of fetal ascites due to erythroblastosis. Although the placenta is not well shown in this scan, it is large and edematous. This is another indication of the disease.

With the use of prophylactic antiserum (RhoGam), erythroblastosis, or "Rh disease," as it is more commonly known, is becoming a rare occurrence. Occasional cases do crop up, however, and the ultrasonographer must be aware of the signs of fetal involvement. Fetal ascites, as demonstrated in Figure 120–A, is considered a late sign and indicates severe disease, according to most authors. In our limited experience it has been one of the earliest and most easily demonstrated findings. The placenta becomes enlarged and edematous, with an irregular, "meaty" appearance (Figure 120–B). There can be widespread fetal edema, most easily recognized in the scalp (Figure 120–C) and the abdominal wall (Figure 120–D).

The treatment for advanced cases is intrauterine fetal transfusion. Ultrasound has proved invaluable for localizing the fetal abdomen for this procedure.

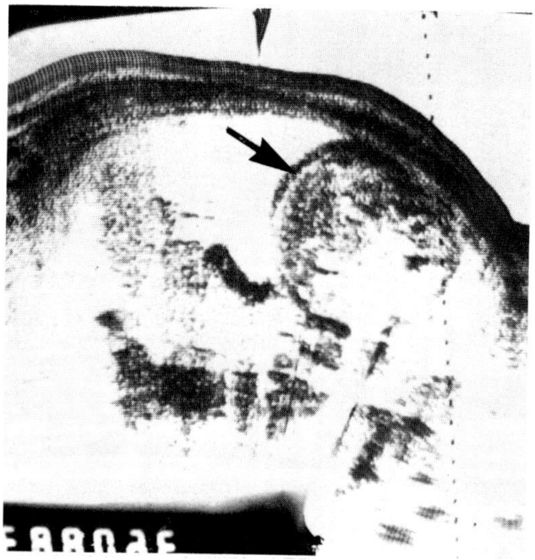

Figure 120–A. Longitudinal section 2L. Ascites in erythroblastosis.

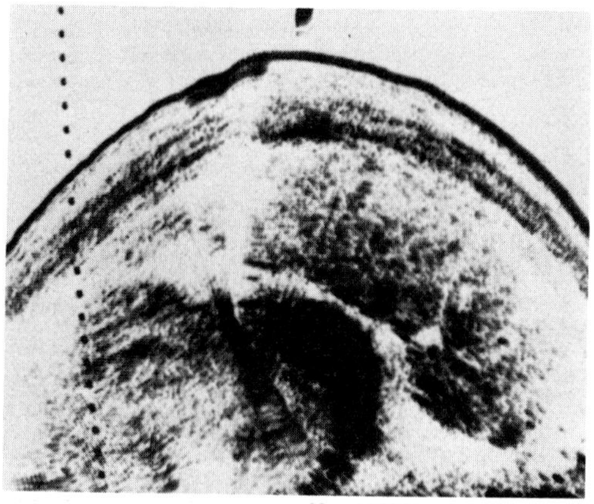

Figure 120–B. Transverse section –8. Edematous placenta in erythroblastosis.

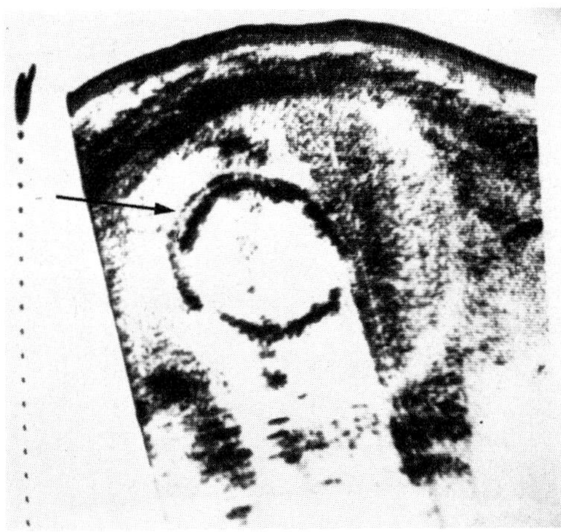

Figure 120–C. Transverse section −3. Scalp edema in erythroblastosis.

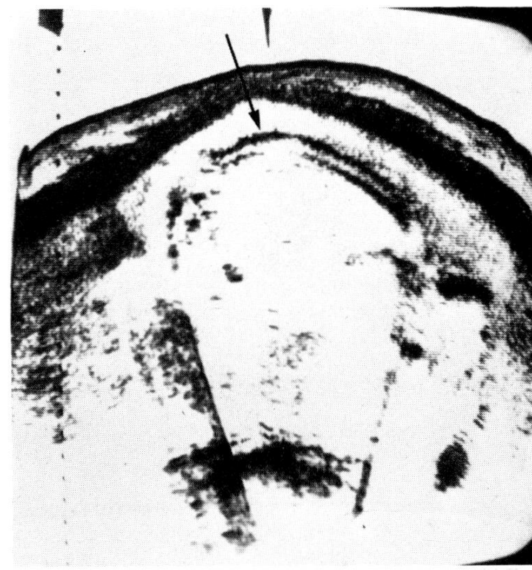

Figure 120–D. Transverse section +4. Abdominal wall edema in erythroblastosis.

REFERENCES

Ghorashi B, Gottesfeld KR: Recognition of the hydropic fetus by gray scale ultrasound. J Clin Ultra 4:193 (1976)

Jones MR: Ultrasonic B-scanning in rhesus incompatibility. J Clin Ultra 2:185 (1974)

Case 121

CLINICAL INFORMATION

This 43-year-old woman was well until 14 hours ago when she was awakened by a severe, colicky pain centered near the umbilicus but radiating into both legs. The pain was accompanied by nausea and vomiting and has increased in severity until it is now constant. Her only previous hospitalization was for a cholecystectomy three years previously: she has been well in the interval. She does not think her appendix has been removed. Physical examination shows a rigid, guarded abdomen, with only mild rebound tenderness and no areas of focal pain. There are some quiet bowel sounds but no rushes. The flanks are not tender nor is there excess tympany in the abdomen. The hematocrit is 42, the white cell count 10,300; serum enzymes and liver functions are normal, as is the creatinine and urinalysis. A plain abdominal film shows a bowel gas pattern interpreted as "partial proximal small bowel obstruction." An IVP is normal. Figures 121–A and 121–B are cross-sectional and longitudinal scans obtained in the right midabdomen.

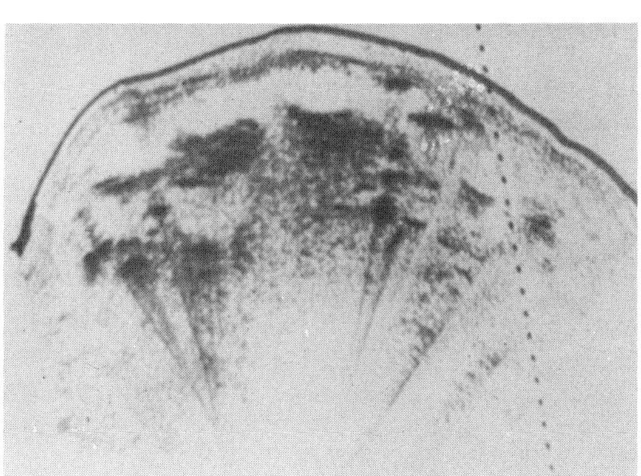

Figure 121–A. Transverse section +4.

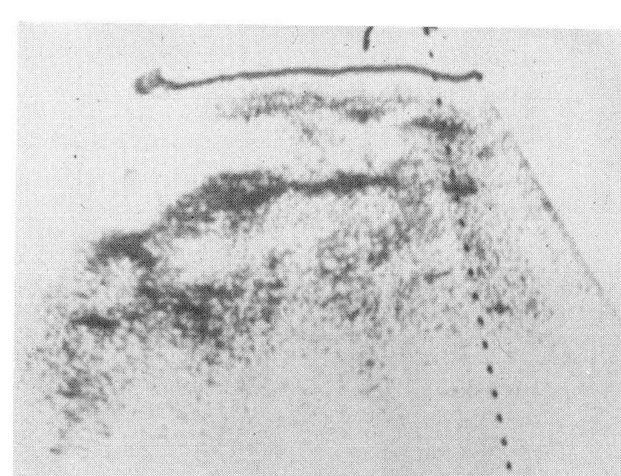

Figure 121–B. Longitudinal section 9R.

DIAGNOSIS

Obstructed, necrotic small bowel secondary to internal hernia.

ULTRASOUND FINDINGS

There is an irregular tubular, relatively echo-free structure in the anterior portion of the abdomen, which is fluid-filled. These scans were obtained in the midabdomen, below the level of the liver, and thus we are looking at either bowel or a pathologic intra-abdominal mass. The latter appears to be unlikely since the patient is relatively young and the disease is acute. Moreover, we cannot feel any mass. We are left with a diagnosis of "dilated fluid-filled bowel." A clue to the correct interpretation is the fact that the echo-free mass is "tubular" or elongated as one might expect a dilated portion of bowel to appear. This fits in well with the remainder of the clinical information and confirms the suspicion of an acute bowel obstruction. At surgery a long loop of small intestine, trapped in an internal hernia, was obstructed and necrotic.

This is another case in which the ultrasound scan provides useful information when integrated into the entire clinical setting, but is rather nonspecific if viewed *in vacuo.* Figures 121–C and 121–D are cross-sectional and longitudinal scans in the left midabdomen in a different patient. The findings are remarkably similar to Figures 121–A and 121–B, yet this patient has no intra-abdominal pathology — we are seeing a feces-filled descending colon. Figures 121–E and 121–F were obtained in another patient. The partially echo-free area above the right lobe of the liver is a focal abscess from a necrotic, perforated loop of bowel: the patient had an embolus to the superior mesenteric artery with bowel necrosis.

The point to remember here is that the ultrasound scan must be integrated into the total clinical picture if an accurate diagnosis is to be made. Always insist on having a full clinical history before interpreting the scan and, in confusing cases, call the referring physician and discuss the problem with him *while the patient is still on the examining table.* If the diagnosis remains obscure, it is then a simple matter to aspirate the suspicious area and obtain a specimen for pathologic evaluation.

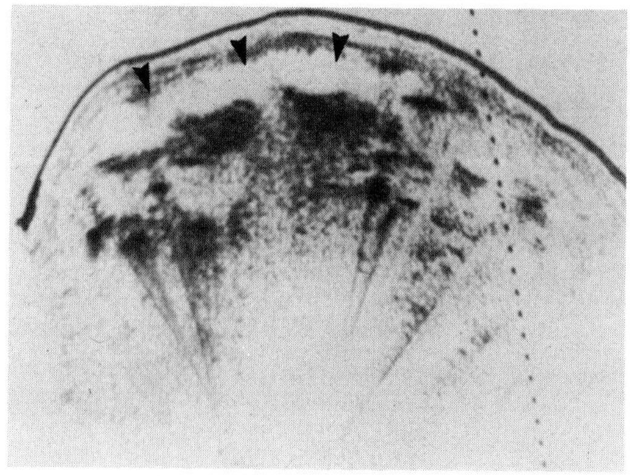

Figure 121–A. **Transverse section +4.** Obstructed small bowel.

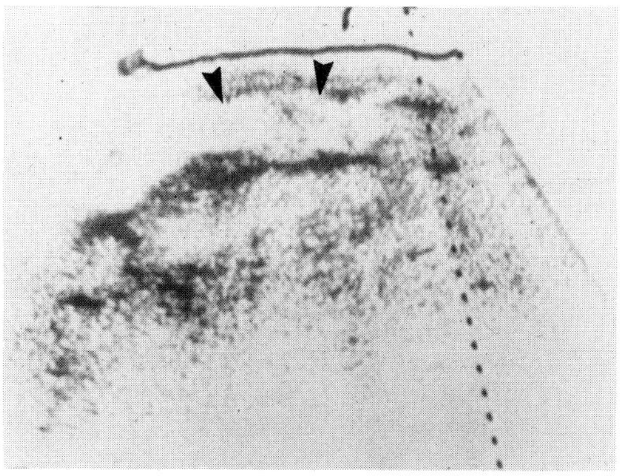

Figure 121–B. Longitudinal section 9R. Obstructed small bowel.

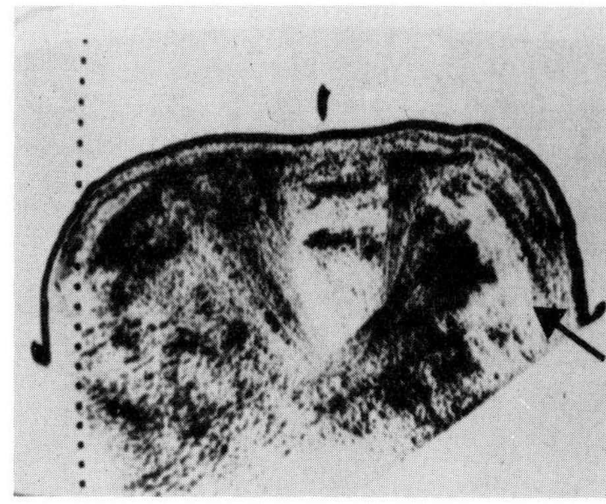

Figure 121–C. Transverse section +1. Feces-filled colon (normal).

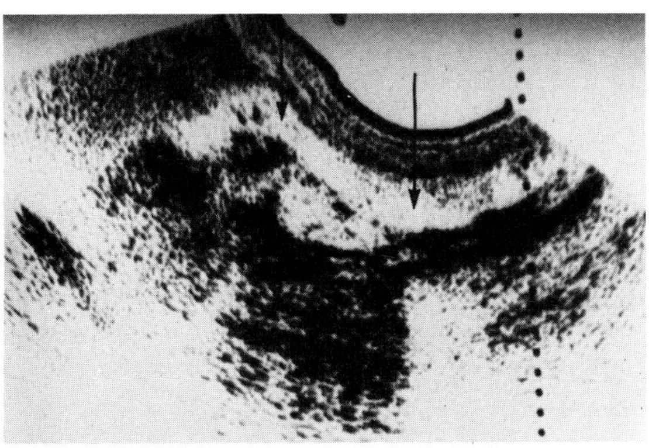

Figure 121–D. Longitudinal section 11L. Feces-filled colon (normal).

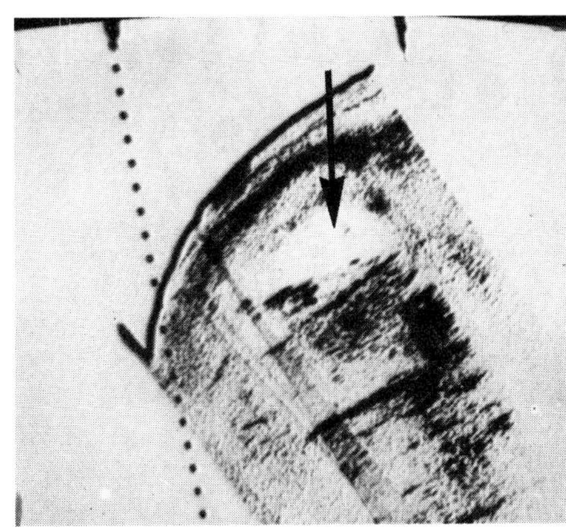

Figure 121–E. Transverse section +4. Necrotic bowel.

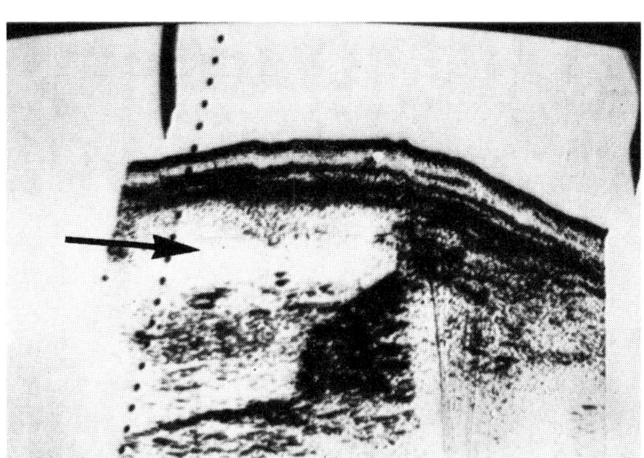

Figure 121–F. Longitudinal section 5R. Necrotic bowel.

Case 122

CLINICAL INFORMATION

This 76-year-old man is complaining of severe pain in his lower abdomen, which has been increasing in severity over the past two days. There has been some nausea associated with the pain, but he has not vomited nor had a bowel movement. Physical examination reveals focal tenderness in the left lower quadrant, and the bowel sounds are diminished but present. The rectal examination is normal but there is occult blood in the stool. His temperature is 102 degrees. The hematocrit is 41, and the white cell count 11,800 with a slight left shift. The remainder of the routine laboratory tests and urinalysis are normal.

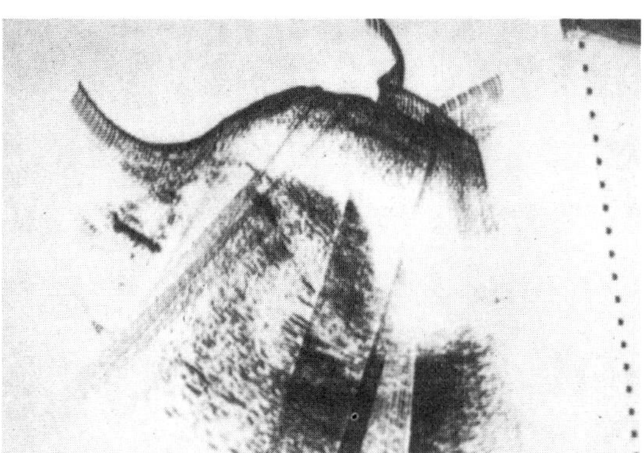

Figure 122–A. Longitudinal section 3L. Scan of lower abdomen.

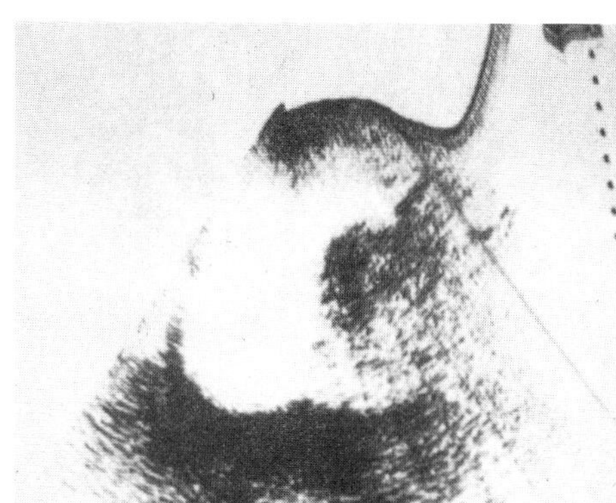

Figure 122–B. Transverse section –6.

DIAGNOSIS

Pelvic abscess (diverticulitis).

ULTRASOUND FINDINGS

There is a mass in the pelvis which is elevating the bladder (Figure 122–A) and displacing it to the right (Figure 122–B). The mass contains echoes of many different intensities and may even have a small echo-free area. The standard ultrasound nomenclature for such a mass is "complex." Unfortunately innumerable different organs and lesions can give rise to such a "complex" mass, so that recognizing this pattern is not of much clinical use: what we really mean when we say a mass is "complex" is that it is not fluid-filled and its components have different densities. Although there is nothing characteristic about the mass itself in this patient, the clinical history is quite suggestive of a pelvic abscess from diverticulitis, and the findings on the scan support this diagnosis. Abscesses can vary in appearance from echogenic, as in this case, to nearly echo-free. The diagnosis of diverticulitis was confirmed by a barium enema.

Could this abscess have been due to a carcinoma of the colon which had perforated? Of course: ultrasound cannot distinguish between a benign and malignant abscess.

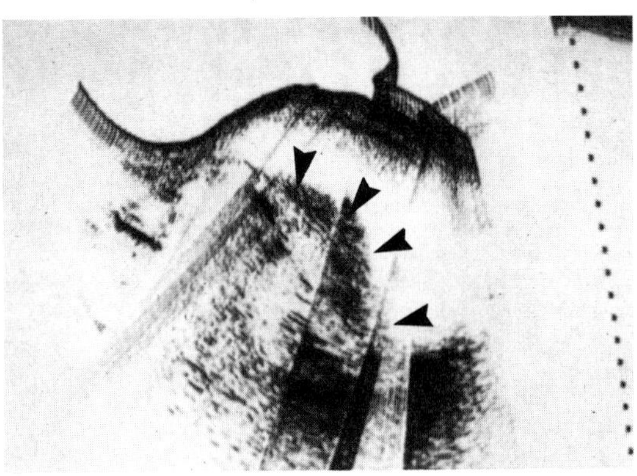

Figure 122–A. **Longitudinal section 3L.** Pelvic abscess.

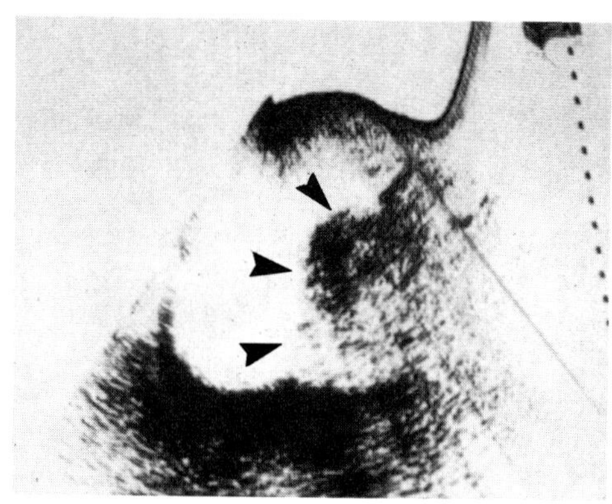

Figure 122–B. **Transverse section −6.** Pelvic abscess.

REFERENCES

Doust BD, Quiroz F, Stewart JM: Ultrasonic distinction of abscesses from other intra-abdominal fluid collections. Radiology 125:213 (1977)

Doust BD: Abscesses, hematomas, and other fluid collections, Chapter 9, in *Abdominal Gray Scale Ultrasonography*, BB Goldberg, editor. John Wiley & Sons, New York (1977)

Case 123

CLINICAL INFORMATION

This is a supine cross-sectional scan of the left side of the abdomen of a 74-year-old man who was admitted to the hospital with a stroke. The patient has developed acute pain in the left flank on the sixth hospital day. The stubby arrow points to the left kidney.

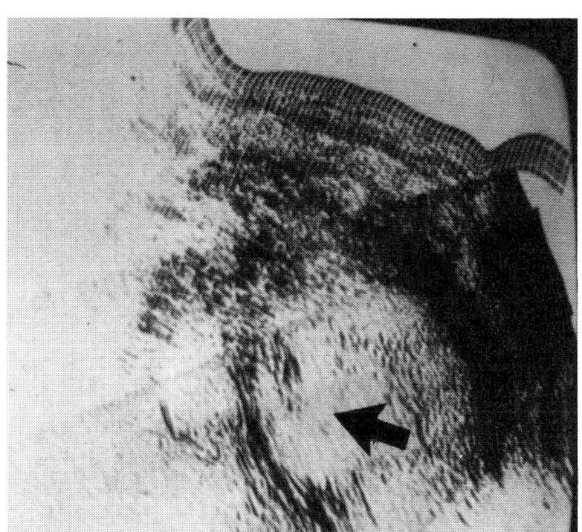

Figure 123–A. Transverse section +6.

DIAGNOSIS

Retroperitoneal hematoma.

ULTRASOUND FINDINGS

The left kidney is displaced medially against the spine by a large, echo-producing retroperitoneal mass. The differential diagnosis includes hypernephroma, retroperitoneal tumor (sarcoma, lymphoma, or metastasis), abscess, and hematoma. There is no way to separate these on the basis of the scan picture alone since they all can appear as an echo-producing mass. The clinical setting in this case leads us to the correct diagnosis of hematoma. Perhaps you need this clue: the patient was on anticoagulants for his stroke and pain came on coincidentally with the appearance of a mass and drop in hematocrit. If the clinical picture were less characteristic we would have performed biopsy of the mass to obtain the specific diagnosis.

Hematomas vary in appearance, depending on their age. When fresh they may be nearly echo-free but, as clotting and internal organization occur, they may acquire internal echoes and become "complex." As resolution and absorption progress, they shrink in size and become more homogeneous.

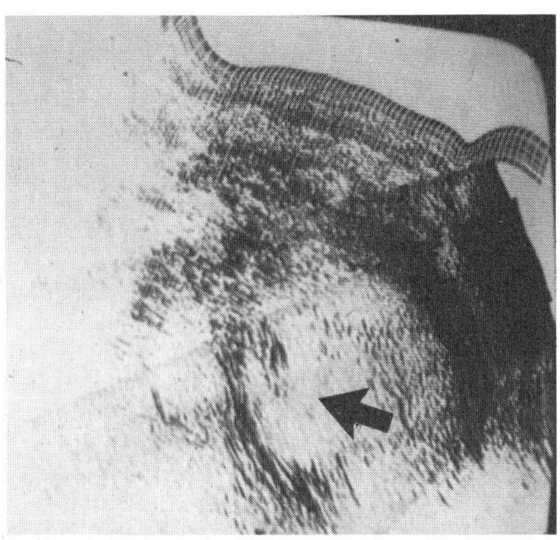

Figure 123–A. **Transverse section +6.** Retroperitoneal hematoma.

REFERENCES

Kaplan GN, Sanders RC: B-scan ultrasound in the management of patients with occult abdominal hematomas. J Clin Ultra 1:5 (1973)

Case 124

CLINICAL INFORMATION

This 31-year-old woman is 33 weeks pregnant by dates and is having an ultrasound examination for gestational dating. She has heard that the baby's sex can be predicted with a sonar examination. Can you tell her? Would you tell her?

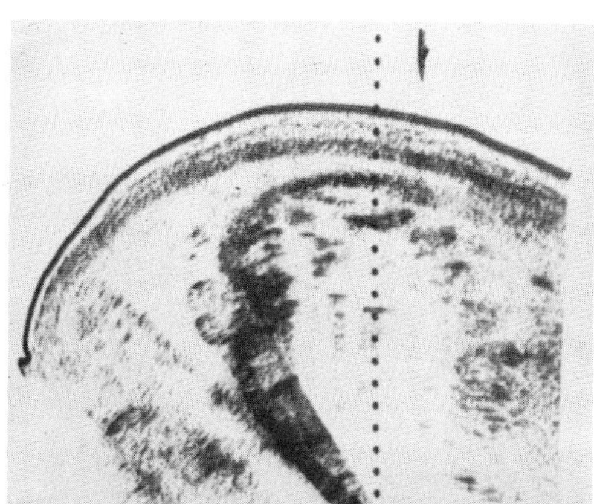

Figure 124–A. Transverse section +8.

DIAGNOSIS

Baby boy.

ULTRASOUND FINDINGS

In order to predict the sex we must make a scan through the fetal perineum. The way to go about this is to first find the fetal abdomen and orient the scan plane so that it is perpendicular to the baby's spine: then make a series of scans at 5 mm intervals, proceeding toward the feet. Soon the fetal bladder should come into view and, shortly after that, the scan plane will move off the abdomen and into the perineum. If there are echoes from the scrotum or penis, we have a boy. (Be careful not to confuse the cord with the penis.) Figure 124–A shows a nice view of the scrotum, while Figure 124–B is a lithotomy-position view of a baby girl.

Should the patient be told the baby's sex? This is a controversial matter, guaranteed to start a fight if discussed by more than three people at one time. Our referring physicians have asked us not to tell the patients, so we do not — in fact we do not even let them know we can tell the sex of the baby. If you do decide to report fetal sex, we caution you to do so only in cases where you are certain; an equivocation is far preferable to a mistake.

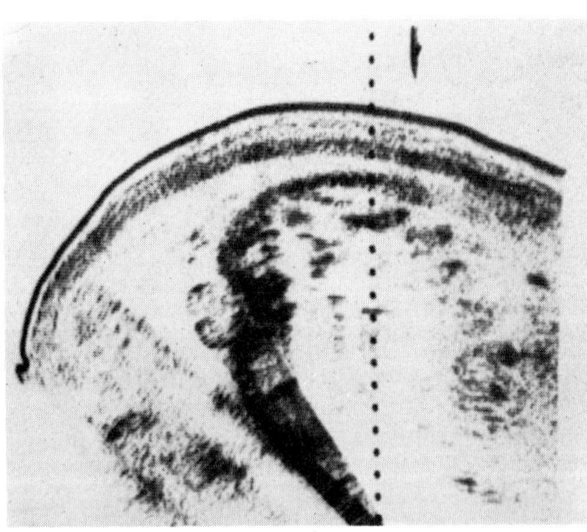

Figure 124–A. Transverse section +8. Baby boy.

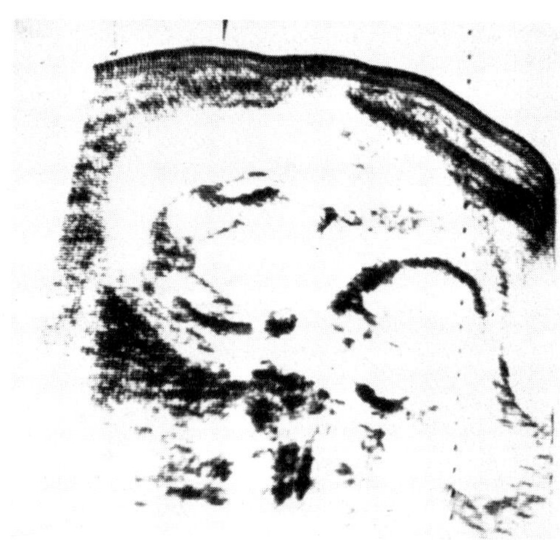

Figure 124–B. Longitudinal section 1L. Baby girl.

Case 125

CLINICAL INFORMATION

This 37-year-old man entered the hospital five days ago with severe abdominal pain associated with nausea and vomiting. The serum amylase on admission was 1220; he was considered to have acute pancreatitis and was treated with nasogastric suction and analgesics. The initial response to treatment has been good, with considerable improvement in the pain and a drop in amylase to 258. Several stool specimens have been positive for occult blood, however, and now that the patient's clinical condition has improved, he is sent for an ultrasound scan and upper GI series to "see what there is to see."

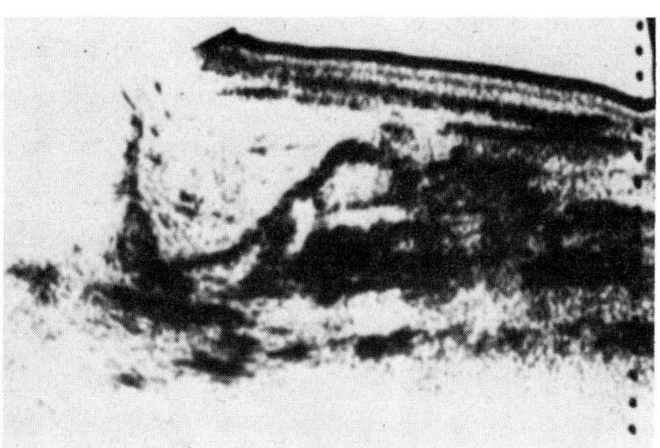

Figure 125–A. Longitudinal section 0L.

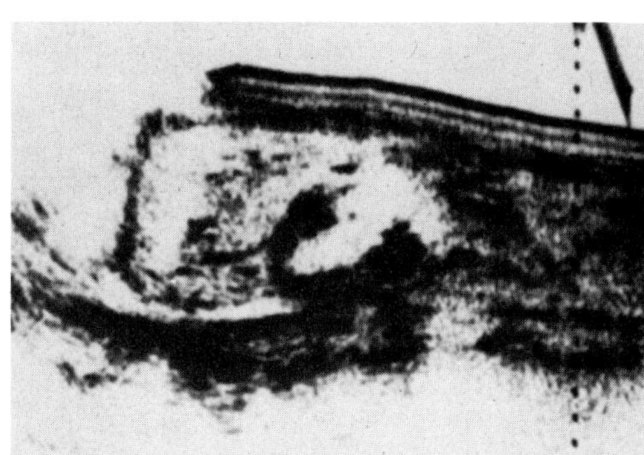

Figure 125–B. Longitudinal section 2R.

DIAGNOSIS

Acute pancreatitis with early pseudocyst formation.

ULTRASOUND FINDINGS

The pancreas is edematous and stands out clearly from the surrounding tissue. In Figure 125–A the splenic vein and superior mesenteric artery define precisely the pancreas cranially and posteriorly, while the undersurface of the liver shows its anterior extent. The normal pancreas produces echoes which are louder (i.e., more intense or darker) than those of the neighboring liver. If the echoes from the pancreas are of the *same or less intensity* than the echoes from the liver, something is amiss. The most common cause of this is pancreatitis, but pancreatic pseudocyst and neoplasms are also responsible for decreased echo production. In this case the diagnosis is no problem, as the patient has clinical pancreatitis. But something else is going on here. In the scan through the head of the pancreas (Figure 125–B) there is a small round area which is totally echo-free. This is a small pseudocyst that has just begun to form in the head of the pancreas. Small cysts like this are not uncommon during acute episodes of pancreatitis; in most patients the cysts will resolve when the pancreatitis subsides, but in some the cysts will continue to enlarge and form the classic pancreatic pseudocyst.

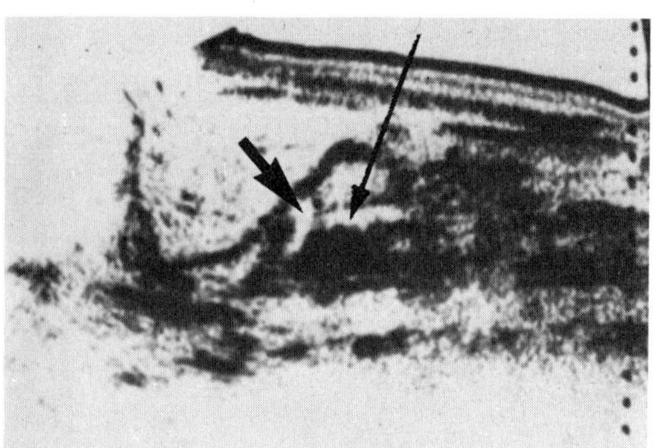

Figure 125–A. **Longitudinal section 0L.** Pancreatitis.

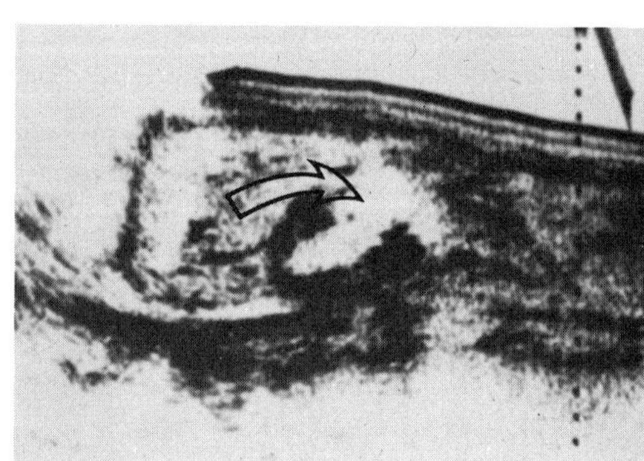

Figure 125–B. **Longitudinal section 2R.** Pancreatitis and small pseudocyst.

REFERENCES

Sarti DA: Rapid development and spontaneous regression of pancreatic pseudocysts documented by ultrasound. Radiology 125:789 (1977)

Case 126

CLINICAL INFORMATION

This case was on your desk when you came to work this morning: apparently the technician did the examination late last evening, after you had left for the day. (Unfortunately she left this morning for a two-week hiking vacation in the White Mountains so that you cannot get any further information.) To make matters worse, this is an outpatient from Rockgap, a village of 37 inhabitants 170 miles to the north. There was no clinical information provided in the requisition slip and there is no answer when you try to call the referring doctor. Despite these handicaps you should be able to make a diagnosis in this case.

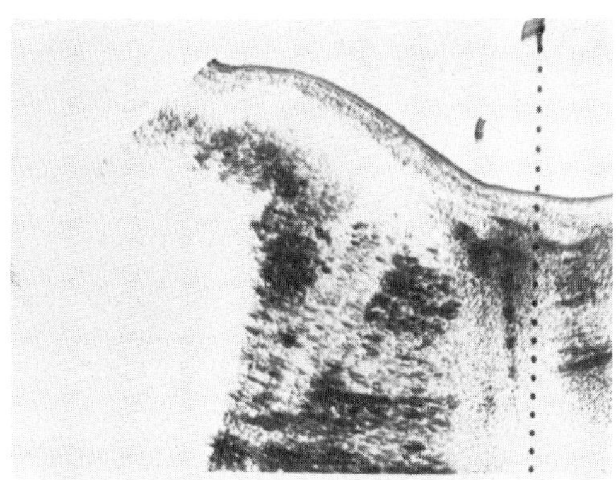

Figure 126–A. Longitudinal section 6R.

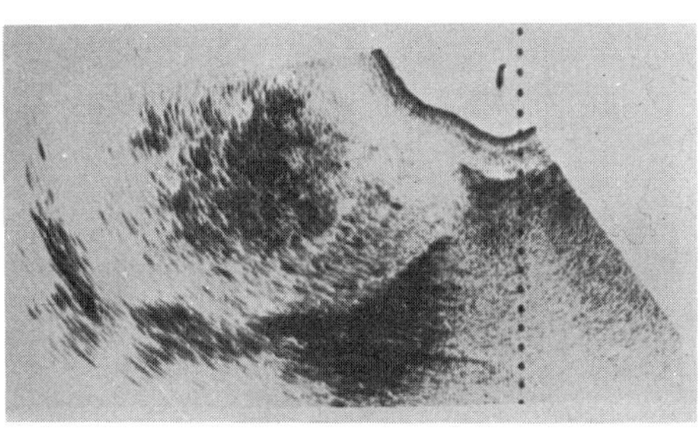

Figure 126–B. Longitudinal section 10R.

DIAGNOSIS

Having made it all the way through the book, you will have no trouble with this case, and we will not insult your ultrasonic acumen by giving an "answer." If you want to check your interpretation, send us a proven case from your practice—make it a tough or an unusual one—and we will send you our diagnosis and comments on this patient.

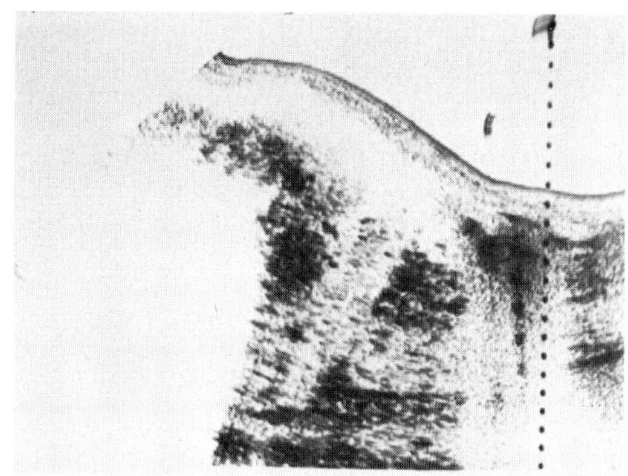

Figure 126–A. Longitudinal section 6R.

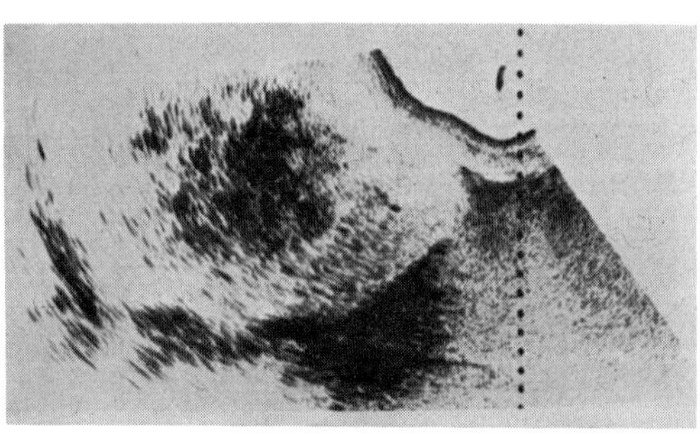

Figure 126–B. Longitudinal section 10R.

Index

Page numbers in *italics* refer to illustrations.